OBJECTS OF HIGH REDSHIFT

Lemaître and Einstein at Pasadena (1933)

ABBÉ GEORGES ÉDOUARD LEMAÎTRE (1894–1966)

"The evolution of the world can be compared to a display of fireworks that has just ended: some few red wisps, ashes and smoke. Standing on a well-chilled cinder, we see the slow fading of the suns, and we try to recall the vanished brilliance of the origin of the worlds." (From "L'expansion de l'espace" in the *Revue des Questions Scientifiques,* November, 1931.)

INTERNATIONAL ASTRONOMICAL UNION
UNION ASTRONOMIQUE INTERNATIONALE

SYMPOSIUM No. 92
HELD IN LOS ANGELES, U.S.A., AUGUST 28–31, 1979

OBJECTS OF
HIGH REDSHIFT

EDITED BY

G. O. ABELL
University of California, Los Angeles

and

P. J. E. PEEBLES
Princeton University

SPRINGER-SCIENCE+BUSINESS MEDIA, B.V.

Library of Congress Cataloging in Publication Data

Main entry under title:

Objects of high redshift.

    (Symposium – International Astronomical Union ; no. 92)
    Includes index.
    1. Redshift–Congresses.   2.   Galaxies–Congresses.
I. Abell, George Ogden, 1927–   II.   Peebles, Phillip James Edwin.
III.   Series: International Astronomical Union. Symposium ; no. 92.
QB465.O24   523.01'584   80–16212
ISBN 978-90-277-1119-9   ISBN 978-94-009-9040-1 (eBook)
DOI 10.1007/ 978-94-009-9040-1

TABLE OF CONTENTS

SCIENTIFIC ORGANIZING COMMITTEE

J.G. Bolton
M.S. Longair
P.J.E. Peebles
L.W. Sargent
G. Setti
H. Spinrad
G.A. Tammann
Ya.B. Zel'dovich
G.O. Abell (Chairman)

PARTICIPANTS

ABELL, G. O., University of California, Los Angeles
ALLER, L. H., University of California, Los Angeles
ANDERSON, S., University of California, Los Angeles
ARP, H. C., Hale Observatories
BAHCALL, J. N., Institute for Advanced Study
BAHCALL, N., Princeton University
BARBON, R., Universita di Padova
BARTKO, F., Martin Marietta Corporation, Denver
BAUM, W. A., Lowell Observatory
BECKER, J., University of California, Los Angeles
BECKLIN, E., University of Hawaii
BEICHMAN, C., California Institute of Technology
BHAVSAR, S. P., Northwestern University
BIRKINSHAW, M., Cavendish Laboratory
BOLDT, E., Goddard Space Flight Center
BOYNTON, P., University of Washington
BRIGGS, F. H., University of California, San Diego
BRUZUAL, G., University of California, Berkeley
BURBIDGE, G., Kitt Peak National Observatory
BURBIDGE, E. M., University of California, San Diego
BUTCHER, H., Kitt Peak National Observatory
CHAFFEE, F. H. JR., University of Arizona
CIARDULLO, R., University of California, Los Angeles
CLAVEL, J., ESA Villafranca Satellite Tracking Station, Madrid
COHEN, M., California Institute of Technology
DAVIDSEN, A. F., Johns Hopkins University
DAVIS, M., Harvard University
DE ZOTTI, G., Osservatorio Astronomico, Padova
EASON, E., University of California, Los Angeles
ELLIS, R. S., University of Durham
EPSTEIN, E., Aerospace Corporation, Los Angeles
EWING, T., Michigan State University, East Lansing
GIOVANNINI-FANTI, C., Laboratorio di Radioastronomia, Bologna
FANTI, C. G., Laboratorio di Radioastronomia, Bologna
FORD, H. C., University of California, Los Angeles
FONG, R., University of Durham
FRENCH, H. B., Hale Observatories
GAISER, B. D., Standford University
GASKELL, C. M., University of California, Santa Cruz
GIACCONI, R., Harvard University
GRANDI, S., University of California, Los Angeles
GRASDALEN, G., University of Wyoming

GREEN, R., California Institute of Technology
GROTH, E. J., Princeton University
GRUENWALD, R. B., Université de São Paulo
GULL, S. F., Cavendish Laboratory
HARMS, R., University of California, San Diego
HAWKINS, M. R. S., Royal Observatory, Edinburgh
HEESCHEN, D. S., National Radio Astronomy Observatory
HUNSTEAD, R. W., University of Sydney
JAFFE, W. J., National Radio Astronomy Observatory
JURA, M. A., University of California, Los Angeles
JUGAKU, J., Tokyo Astronomical Observatory
JANKKARINEN, V., University of California, San Diego
KATGERT, P., Huygens Laboratorium, Leiden
KATGERT-MERKELIJN, J. K., Huygens Laboratorium, Leiden
KENNICUTT, R. C., Hale Observatories
KOCHHAR, R. K., Ecole d'ete de Physique Theorique, Les Houches
KOO, D., University of California, Berkeley
KOSKI, A., University of California, San Diego
KRON, R., University of Chicago
LAKE, G. R., University of California, Berkeley
LARI, C., Laboratorio di Radioastronomia, Bologna
LAWRIE, D. G., University of California, Los Angeles
LEBOFSKY, M., University of Arizona
LESSER, M., University of California, Los Angeles
LIANG, E. P., Stanford University
LO, K. Y., California Institute of Technology
LOCKHART, T., California Institute of Technology
LOEWENSTEIN, M., University of California, Los Angeles
LONGAIR, M. S., Cavendish Laboratory
MANDOLESI, N., Consiglio Nazionale delle Ricerche, Bologna
MARGON, B., University of California, Los Angeles
MARSCHER, A., University of California, San Diego
MASSON, C., California Institute of Technology
MATHEZ, G., Observatoire de Paris-Meudon
MILLER, J. S., University of California, Berkeley
MORABITO, D., Jet Propulsion Laboratory, Pasadena
MURDOCH, H. S., University of Sydney
NARLIKAR, J. V., Tata Institute of Fundamental Research, Bombay
NICHOLSON, G. D., National Institute for Telecommunications Research,
    Johannesburg
NIETO, J. L., l'Observatoire du Pic-du-Midi
NOLTHENIUS, R., University of California, Los Angeles
NORTHOVER, K. J. E., Cavendish Laboratory
O'DELL, S. L., Virginia Polytechnic Institute and State University
OEMLER, A. Jr., Yale University
OKE, J. B., Hale Observatories
OSMER, P. S., Cerro Tololo Inter-American Observatory, La Serena
PEARSON, T. J., California Institute of Technology
PEEBLES, P. J. E., Princeton University
PENSTON, M. V., European Space Agency, Madrid
PENZIAS, A. A. Bell Laboratories, Holmdel, New Jersey

PERRENOD, S. C., Kitt Peak National Observatory
PORNCHAI, P., University of California, Los Angeles
PRESTON, R., Jet Propulsion Laboratory, Pasadena
RAINEY, G., California State University, Northridge
READHEAD, A. C. S., California Institute of Technology
REAVES, G., University of Southern California
REES, M., Institute of Astronomy, Cambridge University
RICHARDS, P., University of California, Berkeley
RICKER, G., Massachusetts Institute of Technology
RIEKE, G., University of Arizona
ROBERTS, D. H., Massachusetts Institute of Technology
ROBERTS, J., University of California, Berkeley
SAKIMOTO, P., University of California, Los Angeles
SCHEUER, P. A. G., Cavendish Laboratory
SCHILD, R. E., Harvard University
SCHMIDT, M., Hale Observatories
SCOTT, E. L., University of California, Berkeley
SEGAL, I. E., Massachusetts Institute of Technology
SEIDOV, Z. F., Shemakha Astrophysical Observatory, Azerbaidzhau, U.S.S.R.
SELDNER, M., Princeton University
SEIELSTAD, G. A., Owens Valley Radio Observatory
SHAFTER, A., University of California, Los Angeles
SILK, J. I., University of California, Berkeley
SIMON, R. S., Owens Valley Radio Observatory
SMITH, H. E., University of California, San Diego
SMOOT, G. F., University of California, Berkeley
SOBEL, H., University of California, Los Angeles
SOIFER, B. T., California Institute of Technology
SONEIRA, R., Institute for Advanced Study, Princeton
SPINRAD, H., University of California, Berkeley
STOCKTON, A. N., University of Hawaii
TAMMANN, G., Astronomisches Institut der Universität Basel
TARENGHI, M., European Southern Observatory-CERN
TOMCZYK, S., University of California, Los Angeles
TRIMBLE, V. L., University of California, Irvine
TSURUTA, S., Max-Planck-Institut für Physik und Astrophysik, München
TURNER, E. L., Princeton University
TYSON, J. A., Bell Laboratories, Holmdel
ULRICH, R., University of California, Los Angeles
UOMOTO, A., University of Texas
VERON, M., European Southern Observatory-CERN
VERON, P., European Southern Observatory-CERN
WALL, J. V., Cavendish Laboratory
WAMPLER, E. J., University of California, Santa Cruz
WEHINGER, P., Max-Planck-Institut für Astronomie, Heidelberg
WEILER, K. W., Max-Planck-Institut für Radioastronomie, Bonn
WEYMANN, R. J., University of Arizona
WILLS, B. J., University of Texas
WILLS, D., University of Texas
WILSON, A. S., University of Maryland

WILSON, M. L., University of California, Berkeley
WOLFE, A. M., University of California, San Diego
WOODY, D., California Institute of Technology
WRIGHT, J. P., National Science Foundation

PREFACE

In 1977 there was a very successful conference in Tallinn on the Large-Scale Structure of the Universe (IAU Symposium 79). Since then a number of developments have greatly increased the body of observational data of cosmological significance.

The Einstein X-ray telescope, launched in late 1978, has shown that the diffuse X-ray background is almost certainly due to quasars. Several independent investigations of the apparent magnitude distribution of faint galaxies have placed new limits on the scale of inhomogeneities in the universe, as well as on the role of galaxy evolution. Ever more remote clusters of galaxies are being discovered and redshifts measured for them that have completely overwhelmed Minkowski's remarkable 1970 achievement of observing z = 0.46 for 3C295 (yet the largest measured redshift for a quasar remains at z = 3.53). The microwave background radiation has still frustrated our attempts to find small-scale anisotropies, but there is a new question on the interpretation of the spectrum, and a very exciting large-scale anisotropy indicating the global peculiar velocity of our galaxy now seems to be well established.

With so many new findings, and in some cases lack of findings where they were expected, it seemed highly appropriate to schedule a conference on these observations dealing with objects of cosmologically interesting redshifts. At the original suggestion of Dr. Malcolm Longair, it seemed particularly fitting to hold the conference in California, where the famous Mount Wilson telescopes as well as the Lick Observatory 3-m and Palomar 5-m telescopes have played such historical roles in optical cosmology, and are still very active today.

The conference (IAU Symposium 92), held at the University of California, Los Angeles, 28-31 August 1979, was attended by 146 participants from 15 countries. There were 33 invited papers and 15 short contributions. The social events included an opening reception, a Hollywood Bowl concert, a Los Angeles Dodgers baseball game, a trip to Disneyland, and, in lieu of a closing banquet, a steak barbecue on Mount Wilson, at a location a short walk from the Observatory, which held open house in the telescope domes. Through the scientific sessions and lengthy conversations at social gatherings, the participants had an opportunity to bring each other up to date on problems of observational cosmology. The purpose of this volume is to share this knowledge with the broader astronomical community.

Following the current procedure for publication of IAU Symposium proceedings, the papers are all reproduced from camera-ready copy submitted by the authors. Invited papers are reproduced in full, and the

short contributions by abstracts.   The one exception is a contribution
by I.D. Karachentsev, who was invited to the Symposium and originally
had intended to attend, but in the end was unable to do so.   The
editors decided that his paper is of sufficient interest and cogency
to the other papers actually presented at the Symposium to warrant
inclusion in the present volume.

The discussions of the papers have been typed from written versions
of questions and comments prepared by questioners immediately following
their remarks after each paper, and from written versions of the
responses prepared by each speaker.   As is inevitable (we suppose) a
small fraction of these written statements never were returned to the
editors.   However, the published discussion is (by actual count) 82
percent complete.

We are most grateful to the majority of the authors who were
prompt in delivering their camera-ready manuscripts.   Only one paper
has not been received in time for inclusion in the Proceedings, a
regrettable omission, but, we think, not a crippling one.   We also owe
thanks to the many graduate students at UCLA who assisted greatly in
providing transportation and in clerical tasks during the scientific
sessions, and we owe special thanks to Mrs. Edna Ford for her splendid
service in helping with the organization of the Symposium itself.
Finally, we thank Mr. Robert O'Daniel, who has done such a fine job in
typing the discussion and in retyping a few "problem" manuscripts.

                            G.O. Abell
                            P.J.E. Peebles, Editors

AUTOMATED FAINT GALAXY COUNTS AND GALACTIC EVOLUTION

J. Anthony Tyson and John F. Jarvis
Bell Labs

Detection and classification of faint images by eye has traditionally encountered systematic errors faintwards of 20th mag on Schmidt plates and 22nd mag on 4-meter plates. Automated classification of Schmidt plate images has pushed the classification limit to 22 mag (Kibblewhite, et al., 1975). Automated detection and classification of faint 4-meter limit plate images has recently led to statistical studies of galaxy numbers and clustering at redshifts where cosmology and galactic evolution dominate over local effects. Here we report on some aspects of the FOCAS (Faint Object Classification and Analysis System) automated classifier (Tyson and Jarvis, 1979) and compare our results of number counts in SA57 with those of Kron, 1979. Differential galaxy counts in six high latitude fields and evidence for galaxy evolution are briefly discussed.

The interactive feature of FOCAS (Jarvis and Tyson, 1979) allows tests of many classification schemes. We have found that for images fainter than 22 mag more than three measured parameters (or features) are required for reliable classification. The data presented here are based on a 6-dimensional feature space. In designing our detection and classification algorithms we have attempted to obtain reliable operation over as wide a dynamic range of magnitudes and morphologies as possible. Six surface luminosity moments about the centroids of images are used as input to a clustering algorithm operating in the 6-dimensional decision space. Figure 1 shows peak intensity vs. magnitude for a subset of objects. The stars form a tight one-parameter cluster and the division between stars and galaxies can be seen visually down to ∿22 mag. First a line is drawn separating these two clusters down to 22 mag, as a zeroth order approximation. This initial line is generally not the optimal separator in other two-dimensional subspaces (see Fig. 2). The classifier then operates in the entire 6-dimensional feature space and evolves a hyperellipsoid decision surface which best separates stars and galaxies. The results of this procedure are checked with duplicate or co-added plates and deep CCD images of parts of the same fields. Classifier thresholds are set on the basis of these checks. Photographic density of sky is well into the linear portion of the gamma curve. Magnitudes are photoelectrically determined through a photoelectric sequence for each area studied.

G.O. Abell and P. J. E. Peebles (eds.), Objects of High Redshift, 1–8.
Copyright © 1980 by the IAU.

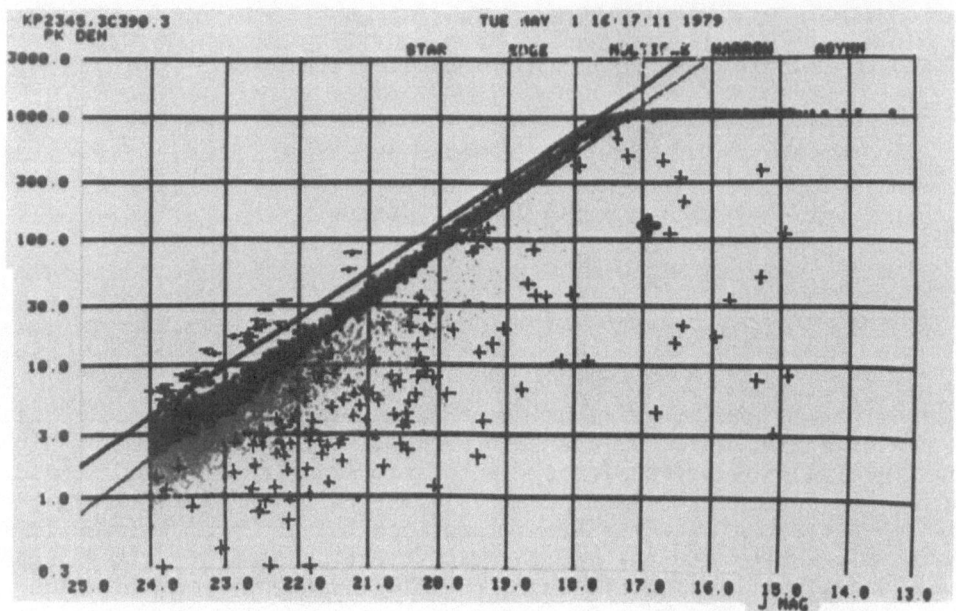

Fig. 1 Peak intensity vs. J. magnitude for a subset of objects. Dark
points are stars, points with horiz. bars are noise, light points are
galaxies, and crosses are multiple overlapping objects. The lines are
the zeroth order intersection of the decision surface. Plate saturation
for stars occurs at 17.5 mag.

Stars and galaxies form distinctly separate clusters in the 6-d
decision space, as can be seen in Fig. 3 which is a histogram of
distances in the 6-d space from the decision surface for a large number
of objects. Stars of all magnitudes cluster around the same 6-d point.

Fainter galaxies tend to cluster closer to the star cluster but are
reliably distinguished statistically even at 24th magnitude. The majority of galaxies in the range 19-20 mag and many in the range 23-24 mag
are off the left boundary of the histogram. Since galaxies greatly out-
number stars at 24th mag, misclassification error will not significantly
affect galaxy differential number counts but could adversely affect the
star counts at the faint end.

An example of the breakup of multiple or overlapping objects
(crosses in Figs. 1 and 2) is shown in Figures 4a,b. (The original display shows color-coded squares for each type of object classified.)

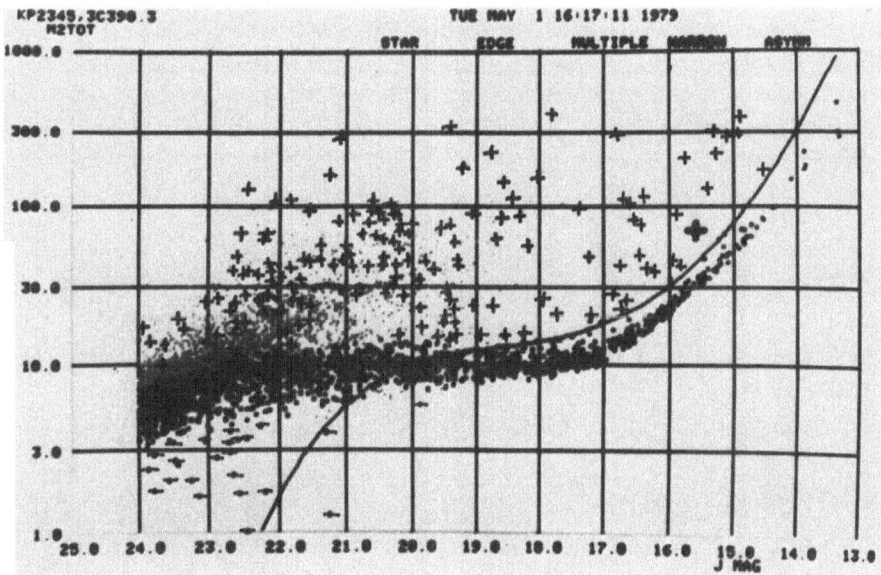

Fig. 2   Second moment M_2 vs. J magnitude for a subset of objects.   The line is the intersection of the zeroth order try at a decision surface.

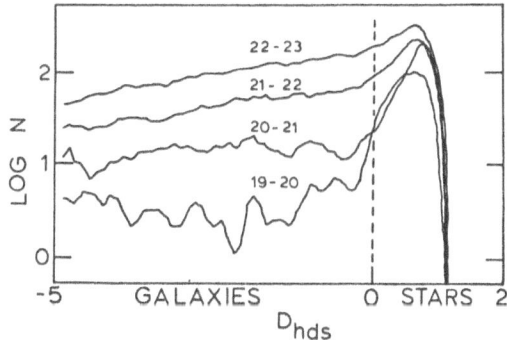

Fig. 3   Clustering in decision space. The region near the decision surface is shown. The number of objects found in several magnitude ranges are plotted vs. their distance from the hyperellipsoid decision surface.

Comparison of Kron's very different technique (Kron, 1978,9) and ours is shown in Fig. 5 where we have analyzed the same raw data plate. Star counts also agree well; this is somewhat surprising since the classification methods are so different, and star counts are expected to be more sensitive to classification error at the faint limit.

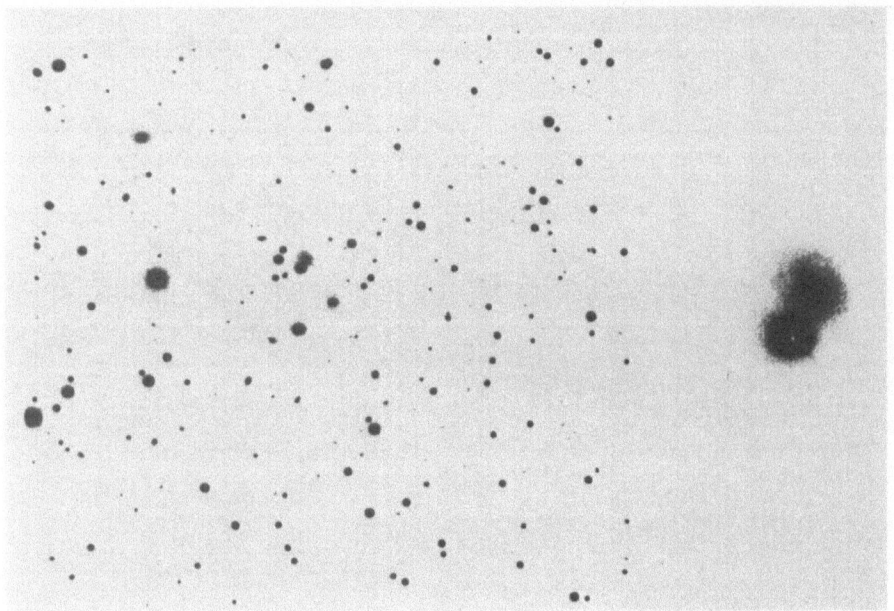

Fig. 4a,b Above (4a) is shown a part of one digitized plate 12 arc min across containing a multiple object shown at full resolution on the right. In Fig. 4b (below) is shown the breakup of the multiple into a final classification of a galaxy and star.

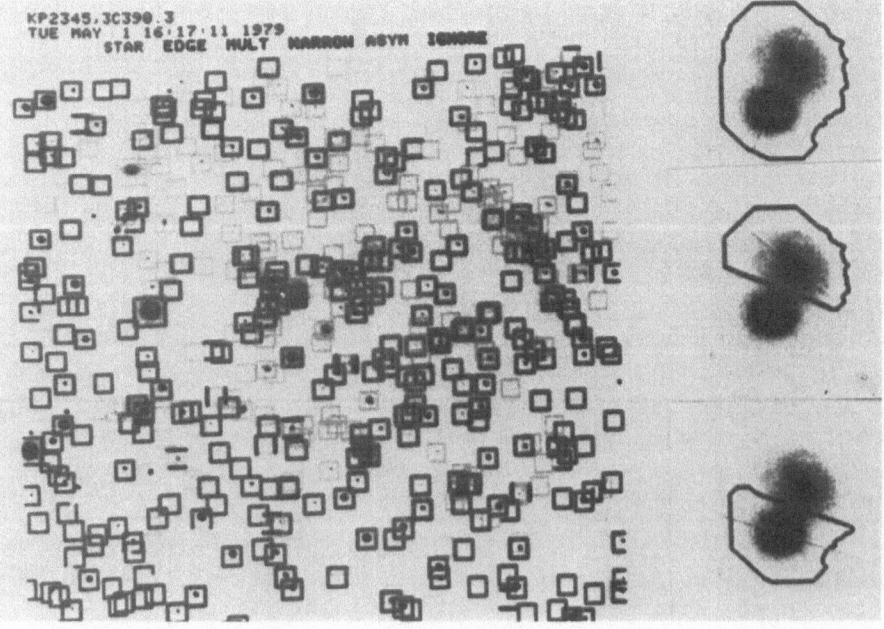

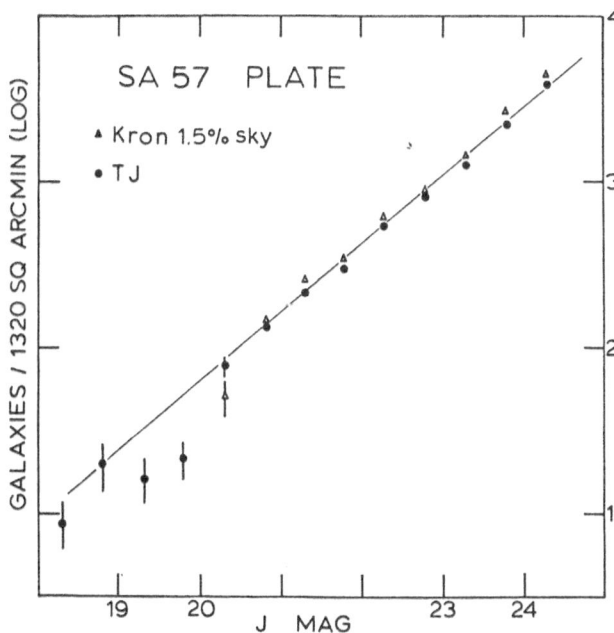

Fig. 5 Differential numbers of galaxies per half-magnitude bin counted on the same plate of SA57 by Tyson-Jarvis (filled circles) and Kron (triangles). A re-binning correction (Kron 1979, Fig. 6) has been applied to Kron's data to convert it to magnitudes integrated out to 1.5% of night sky surface luminosity. The line has slope 0.41.

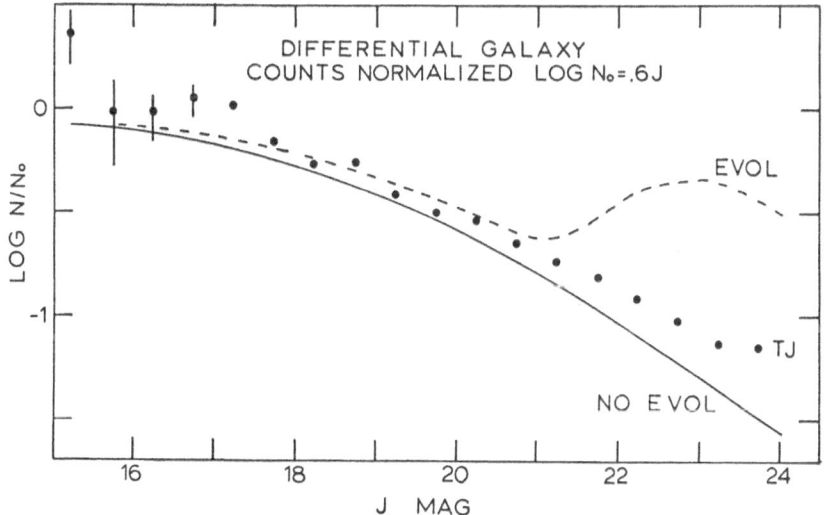

Fig. 6 Observed and theoretical differential galaxy counts. TJ are our data from the average of six high north latitude fields. Errors are smaller than the size of the points, except for the four brighter than 17th mag. The two model curves are from Tinsley, 1979. N (J) has been normalized by the Euclidean $N_o(J)$ where Log N_o = 0.6 J - 9.

We have examined eleven fields in the north with the FOCAS classifier so far, resulting in the classification of approximately 60,000 galaxies in these fields down to J = 24 mag. Although data down to J = 24.5 mag is kept and classified, we currently trust number counts for magnitudes brighter than 24th. Fig. 6 shows the sum of 6 high north galactic latitude fields. A total of 34,241 galaxies are found in these 6 fields to J = 24 mag. At high latitude we find 17,100 ± 800 galaxies per square degree to 24th mag. The dispersion in this number between individual high-latitude fields is 2500 per square degree. This is more than 10 times larger than the expected errors, due to high latitude clumped extinction. Local clustering effects are also seen, mostly around J = 17th mag. It appears that the "local" supercluster is considerably larger than we had thought. None of our fields are near any known rich clusters brighter than 18 mag per galaxy.

Our data for 18 < J < 24 mag are approximated by the relation Log Ntot = 0.41 (± .004) J - 5.63. The differential galaxy number counts shown in Fig. 6 (TJ) are compared with theoretical models of Tinsley (1979) for no evolution and for somewhat conservative evolution. These models are constrained to give present galaxy colors. The EVOL model has galaxy formation taking place at a redshift of 5. Although the data show evidence for some evolution, it is clear that there has been considerably less evolution over the look-back times involved in a magnitude-limited sample at J = 24.5 than had been suspected. Although the theory contains several adjustable parameters, perhaps the simplest is to move galaxy formation to earlier epochs. A redshift of formation greater than 10 would give agreement with the data. A larger sample of data will be discussed in greater detail in an article in preparation. In brief:

1. The bright flash of initial stellar burning in E's and SO's is not seen, either because galaxies formed earlier than Z ⌣ 10 or perhaps because gas and dust then surrounding the galaxies prevents us from seeing this flash.

2. High latitude clumped extinction averaging ⌣0.4 mag exists.

3. There is some evidence for a local supercluster to the north, containing galaxies at J ⌣ 17 mag.

REFERENCES

Jarvis, J.F. and Tyson, J.A.: 1979, SPIE Proc. Instrumentation in
      Astron. III, 172, pp. 422-428.
Kibblewhite, E.J., Bridgeland, M.T., Hooley, T., and Horne, D.: 1975, in
      "Image Processing Techniques in Astronomy" (eds. C. de Jager and
      H. Nieuwenhuijzen), D. Reidel, Dordrecht, p. 245.
Kron, R.G.: 1978, dissertation, University of California, Berkeley.
Kron, R.G.: 1979, Physica Scripta (Sweden) November; 1979, Astrophys. J.
      Suppl., in press.
Tyson, J.A. and Jarvis, J.F.: 1979, Astrophys. J. Lett. 230, pp. L153-L156.
Tinsley, B.M.: 1979, Preprint.

DISCUSSION

Fong: In your star/galaxy separation plots, there is a clear separation between stars and galaxies down to $20^{m}0$ or $21^{m}0$, but, going fainter, the objects overlap and become just a uniform spread. How do you decide which are stars and which are galaxies?

Tyson: On scatter plots which are not saturated by thousands of points, there is a clear separation down to 22 mag–23 mag. However, several (we currently use 8) classifier features are necessary in order to separate stars and galaxies to 24th mag. In the resulting multi-dimensional feature space, clustering is well-defined down to our faint limit. Fainter than 24th mag, the clusters which define stars and galaxies merge. This can be seen in our histogram of numer as a function of distance from the decision surface in the feature space. Obviously, such a procedure must be checked: we use CCD deep exposures to examine and verify our classifier decisions for a small subset of the data.

Baum: Could you comment further on the models with which your galaxy counts are being compared? In particular, if a substantially different luminosity function were considered, could you disentangle the effect of the luminosity from the effect of evolution?

Tyson: If the luminosity function had an enhanced faint tail, you might expect to see an excess of faint objects locally. This is not seen. Accordingly, Tinsley has used the Schechter Luminosity function in constructing her models. The dashed line EVOL shows Tinsley's estimated number counts for a formation redshift of 5 in a model constrained to give present epoch colors and in which star formation in spirals decays exponentially without a bright early burst, and ellipticals and S0's form stars up to 10^{9} yr. There are many model adjustable parameters which could be modified to fit the data. Early formation redshift is the simplest but not the only possibility.

Hawkins: How secure is your magnitude scale at the faint end, and what comparison do you have between photoelectric magnitudes? Do C.C.D. magnitudes have systematic errors?

Tyson: After field flattening there is no evidence for systematic error in C.C.D. magnitudes -- the C.C.D.'s have a very large dynamic range. (R. Lynds recently found nonlinearity in the edge of the field in the JPL C.C.D. due to post-C.C.D. electronics.) In addition to spot C.C.D. tests of our FOCAS-assigned faint magnitudes, we have obtained photoelectric sequences well into the linear portion of the γ-curve of our plates for each of our fields. It is crucial to obtain accurate sky subtraction at the faint end, since any systematic error changes the N(J) slope htere. In that sense, the algorithm for sky determination is what is being tested by various photoelectric tests.

Tarenghi: What do you mean by "supercluster," the local supercluster
               or a more distant one?

Tyson: In addition to the excess counts to 16th mag in the north
               recently observed by Kirshner, Oemler, and Schechter, we
find a significant excess in the form of two clumps centered at 15.5 mag
and 17 mag. We have checked the PSS plates for clusters and we are near
none. If this "local" superclustering is all part of one cluster, its
size is probably larger than 500 Mpc.

COUNTS AND COLORS OF FAINT GALAXIES

Richard G. Kron
Yerkes Observatory, University of Chicago

The color distribution of faint galaxies is an observational dimension which has not yet been fully exploited,
despite the important constraints obtainable for galaxy
evolution and cosmology. Number-magnitude counts alone
contain very diluted information about the state of things
because galaxies from a wide range in redshift contribute
to the counts at each magnitude. The most-frequently-seen
type of galaxy depends on the luminosity function and the
relative proportions of galaxies of different spectral
classes. The addition of color as a measured quantity can
thus considerably sharpen the interpretation of galaxy
counts since the apparent color depends on the redshift and
rest-frame spectrum. To a first approximation two colors
for a galaxy can determine a redshift and a spectral class,
because redshift loci in a color-color diagram run roughly
parallel to each other, and roughly perpendicular to the
zero-redshift galaxy "main sequence" (Tinsley 1977a, Pence
1976) for small redshift. This game becomes more and more
uncertain at higher redshift, because the systematics of
galaxy UV spectral energy distributions are not well known,
and what is known is not well understood (Code and Welch
1979). Redshifts for some random sample of faint galaxies
is required to pin down the color-redshift relations; steps
in this direction have already been taken by E. Turner.
The reason for stressing colors, as opposed to redshifts,
is that colors can be obtained relatively easily for large
samples of faint galaxies if panoramic detectors are used:
indeed, colors are not much more difficult to obtain than
magnitudes.

Granting, however optimistically, that colors can be
used as statistical redshift indicators, galaxies could
then be counted to successive limits in redshift, rather
than magnitude, which would sample the volume element
directly. A full discussion of the use of colors as a

G.O. Abell and P. J. E. Peebles (eds.), Objects of High Redshift, 9–15.
Copyright © 1980 by the IAU.

test for cosmological models and as a test for evolution
will be found in Bruzual and Kron (1979).

Besides the broader context of the color distribution
of a complete sample of galaxies, there are the specific
applications of using colors to isolate particular classes
of objects. For instance, galaxies at very high redshift,
z ≳ 4, would appear to be very red and might perhaps be
identified by this property alone (Davis and Wilkinson
1974). Very blue galaxies, on the other hand, are presumably galaxies at lower redshift with active star formation. Studying the characteristics of these objects as a
function of redshift might reveal important aspects of the
history of star formation in galaxies. For example, a
natural extension of the work of Butcher and Oemler (1978)
would be to look at the distribution of galaxies in remote
clusters in a color-color diagram, and see whether deduc-
tions about the star formation history can be made, à la
Larson and Tinsley (1978) and Huchra (1977). The colors
of radio galaxies probably relate both to recent star formation (van den Bergh 1978) and to the existence of a non-
thermal (evolving?) contribution to the light (Smith 1977,
Yee and Oke 1978). Many of these topics will be examined
in detail later in the symposium and so will not be discussed here.

Figure 1 displays some examples of empirical determinations of the color distribution of faint galaxies.
Figure 1a gives the photo-
electric data used by Kron and
Shane (1974) to calibrate the
depth of the Shane and Wirtanen galaxy count survey; the
galaxies should be representative of the survey limit. Figure 1b shows a section of
Rainey's (1977) photographic
data for Selected Area 57,
chosen here to correspond
roughly in magnitude with the
Kron and Shane photometry.
Finally, Figure 1c reproduces
Butcher and Oemler's (1978)
faint ISIT photometry of a region ∿100 arc sec away from
3C 295, which is taken to represent the field.

The various data have
qualitative differences due to
(1) different bandpasses;

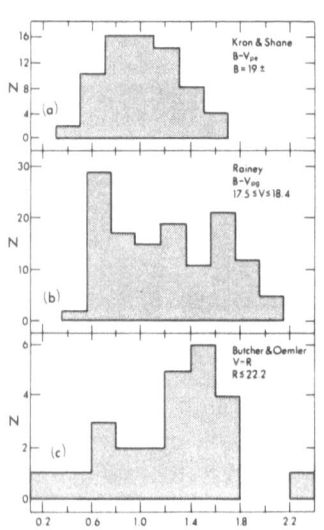

Fig. 1 - Color distribu-
tions for faint galaxies.

(2) different depths represented, at least in the case of
Butcher and Oemler; (3) different random errors; (4) different selection rules defining the sample; and (5) completely
different photometric methods: large-aperture photoelectric
photometry, iris photometry, and digital surface photometry,
for Figures 1a, 1b, and 1c, respectively. Because of this
diversity, any effort to rationalize the various data would
be difficult; a better approach would be to build a numerical model for the expected color distribution explicitly for
a specific set of bandpasses and magnitude range. One example of such a model (for B-V) has been presented by Pence
(1976), which does agree, within the uncertainties in the
comparison, with Figure 1a, and to a lesser extent with Figure 1b. The Butcher and Oemler color distribution, Figure
1c, appears to be similar to the distant cluster color distributions. This is particularly interesting because numerical models by several authors (e.g., Pence 1976, Tinsley
1977b, 1978, 1979, Bruzual and Kron 1979) have suggested
that the median redshift for field galaxies with $m_R \sim 22$ is
about the same as the redshift of 3C 295. The Shane and
Wirtanen survey has a depth corresponding to a median redshift near 0.13, so that in terms of the placement of the
$\lambda 4000$ feature with respect to the bandpasses, the Shane and
Wirtanen depth viewed with B-V should be similar to the
Butcher and Oemler depth viewed with V-R.

Anyway, the main point of Figure 1 is to present what
is currently known about the color distribution of faint
galaxies and to show that the colors are not really exceptional—there do not appear to be large numbers of blue
galaxies or large numbers of very red galaxies, and the
colors are not too different from those encountered at z = 0.
This stability of color is the result of two competing effects: all types of galaxies, even the bluest, become apparently redder with increasing redshift (Wells 1972); but on
the other hand the intrinsically bluer galaxies get dimmer
less rapidly with redshift, so that the most-frequently-seen
type of galaxy at a given magnitude shifts blueward with increasing redshift (Greenstein 1938). Figure 2 shows how
these effects also depend on galaxy evolution and, because
of the cosmological sensitivity of the time-redshift relation, on the choice for q_0 and t_0. The curves were computed
using a particular galaxy spectral-energy distribution model
by Bruzual (1979), which adopts a stellar synthesis for the
galaxy light, with standard assumptions about the mass function and the star formation rate. The J and J-F system refer to the photographic IIIaJ and IIIaF emultions. Figure
2a gives the predicted color vs. redshift curves for a galaxy type which, with $J\text{-}F \sim 1.05$ at z = 0, corresponds to
something like morphological type Sa. The no evolution
curves differ from each other because t_0 was chosen to be

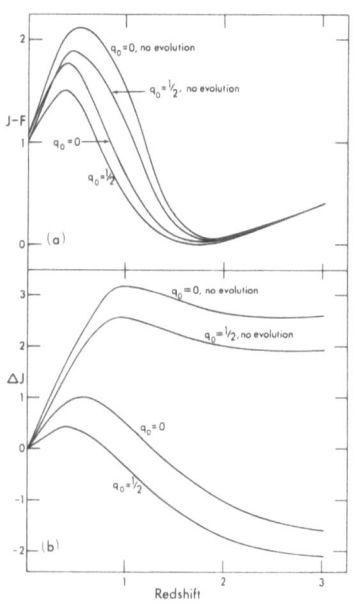

Fig. 2 - Color-redshift and
Δ mag-redshift curves for a
particular galaxy energy
distribution model.

16 b.y. for the q_0 = 0 example
and 13 b.y. for the q_0 = 1/2 example. The curves with evolution
differ partly because of this,
and also because for each redshift the look-back time is
different for the two values of
the deceleration parameter (the
galaxy spectral evolution model
is specified as a function of
time, not redshift, because it
is based directly on the component stellar evolutionary
tracks). The color-redshift
curves evidently depend quite
strongly on q_0 and t_0, at least
for this particular form of the
star formation rate. Other energy distribution models, with
either redder or bluer present-
day colors, generally behave
similarly to the particular
case which is illustrated, but
some galaxy types may not
evolve strongly in color, e.g.,
ellipticals in the recent past
(Tinsley 1972). Still, if the
star formation rate were known
as a function of time, the
color-redshift relation would be a cosmological test in its
own right, essentially reflecting the time-redshift relation.
More to the point, the number of galaxies seen at each color
depends on q_0 both via the time-redshift relation and via the
volume element.

Figure 2b gives, for the same energy distribution model
used in Figure 2a, the redshift dimming ΔJ (positive = fainter) in excess of the bolometric cosmological dimming. For no
evolution, ΔJ is just the k correction, with the two top
curves differing as before only in the present-day ages which
were assumed. When the energy distribution is allowed to
evolve, ΔJ incorporates both the k correction and the evolution. The calculation of the number-color relation for this
galaxy type would then use Figure 2a, and the volume element
as a function of redshift, to compute the number of such
galaxies at each color, but would need also Figure 2b to
specify the relative contributions brighter than some magnitude limit, which is what is required for a practical application. Note that all of the curves in Figure 2a tend to
the same color at high redshift, but whether or not these
are actually seen depends strongly on evolution, as evident

from Figure 2b. As an extreme example, an elliptical galaxy, or more precisely speaking, a galaxy which has produced few hot stars for the past few times 10^9 years, would become very red with increasing redshift, but would however not be seen (in a blue band) for the same reason: the k correction would be very large. Nevertheless, there must have been some epoch when the star formation rate was highest, and depending on how high this rate was, young ellipticals could have been very luminous and either blue, if $z \lesssim 3$, corresponding to that epoch, or red, if $z \gtrsim 4$.

Any model for the evolution of galaxies must be consistent with the counts as a function of both magnitude and color. Figure 3 gives an example of such a comparison. The model for the color distribution was derived again from Bruzual's synthetic energy distributions, with a standard assumed galaxy luminosity function and a count zero point which agrees with the bright data of Kirshner et al. (1978). The numerical model includes individual evolving energy distributions such as that used in Figure 2, as well as several others with different colors at $z = 0$ so as to reproduce the color distribution of Kirshner et al. The dashed curve in Figure 3a was computed for $q_0 = 1/2$, $t_0 = 13$ b.y., and the solid curve for $q_0 = 0$, $t_0 = 16$ b.y. The lower set of curves in Figure 3a is for the magnitude interval $21 \leq (J+F)/2 < 22$, and the upper set is for the magnitude interval $22 \leq (J+F)/2 < 23$. Figure 3b reproduces the equivalent photographic data from Kron (1978) for Selected Area 57. Note that Figure 3a gives not only the expected color distribution, as in Figure 1, but also the expected counts themselves in the area relevant to Figure 3b. The random errors for the photometry were determined from comparison of pairs of plates, and an analytic approximation to these errors has been included in the model of Figure 3a. The main conclusion here is that

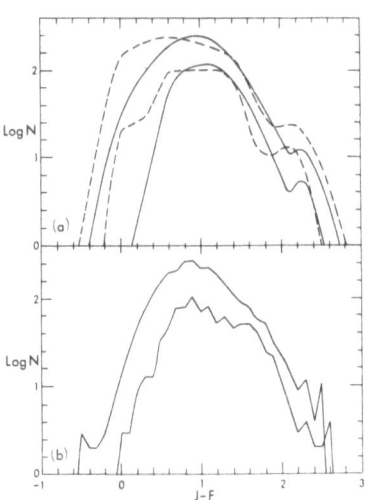

Fig. 3 - Theoretical (top) and observational (bottom) color distributions. Dashed curves computed for $q_0 = 1/2$, solid curves for $q_0 = 0$. Upper and lower sets of curves in each panel refer to $V \sim 22.5$ and $V \sim 21.5$, respectively.

the q_0 = 1/2 model predicts more blue galaxies than are observed, which is reflected in the behavior of the one component to the numerical model that is shown in Figure 2. The next step is to check that the assumed star formation rates are realistic, and this could be done either by getting redshifts for a relatively small sample, or by obtaining more than one color for each galaxy, a program which is currently under way by D. Koo and co-workers and by J. A. Tyson and J. Jarvis.

I thank Gustavo Bruzual for allowing me to use and to present some aspects of his galaxy spectral synthesis models in advance of publication.

REFERENCES

Bruzual-A., G.: 1979 (in preparation).
Bruzual-A., G., and Kron, R. G.: 1979 (in preparation).
Butcher, H., and Oemler, A.: 1978, Astrophys. J. 219,
      pp. 18-30.
Code, A. D., and Welch, G. A.: 1979, Astrophys. J. 228,
      pp. 95-104.
Davis, M., and Wilkinson, D. T.: 1974, Astrophys. J. 192,
      pp. 251-259.
Greenstein, J. L.: 1938, Astrophys. J. 88, pp. 605-617.
Huchra, J. P.: 1977, Astrophys. J. 217, pp. 928-939.
Kirshner, R. P., Oemler, A., and Schechter, P. L.: 1978,
      Astron. J. 83, pp. 1549-1563.
Kron, R. G.: 1978, dissertation, University of California,
      Berkeley.
Kron, G. E., and Shane, C. D.: 1974, Astrophys. Space Sci.
      30, pp. 127-134.
Larson, R. B., and Tinsley, B. M.: 1978, Astrophys. J.
      219, pp. 46-59.
Pence, W.L. 1976, Astrophys. J. 203, pp. 39-51.
Rainey, G. W.: 1977, dissertation, University of
      California, Los Angeles.
Smith, H. E.: 1977, in D. L. Jauncey (ed.), "Radio
      Astronomy and Cosmology", IAU Symp. No. 74, D. Reidel,
      Dordrecht, pp. 279-293.
Tinsley, B. M.: 1972, Astrophys. J. 178, pp. 319-336.
Tinsley, B. M.: 1977a, Publ. Astron. Soc. Pacific 89,
      pp. 245-250.
Tinsley, B. M.: 1977b, Astrophys. J. 211, pp. 621-637.
Tinsley, B. M.: 1978, Astrophys. J. 220, pp. 816-821.
Tinsley, B. M.: 1979 (preprint).
van den Bergh, S.: 1978, Vistas Astron. 22, pp. 307-319.
Wells, D. C.: 1972, dissertation, University of Texas,
      Austin.
Yee, H. K. C., and Oke, J. B.: 1978, Astrophys. J. 226,
      pp. 753-769.

DISCUSSION

Hawkins: I think that the method of distinguishing stars from galaxies
using the colour/colour plot is a much sounder approach than
the image structure approach described by Tyson.

NUMBER COUNTS OF GALAXIES TO B = 25*

I.D. Karachentsev
Special Astrophysical Observatory, the USSR Academy of
Sciences

ABSTRACT. Results are presented of counts of galaxies over the apparent magnitude range $21 \leq B \leq 25$, which have been obtained at the 6-meter telescope using McMullan's electronographic camera. These data do not indicate evidence for rapid luminosity evolution of galaxies. It is noted that the faintest galaxies, on the whole, have bluish color and a small correlation function amplitude.

Counts of faint galaxies at large telescopes make it possible to investigate a luminosity evolution of galaxies. The most valid information derives from counts log $N(m)$, which are performed in different colors simultaneously. A determination of the amplitude and the timescale of the luminosity evolution of galaxies must be treated, obviously, as a necessary prelude to the precise determination of the cosmological deceleration parameter, q_0.

The first counts of galaxies fainter than 23rd magnitude were carried out by Karachentsev and Kopylov (1977) and by Kron (1978). The fundamental work of Kron contains a detailed analysis of theoretical and observational aspects of the problem of counting galaxies. There is an appreciable disagreement, however, between the data of these authors in the sense that the counts of Kron show a relative number excess, log $N(B = 23.5) \approx 0.4$. To understand the reason for this disagreement, we have undertaken new observations, the results of which are presented here.

Observations were made in Selected Area 57 (1950: $\alpha = 13^h05^m8$; $\delta = +29°35'$) near the north galactic pole with deep photoelectric standards. This is one of the two fields investigated by Kron. On April 30, 1979, the author obtained films of the field in three colors (B, V, and R) with McMullan's electronographic camera installed in the 6-meter telescope cage. The camera has a high-quantum-efficiency photocathode

---

*This paper arrived after the Symposium, but has been included here (with editing by G.O. Abell) because of its interest and cogency.

G.O. Abell and P. J. E. Peebles (eds.), Objects of High Redshift, 17–21.
Copyright © 1980 by the IAU.

of 44 mm diameter. The linearity of the system response, a wide
dynamic range, and its low noise make the electronographic camera a
valuable instrument for a photometry of extremely faint objects. The
electronograms were obtained from a 35-minute exposure in each color,
with a mean seeing of 1".8. The films have sky background densities of
0.6, 1.2, and 1.4 in B, V, and R, respectively.

At present, we restrict ourselves to photometry of the field in
the blue color only. In an area of 30 (arc minutes)^2 were about 400
objects to B = 25. Stars and galaxies are distinguished reliably to
B = 23. Among the faint images, stars contribute no more than 10%.
The numbers of galaxies in the field, and their number per square degree,
brighter than apparent magnitude B are presented in Table 1 over the
range 21 ≤ B ≤ 25.

Table 1. Galaxy Counts in SA 57

m_B	Number in the field	log N(m) per degree
21.0	6	2.85:
21.5	12	3.15
22.0	17	3.30
22.5	28	3.52
23.0	47	3.74
23.5	98	4.06
24.0	166	4.29
24.5	240	4.45
25.0	380	4.65
26.2:	930	5.05:

The small-scale noise of the electronogram is rather low, amount-
ing to 0.5% of the sky background at 1 (arc sec)^2, or about 3 times
smaller than IIIa-J emulsion noise (Kron 1978). For the given noise
level a galaxy of the 25th apparent magnitude has a signal-to-noise
ratio, S/N = 3, with a typical image square of about 10 arc sec^2.

At a still lower detection level, down to S/N ≃ 1, or B = 26.2,
there are also visible extremely faint images; their log number is
given in the last line of Table 1. The reliability of this last entry
is certainly rather poor, because, on the one hand, an unknown fraction
of the galaxies could have been missed, and on the other, the grain
noise introduces some false images. This estimation needs confirmation
by repeated plates of the field.

Recently, data on number-magnitude counts of faint galaxies have
been enriched by new work (Peterson et al. 1979, Tyson and Jarvis 1979).
A summary of the deep counts is presented in Figure 1. The differential
counts by Kron and by Peterson et al. have been transformed into integral ones and normalized at B = 20-21. Note that both counts have been

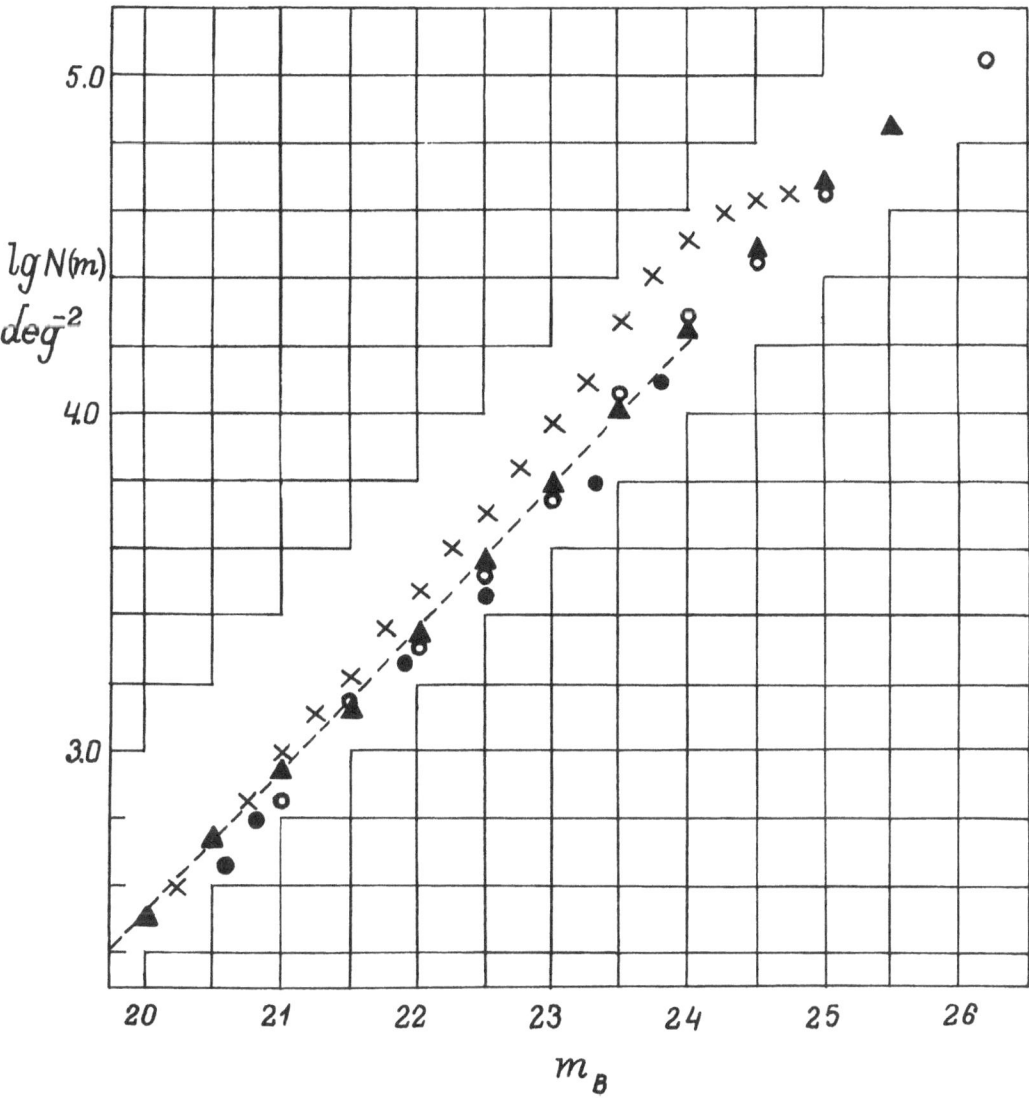

Figure 1. Integral number of galaxies per square degree brighter than apparent magnitude m_B. The data by different authors are indicated by the symbols: dots -- Karachentsev and Kopylov (1977); open circles -- the present paper; crosses -- Kron (1978); triangles -- Peterson et al. (1979); dashed line -- Tyson and Jarvis (1979). A 45° line in the figure corresponds to the expected number distribution slope for the case of no luminosity evolution.

carried out in the J system. Transformation from J to B requires a correction of about + 0ᵐ2, which is slightly dependent on the mean color of galaxies over the magnitude range under consideration.

As seen in Figure 1, our new counts show slightly more galaxies than our previous ones. This may be caused by an underestimation of the light of faint outlying parts of galaxies in our former counts, which were made mostly on IIa-O plates. Nevertheless, a significant difference exists between our present counts and those of Kron. The discrepancy is at a maximum near B = 24. The humpbacked shape of log N(m)-relation by Kron, which is interpreted by him as due to evolution, is not consistent with the present data. Considering also the data of other authors, we suggest that the excess of faint galaxies in Kron's counts may be due to his specific reduction procedures and/or selection of faint images.

The simulation of measurement errors and selection effects is a rather difficult task, the principal difficulty being the lack of advance knowledge of the shape of the log N(m) curve for extremely faint magnitudes. A bias in log N(m) is quite sensitive to the emulsion contrast γ, together with the seeing factor. In our experience, the linearity ($\gamma = 1$) and the small noise of electronograms minimizes the role of systematic measurement errors near the detection threshold.

The results of our counts of galaxies agree with the data of the Anglo-Australian group (Peterson et al. 1979). It is interesting that the agreement extends beyond B = 24.5, which is the completeness limit of the counts by Peterson et al. (1979). The last points in Figure 1 do not deviate strongly from the linear relation, log N(m) = constant + 0.41 m, founded by Tyson and Jarvis (1979). This excellent agreement of independent observers' data, obtained under different conditions, provides hope of obtaining a reliable basis for theoretical analysis of evolution effects.

Note that counts near B = 25-26 correspond to galaxies whose redshifts reach or exceed z = 1. Therefore, the statistics of QSO absorption lines may provide an independent confirmation of the counts, if intervening galaxies cause the absorption lines.

In the field investigated we do not find a large variation in the surface density of galaxies. The amplitude of two-point correlation function, w(Θ) $\leq$ 0.04, at scales of Θ = 10" to 60". Assuming that our sample is 10 times deeper than the survey by Shane and Wirtanen, we estimate the value of w(Θ) by using the scaling method of Groth and Peebles (1977). The amplitude of w (60") agrees satisfactorily with the expected one; however, the observed w (10") is 2-3 times lower than the calculated value. This may mean that small-scale clustering of galaxies ($\leq$ 100 kpc) has not developed yet at z = 1. Another explanation could be a selection of galaxies according to their colors, because blue spiral galaxies, with a lower space concentration, are more easily visible at large distances than elliptical galaxies. A larger sample of galaxies is needed to check this result.

Even a passing glance on the electronograms confirms Kron's conclusion that remote galaxies are predominantly blue. Most faint groups are more distinct on the blue plates than in the visual or red ones. The relative numbers of galaxies to the film limit in B, V, R are in the ratios $N(B \leq 25.0): N(V \leq 24.5): N(R \leq 23.5) = 1:0.79:0.76$. Our measurements of a dozen galaxies with $<B + V>/2 \simeq 23$ give the mean colors $<B - V> = +0.52 \pm 0.08$ and $<B - R> = +1.04 \pm 0.24$.

Considering the preliminary character of the present photometric data, we shall not discuss evolutionary effects in detail here. We do note, however, that there is good agreement among different authors on the expected slope of the curve, $\alpha = d \log N/dm$, which has been calculated under the assumption of no luminosity evolution, over the range of 21-24 mag; namely: $\alpha = 0.39$ (Karachentsev and Kopylov 1977); $\alpha = 0.39$ (Kron 1978); and $\alpha = 0.42$ (Peterson et al. 1979). The observed data points are only a little steeper than the mean slope $\alpha = 0.40$ (the diagonal in Figure 1). A moderate effect of galaxy evolution probably exists. However, it is necessary to know the precise energy distribution in the ultraviolet for a large sample of galaxies of different morphological types before we can confidently interpret a small excess of faint galaxies as a product of galaxy evolution.

The author thanks the Director of the Royal Greenwich Observatory for the use of the RGO electronographic camera. The author also gratefully acknowledges Dr. D. McMullan, Dr. J. Powell and V.L. Afanasjev for their help in solving the many problems of installing the camera in the 6-meter telescope cage. Finally, the author is indebted to Dr. I. King for providing his unpublished photometric standards in SA 57.

REFERENCES

Groth, E.J., and Peebles, P.J.E.: 1977, Astrophys. J., 217, 385.
Karachentsev, I.D., and Kopylov, A.I.: 1977, Sov. Astr. Lett., 3, 130.
Kron, R.: 1978, Ph.D. thesis, University of California, Berkeley.
McMullan, D., and Powell, J.R.: 1976, Proceedings of IAU Colloquium
    Number 40, Meudon.
Peterson, B.A., Ellis, R.S., Kibblewhite, E.J., Bridgeland, M.T.,
    Hooley, T., and Horne, D.: 1979, Epping, 13, Anglo-Australian
    Observatory, preprint.
Tyson, J.A., and Jarvis, J.F.: 1979, Astrophys. J., 230, L153.

ANALYSES OF DEEP GALAXY SAMPLES

Richard S. Ellis
Durham University, England

A.   INTRODUCTION

The combination of deep photography using fine-grain emulsions and fast automated plate-measuring machines is proving to be a valuable tool in studying galaxy evolution. Until recently, the favoured method for monitoring evolution was the spectroscopic study of one type of galaxy at various redshifts (see Spinrad 1977 for a review). It is considerably more economical, however, to derive evolutionary information from the statistical properties of the numerous faint images detectable on deep plates. The drawbacks are that the data at each apparent magnitude involves galaxies of different types seen over large redshift ranges. Also to interpret the statistics, we need to know the properties of large numbers of galaxies, for example their luminosity function, ultraviolet spectra and, particularly, the morphological variations in such properties.

Photography provides us with number-magnitude, colour-magnitude and colour-colour data. These topics have been discussed at this symposium by both Tyson and Kron. As Kron (1979) has emphasised, the automatic measurement and detection of faint images is a young subject and it is not clear which of the many procedures adopted by the numerous research groups active in this field is the most appropriate. The technical side of the issue has been discussed by Kron in some detail. A topic which has received less attention and which is even more important is concerned with the difficulties in interpreting the counts and colours, regardless of how they are measured, because of the poor knowledge of the fundamental properties of galaxies.

In addition to demonstrating the effects of these uncertainties, I shall outline a further use of deep galaxy data, namely studies of galaxy clustering. Until now, correlations of galaxy positions have been studied only using samples taken from Schmidt plates. This is because very large areas are required to avoid "fair sample" problems. The conclusions I reach will hopefully stress the need to continue this aspect of galaxy analysis with 4-metre plates and eventually with the Space Telescope.

G.O. Abell and P. J. E. Peebles (eds.), Objects of High Redshift, 23–30.

B. INTERPRETATION OF NUMBER-MAGNITUDE COUNTS

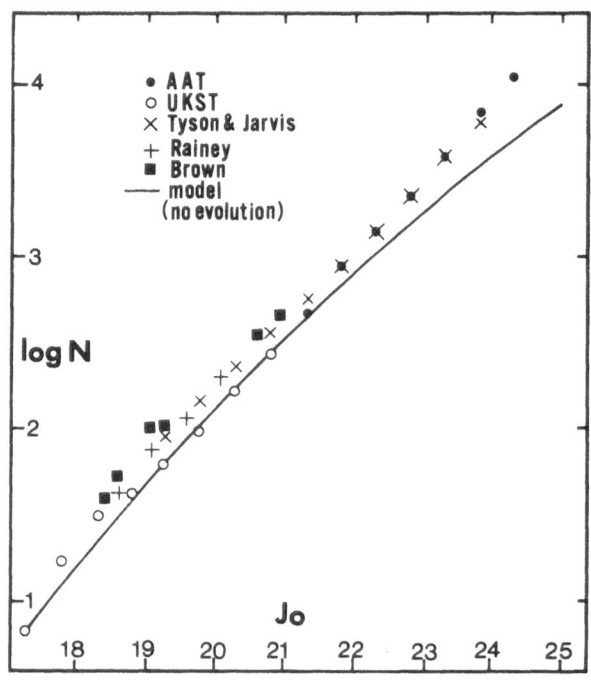

Figure 1. Differential galaxy counts per deg^2 per 0.5 mag interval
in a isophotal J (IIIaJ plus Schott GG385) system.

        Figure 1 shows counts from some deep 4-metre and Schmidt plates.
AAT/UKST refers to work described by Peterson, Ellis and Kibblewhite
(1979).   These counts agree well with those of Tyson and Jarvis (1979)
and also with those of Kron when allowance is made for his different
photometric techniques. To derive evolutionary information from such
counts, one must first predict the counts in the absence of any
evolution. This requires accurate K-corrections, luminosity functions
and the variations in these properties from one type of galaxy to
another.
        Consider first the K-correction. When comparing data with models
the crucial factor in assessing the presence of evolution is the count
slope. As galaxies are being lost from these samples typically over
redshifts 0.3 < z < 1.0, reliable galaxy fluxes are required for wavelengths around 2500 Å for work involving the J band. It is well-known
that ultraviolet galaxy measurements when compared are often discrepant.
Even with one detector, Code and Welsh (1979) have shown the variation
within one galaxy type displaying similar optical colours may be
∿ 2 mag at 2500 Å. Ultraviolet data on normal galaxies, particularly
spirals, is very much in short supply. This intrinsic scatter means
that a large number of each type will have to be studied before
reliable K-corrections can be provided.

What effect does the present scatter in the UV fluxes have on
our interpretation of the number counts? Figure 2 shows the effect
of using K-corrections as determined from the fluxes of some early-
type galaxies. This difference represents the minimum effect we can
expect.

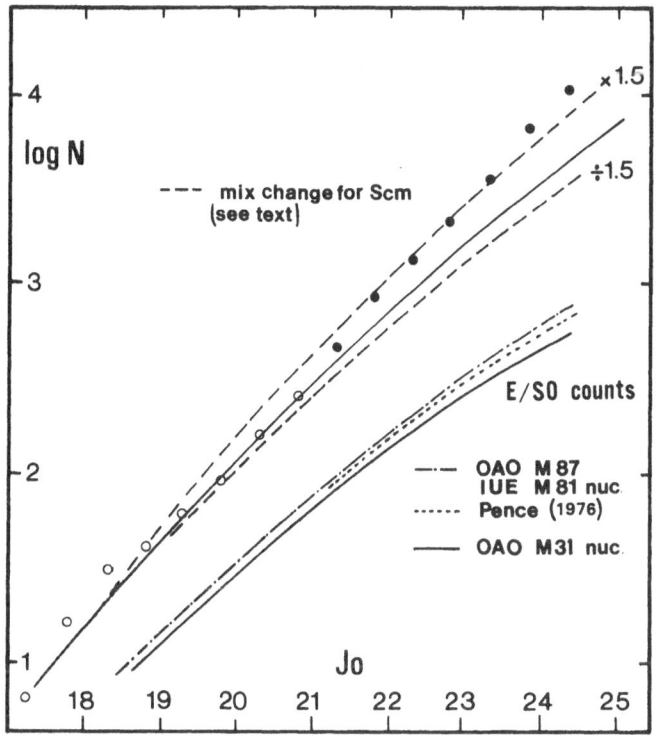

Figure 2. The effect of various uncertainties on the predicted number
magnitude counts for the J band. The lower curves represent predictions
for E/SO galaxies only.

For the later types, which form a less homogenous sample in the visible,
the scatter is likely to be much larger and their contribution to the
counts (see below) is more important. At present UV measurements of
spirals are so few we cannot even estimate the scatter.
     This imbalance of knowledge between early and late types is
disastrous, for by far the greatest proportion of galaxies seen on
deep plates are late-type spirals which have small K-corrections
(Figure 3).

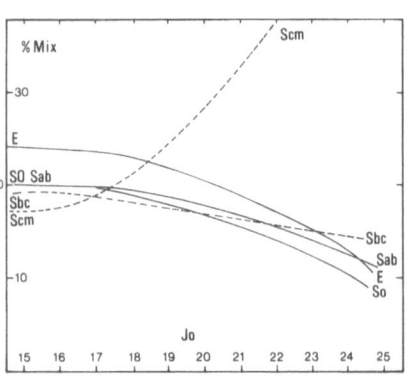

It is vital, therefore, to know their true mix and K-corrections as accurately as possible. A 50% perturbation on their true mix produces the dashed lines in Figure 2 thus drastically changing our interpretation of the data. The apparent mix would be modified in the reverse sense if there were evolution of the kind envisaged by Tinsley (1978) whereby early types were substantially brighter at moderate redshifts. Ironically such evolution cannot be confirmed until the no-evolution prediction is accurately defined.

Figure 3.   Apparent galaxy mix at various limiting J magnitudes.

The model presented together with the data in Figure 1 utilises the K-corrections of Pence (1976) and a galaxy mix consistent with the sample of Kirschner, Oemler and Schechter (1979) limited at 15.5 in their J system. The galaxy luminosity functions for each of 6 types are of the Schechter (1976) form adopting Felten's (1977) overall normalisation. If this model is correct, the data shows clear evidence for some luminosity evolution. On the pessimistic side, Kron (1979) has argued that technical problems such as systematic and random errors could mimic the effect of evolution by artificially steepening the count slope. In any case the above discussion makes further interpretation of Figure 1 rather pointless. What can perhaps be said (see Ellis 1979) is that the counts are unlikely to be consistent with the very strong evolutionary models where all galaxy types were substantially brighter in the past as a result of a rapidly declining star formation rate (Tinsley 1977).

On the optimistic side, some effort is now being directed towards filling in the missing details on galaxy properties. Far more effort is needed however. The literature is full of information on peculiar objects; we tend to forget the universe is populated primarily by normal galaxies about which we know very little indeed! Several groups are constructing larger redshift samples from which a better understanding of the galaxy luminosity function and its morphological variations should follow. Ultraviolet spectra of spirals with IUE will also soon be available. This will enable us to converge, albeit slowly, to the true K-corrections appropriate to the large samples on the deep plates.

C. GALAXY CORRELATIONS

Statistical measures of galaxy clustering are important not only in describing the distribution of visible matter but also because they can be compared with model predictions starting from various initial conditions (see Fall 1979 for a review).

The two-point correlation function $\xi(r)$ is defined as the excess probability of detecting two galaxies r Mpc apart; an equivalent angular function $W(\theta)$ can be defined for the projected distribution. The relationship between ξ and w involves geometrical projection (Limber 1953, Phillipps et al. 1978). In the gravitational instability picture of galaxy formation with isothermal (matter only) perturbations, $\xi(r)$ is expected to resemble two power-laws over the observable range (David, Groth and Peebles 1977). At small r where ξ is large the clusters are bound and do not evolve in proper space. At large r the function should fall off more rapidly. The junction of the two regimes defines a "feature" whose position is highly Ω-dependent. The feature should not be seen in the observable range $0.5 < r < 50\ h^{-1}$ Mpc unless $\Omega \sim 1$ (Efstathiou 1979).

The observed correlation function is indeed of this two power-law form. Groth and Peebles (1977) find evidence for such a feature in the distribution of the Shane-Wirtanen catalogue at an angle corresponding to $9\ h^{-1}$ Mpc. Its reality is not clear, however, because over such large angles the distribution may be affected by Galactic obscuration and plate-plate variations. The reality can be checked with deeper samples where the corresponding angle is much smaller.

Another important application of the correlations on deep plates is to check whether the amplitudes of the bound regime agree with those expected on the basis of local results. This scaling tests the uniqueness of ξ over large volumes of space. It might also be possible to monitor evolution in the clustering over these look-back times. Such evolution is expected to be small unless galaxies formed very recently, which is unlikely in view of the modest luminosity evolution detected in Figure 1.

Using 5 UKST plates forming two red (IIIaF) - blue (IIIaJ) pairs together with a third J plate, the galaxy distribution has been studied by Shanks et al. (1980) over a total area of $\sim 40\ deg^2$ to limits of J = 21.5 and R = 19.75. The angular correlation functions were fitted by power-laws of index -0.8 over the range $0.005 < \theta < 0.1$ deg. The amplitudes for the ensemble average J samples at 3 limiting magnitudes are compared with the corresponding amplitudes for shallower samples in Figure 4. The data is in good agreement with the virial prediction implying the locally derived distribution holds to very large depths ($\sim 700\ h^{-1}$ Mpc). This is convincing evidence for the isotropy and homogeneity of the Universe on very large scales. With substantially deeper data covering large areas, it might be possible to monitor the evolutionary trends modelled in Figure 4.

On larger scales the correlation functions for all 5 plates show good evidence for a feature (Figure 5). The functions drop rapidly beyond $\theta \sim 0.3$ deg. The effect is not due to noise as the scatter is small. The drop-off is not caused by the self-normalisation of $w(\theta)$ as can be shown from the number densities involved. In Shanks et al. (1980) we compare the entire ensemble-average functions with those for the Shane-Wirtanen and Zwicky catalogues, thereby checking the reproducibility of the feature at various depths. Internally in our data sets the feature is at the same spatial separation, $3\ h^{-1}$ Mpc, to within 30%. Its position is 3 times smaller than that for the S-W catalogue (though in reasonable agreement with that for Zwicky). The important point is that <u>all</u> analyses find the feature. Considering the problems in compiling the shallower catalogues, this discrepancy in position may not be too serious. A complete redshift sample of several hundred galaxies would be a convincing way to pinpoint its precise position.

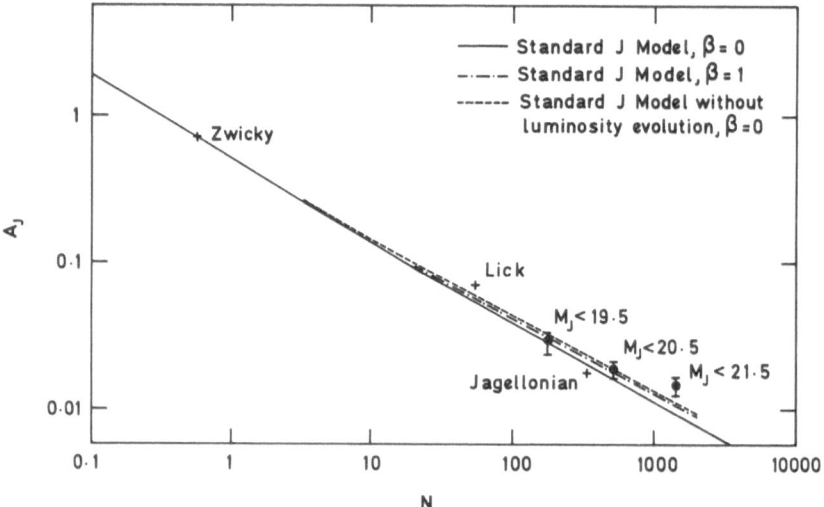

Figure 4. Correlation amplitudes appropriate to 1 degree versus galaxy density deg^{-2}. The solid line represents the expectation for virialised clusters.

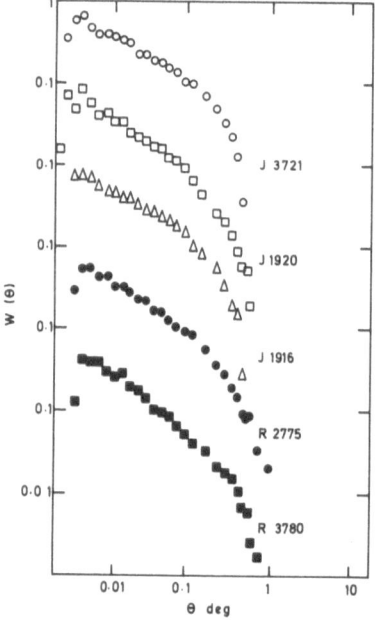

Figure 5.
Angular correlation functions for 5 UKST plates.

REFERENCES

Brown, G.S.; 1974, Ph.D. Thesis, University of Texas at Austin.
Code, A. and Welch, G.; 1979, Astrophys. J., 228, 95.
Davis, M., Groth, E.J. and Peebles, P.J.E.; 1977, Astrophys. J.
    212, L107.
Efstathiou, G.: 1979, M.N.R.A.S., 187, 117.
Ellis, R.S.: 1979, Proc. Roy. Sci., in press.
Fall, S.M.; 1979, Rev. Mod. Phys., 51, 21.
Felten, J.E.: 1977, Astron. J., 203, 39.
Groth, E.J. and Peebles, P.J.E.: 1977, Astrophys. J., 217, 385.
Kirschner, R.P., Oemler, A. and Schechter, P.L.: 1978, Astron. J.,
    83, 1549.
Kron, R.; 1979, Physica Scripta, in press.
Limber, D.N.; 1953, Astrophys. J., 117, 134.
Pence, W.; 1976, Astrophys. J., 203, 39.
Peterson, B.A., Ellis, R.S., Kibblewhite, E.J., Bridgeland, M.T., Hooley,
    T., and Horne, D.: 1979, Astrophys. J., in press.
Phillipps, S., Fong, R., Ellis, R.S., Fall, S.M. and MacGillivray, H.T.;
    1978, M.N.R.A.S., 182, 673.
Rainey, G.W.: 1977, Ph.D. Thesis, University of California at Los
    Angeles.
Schechter, P.L.; 1978, Astron. J., 82, 569.
Shanks, T., Fong, R., Ellis, R.S. and MacGillivray, H.T.: 1980,
    M.N.R.A.S., in press.
Spinrad, H., 1977, Evolution of galaxies and stellar populations,
    p301. (Yale).
Tinsley, B.M.: 1977, Astrophys. J., 211, 621.
Tinsley, B.M.: 1978, Astrophys. J., 220, 816.
Tyson, A. and Jarvis, J.: 1979, Astrophys. J., 230, L153.

DISCUSSION

Davis: How much extinction is indicated by the band of low galactic
density across the Schmidt plate?

Ellis: The band is seen on both J and R plates of this field and is
over 3 degrees in length and 0.5 degrees in width. The extinctions of 0.25 mag in J and 0.12 mag in R are consistent with that produced by an interstellar cloud. The deficiency of galaxies makes
other areas on the field look overdense on large scales. Consequently,
it is only after applying a moving-average filter to this distribution
that we obtain a feature at the position seen in the analysis of the
other plates.

Tyson: 1. Analysis of deep 4-meter plates by J. Jarvis and me shows
significant variation in galaxy number counts in different
high latitude areas, over and above that expected from photometry
errors and known clustering. These variations may be caused by high
latitude clumped extinction, and are similar to those just mentioned
by R. Ellis seen at brighter limiting magnitudes.

2.   We have obtained 2-point correlation functions for our
     11 fields (approximately 50,000 galaxies) and find prelimin-
ary evidence for less clustering at a limit of J = 24th magnitude.

Ellis: 1.   If extinctions of $\gtrsim$ 0.25 mag in J were common at high
            latitudes, it would indeed produce fluctuations of $\gtrsim$ 30% in
deep counts. Structure might not be seen in the smaller areas covered
by 4-meter plates.

           2.   We also analysed the correlations on our single 4-meter
                plate but found large discrepancies which we believe are
related to sampling problems rather than evolution (Ellis 1979: Proc.
Roy. Sci., in press).

Abell: The galaxies were identified with the COSMOS engine at Edin-
           burgh, were they not?

           Since you and Tyson both use J magnitudes, you should agree
           on the counts at the bright end (J $\approx$ 18); do your data also
show the bright excess reported by Tyson?

Ellis: The U.K. Schmidt counts were determined from COSMOS scans of
           three IIIa-J plates. Even with such a large area, the galax-
ies are few in number brighter than J = 17. Over the region 17 $\leq$ J $\leq$
21 becomes noisy at J $\sim$ 16, but I suspect that this relates to sampling problems or individual clusters in the KOS fields (as the authors
commented), rather than a very large supercluster such as Tyson suggested.

STATISTICAL STUDIES OF THE CLUSTERING OF GALAXIES

Edward J. Groth
Physics Department, Princeton University

An investigation of the clustering of galaxies making use of correlation function techniques is discussed. Correlation functions are defined and the relations between the angular and spatial functions are presented. The results obtained from application of the method to three samples of galaxies are described and some conclusions are drawn. Finally, some directions for future research are suggested.

1. INTRODUCTION

The study of the clustering of galaxies is an exciting and active field of research. Many complementary techniques from qualitative examination of photographs to quantitative analyses of measured clustering parameters are being brought to bear on the problems. In the limited space available only a small part of this research can be reviewed. In particular, this paper concentrates on the work done at Princeton in recent years. Clustering is studied with statistical methods making use of correlation function techniques. P.J.E. Peebles is the "prime mover" of this work and, building on the earlier work of Limber (1953) and Neyman, Scott, and Shane (1953), has summarized the mathematical details and assumptions used in the application of correlation function techniques (Peebles 1973). These techniques have been refined through many applications (see Groth and Peebles 1977 and earlier references therein) and can be applied to study the clustering of objects among themselves or the "cross clustering" of different classes of objects (e.g. the clustering of galaxies with radio sources has been investigated by Seldner and Peebles 1978). The clustering of galaxies with galaxies is the subject of this paper.

G.O. Abell and P. J. E. Peebles (eds.), Objects of High Redshift, 31–38.

2. CORRELATION FUNCTIONS

2.1. Definition

The angular two-point correlation function for a sample of objects distributed on the sky may be defined through the joint probability of simultaneously finding an object within solid angle $d\Omega_1$ and a second object within solid angle $d\Omega_2$ at angular distance θ from $d\Omega_1$,

$$dP = N^2[1+w(\theta)]d\Omega_1 d\Omega_2, \tag{1}$$

where N is the surface number density of objects in the sample and $w(\theta)$ is the angular two-point correlation function for separation θ. Given a sample of positions on the sky, one can estimate w; however the quantity of interest is the spatial two-point correlation function, $\xi(r)$, which is similarly defined through the joint probability of finding objects within volume elements dV_1 and dV_2 separated by distance r,

$$dP = n^2[1+\xi(r)]dV_1 dV_2, \tag{2}$$

where n is the volume number density of objects. Note that,

$$dP = n[1+\xi(r)]dV, \tag{3}$$

is the conditional probability of finding an object within dV at distance r from a given object.

The angular and spatial three-point correlation functions are defined through the joint probability of finding a triplet of objects,

$$dP = N^3[1+w(\theta_{12})+w(\theta_{23})+w(\theta_{31})+z(\theta_{12},\theta_{23},\theta_{31})]d\Omega_1 d\Omega_2 d\Omega_3, \tag{4}$$

$$dP = n^3[1+\xi(r_{12})+\xi(r_{23})+\xi(r_{31})+\zeta(r_{12},r_{23},r_{31})]dV_1 dV_2 dV_3, \tag{5}$$

where z and ζ are the angular and spatial three-point correlation functions, respectively.

The above definitions are rather dry and do not provide much insight into what the coorelation functions really measure. The following qualitative interpretations of equations (2) and (3) may help to provide a "feel" for the two-point correlation function. First, equation (2) is equivalent to the statement that ξ is the normalized auto-correlation function of the galaxy density distribution, ρ,

$$\xi = (\rho-\bar{\rho})*(\rho-\bar{\rho})/\bar{\rho}^2. \tag{6}$$

With equation (3), $\xi(r)$ may be interpreted as the average correlated or clustered density (in units of the mean density) of galaxies at distance r from an "average" galaxy. Finally, $\xi(r) \propto \delta\rho/\bar{\rho}$, the average density contrast in a cluster of size r around an "average" galaxy.

2.2. Relations Between the Angular and Spatial Functions

The relation between the angular and spatial two-point correlation functions was derived by Limber (1953); extensions to the relativistic case and relations for the three point functions are given by Groth and Peebles (1977). In brief, the angular functions can be written as integrals of the spatial functions along the line-of-sight,

$$Nd\Omega = \int(\text{Volume Element}) \times (\text{Comoving Density})$$
$$\times(\text{Probability of Seeing an Object}), \qquad (7)$$

$$N^2 w(\theta) d\Omega_1 d\Omega_2 = \int\int(\text{Volume 1}) \times (\text{Volume 2})$$
$$\times(\text{Density 1}) \times (\text{Density 2})$$
$$\times(\text{Probability 1}) \times (\text{Probability 2})$$
$$\times\xi(r_{12}), \qquad (8)$$

with a similar expression for the three-point functions. Note that the probability of seeing an object depends on the luminosity function as well as the method of construction of the sample. The usual procedure for estimating the spatial function given the angular function is to guess a spatial function, choose a luminosity function, plug them into the integral, and iterate until the resulting angular function agrees with that desired. In special cases, the evaluation of the integrals is straightforward; for example, a power law spatial function yields (at small angles) a power law angular function with the power law exponent increased by 1. Fall and Tremaine (1977) have presented a method of solving the integral equation (8).

2.3. Scaling of the Angular Functions with the Depth of the Sample

A characteristic depth, D, may be associated with a magnitude limited sample. For example, D may be the distance at which a galaxy of absolute magnitude M appears at the limiting magnitude of the sample. In the absence of relativistic effects, k corrections, and evolutionary effects, $D \propto \text{dex}(0.2m)$, where m is the limiting magnitude of the sample. A key check that the correlation functions are measuring properties of the galaxy distribution and not local effects is provided by the scaling law relating angular two-point functions for two samples of depths D_1 and D_2,

$$w_2(\theta) = (D_1/D_2)w_1[(D_2/D_1)\theta]. \qquad (9)$$

A similar scaling law applies to the three-point function, The factor D_1/D_2 accounts for the greater overlap of clusters along the line of sight as the depth of the sample increases while the factor D_2/D_1 accounts for the smaller angular size of a given physical structure as the depth of the sample increases. The small relativistic, k, and evolutionary corrections to the simple scaling law are discussed by Groth and Peebles (1977).

2.4. Advantages and Disadvantages of Correlation Function Techniques

Among the advantages of correlation function techniques are the following:
1. The method is straightforward to apply.
2. Only positional information is required.
3. The method is quantitative and in some cases fairly precise.
4. There are well defined relations connecting the angular and spatial functions.
5. The absence of significant local effects (e.g. variable obscuration) can be verified.
6. Correlation functions are easy to compare with theoretical predictions.
Some disadvantages are:
1. A large sample is usually required.
2. The method is insensitive to some features of the galaxy distribution (e.g. holes, filaments, sheets).
3. No specific information concerning any particular cluster or galaxy is obtained.

3. GALAXY CLUSTERING

3.1. Catalogs

Angular correlation functions have been estimated for three magnitude limited samples of galaxies: the Zwicky catalog (Zwicky *et al.* 1961-68), the Lick Catalog (Shane and Wirtanan 1967, Seldner *et al.* 1977), and the Jagellonian field (Rudnicki *et al.* 1973). Properties of these catalogs are summarized in the following table.

Table 1   Catalogs Analysed

Sample	Limiting Magnitude	Resolution Arcmin	Solid Angle Square Degrees	Typical Redshift	Galaxies
Zwicky	14.9	~2	~6000	0.016	~3700
Lick	18.6	10	~11000	0.07	~590000
Jagellonian	20.3	~3	36	0.13	~12000

3.2. Results

The angular two-point correlation functions for the three samples are shown in Figure 1a. When the Lick and Jagellonian samples are scaled (eq. [9]) to the depth of the Zwicky sample, the curves appear as in Figure 1b. From these data and other data on the

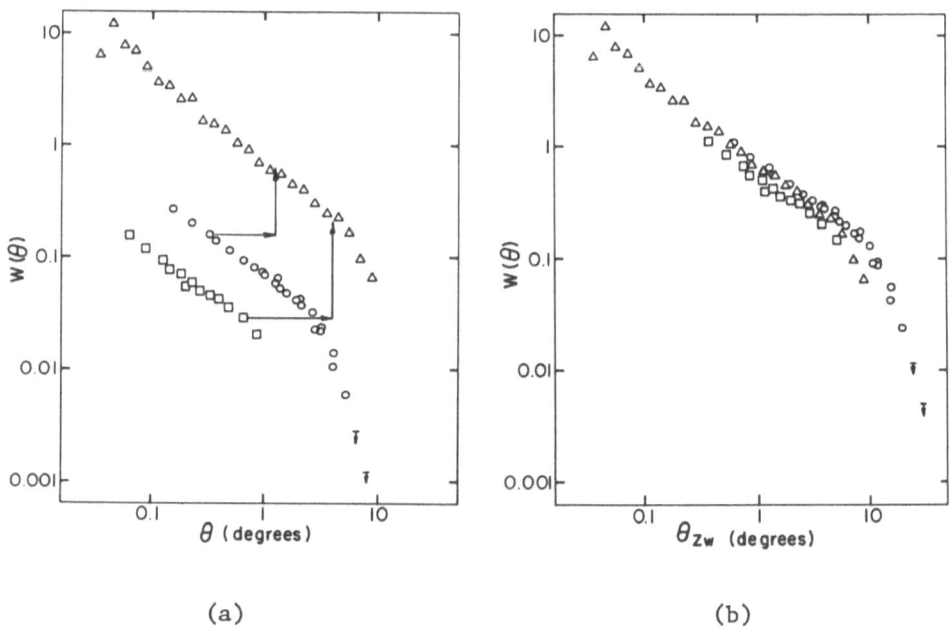

(a)   (b)

Figure 1. (a). The angular two-point correlation functions for the Zwicky (triangles), Lick (circles), and Jagellonian (squares) samples. (b). Same as (a) except the Lick and Jagellonian samples have been scaled to the depth of the Zwicky sample.

three-point function, the following results have been obtained:
  1. The agreement between the scaled functions indicates that spurious correlations due to variable obscuration are not a problem with these data.
  2. Over the range 0.05 Mpc $\leq$ hr $\leq$ 9Mpc the two-point correlation function is well represented by the power law

$$\xi(r) = (r_0/r)^{\gamma}, \quad \gamma = 1.77 \pm 0.04, \quad hr_0 = 4.7\text{Mpc}, \tag{10}$$

where h is Hubble's constant in units of 100 km s^{-1} Mpc^{-1} and r_0 is uncertain by about 50% due to uncertainties in the luminosity function.

3. At hr ~ 9Mpc, the correlation function has a "break". For larger distances, it falls much more rapidly than the power law in equation (10).

4. The three-point correlation function is well represented by the model

$$\zeta(r_1, r_2, r_3) = Q[\xi(r_1)\xi(r_2) + \xi(r_2)\xi(r_3) + \xi(r_3)\xi(r_1)],\tag{11}$$

where $Q = 1.29 \pm 0.21$.

3.3. Inferences

From the results presented above, a number of (sometimes controversial) inferences have been drawn:

1. The power law (eq. [10]) may indicate that the mechanism of cluster formation has no preferred scales. One such mechanism is the "gravitational instability picture" (Peebles 1974).

2. The form of the three-point function (eq. [11]) suggests that galaxies are distributed in a continuous clustering hierarchy (Soneira and Peebles 1978, Groth *et al.* 1977).

3. The gravitational instability picture, assuming white noise density fluctuations at recombination, predicts a power law exponent of 1.8, in good agreement with the observed index (Peebles 1974).

4. The break in the two-point function can be interpreted as the transition between linear (longward of the break) and non-linear growth of density fluctuations. The fact that $\xi \sim 0.3$ at the break may indicate that the density is close to the closure density (Davis, Groth, and Peebles 1977).

4. FUTURE DIRECTIONS

Although much has been learned via correlation function techniques, future applications should continue to add substantially to our understanding of galaxy clustering. An exciting development will be the construction of a redshift sample complete to ~15th magnitude and covering a substantial portion of the galactic polar cap. Such a catalog will allow the determination of the galaxy two-point function in both position *and velocity* space.

However, if the discussion is restricted to what can be learned from positional information only, then deeper samples are required. Questions which might be studied with deeper samples are:

1. Is the Universe statistically homogeneous and isotropic?

2. Does the power law (eq. [10]) extend all the way down to galactic dimensions, or is there a discontinuity or other feature at small scales (Peebles 1974)?

3. Is the break (seen mainly in the Lick sample) real or only an artifact of the analysis?

4. How has galaxy clustering (as measured by the correlation functions) evolved? Are clusters still growing?

Since the angular correlation functions decrease with the depth of the sample due to overlapping of clusters along the line of sight, the correlation functions for a sample significantly deeper than those listed in Table 1 will be very small. This means that extension of correlation function techniques to deeper samples will be difficult and will require great care in the construction of the sample. Nevertheless, the questions to be addressed are so interesting that several investigators have already begun to construct the required samples. Kron (1978) and Tyson and Jarvis (1979) have generated several deep samples using 4m prime focus plates. Phillipps *et al.* (1978, see also Ellis *et al.* 1977) are using 1.2m Schmidt plates. In the past year, I have obtained 4m prime focus plates for two fields of approximately 1°×10°. The sample to be constructed from these plates will be especially useful for verifying the existence of the break.

With ground based photography (to obtain the large field of view required) one may perhaps construct a sample with typical redshift ~0.4. With the Space Telescope, the typical redshift will be ~1. At this redshift the angular size of the break will be ~0°.25, so the small field of view of the ST should not be a major limitation. Thus, the statistical study of galaxy clustering at cosmologically significant distances promises to be an exciting and active field of research in the coming decade.

ACKNOWLEDGMENT

This research was supported in part by the National Science Foundation.

REFERENCES

Davis, M., Groth, E. J., and Peebles, P. J. E.: 1977, *Ap. J. Letters*, 212, p. L107.
Ellis, R. S., Fong, R., and Phillipps, S.: 1977, *M.N.R.A.S.*, 181, p. 163.
Fall, S. M., and Tremaine, S.: 1977, *Ap. J.*, 216, p. 682.
Groth, E. J., and Peebles, P. J. E.: 1977, *Ap. J.*, 217, p. 385.
Groth, E. J., Peebles, P. J. E., Seldner, M., and Soneira, R. M.: 1977, *Scientific American*, 237, No. 5, p. 76.
Kron, R. G.: 1978, Ph.D. Thesis, University of California.
Limber, D. N.: 1953, *Ap. J.*, 117, p. 134.
Neyman, J., Scott, E. L., and Shane, C. D.: 1953, *Ap. J.*, 117, p. 92.
Peebles, P. J. E.: 1973, *Ap. J.*, 185, p. 413.

Peebles, P. J. E.: 1974, *Ap. J. Letters*, 189, p. L51.

Phillipps, S., Fong, R., Ellis, R. S., Fall, S. M., and MacGillivray, H. T.: 1978, *M.N.R.A.S.*, 182, p. 673.

Rudnicki, K., Dworak, T. Z., Flin, P., Baranowski, B., and Sendrakowski, A.: 1973, *Acta Cosmologica*, 1, p. 7.

Seldner, M., and Peebles, P. J. E.: 1978, *Ap. J.*, 225, p. 7.

Seldner, M., Siebers, B., Groth, E. J., and Peebles, P. J. E.: 1977, A. J., 82, p. 249.

Shane, C. D., and Wirtanen, C. A.: 1967, *Pub. Lick Obs.*, vol. 22, Part 1.

Soneira, R. M., and Peebles, P. J. E.: 1978, A. J., 83, p. 845.

Tyson, J. A., and Jarvis, J. F.: 1979, *Ap. J.*, 230, p. L153.

Zwicky, F., Herzog, E., Wild, P., Karpowicz, M., and Kowal, C. T.: 1961-1968, *Catalogue of Galaxies and Clusters of Galaxies*, in 6 vols. (Pasadena: California Institute of Technology).

DISCUSSION

Rees: Do you think that galaxy correlation studies will eventually be able to detect "linear" fluctuations ($\lesssim 10\%$, say) on scales greater than 100 Mpc; or can this be done better by studying the distribution of rarer classes of objects such as X-ray sources or radio sources?

Groth: I can't give a definite answer to this question. However, to set the scale, note that the typical depth of the Lick sample is ~ 200 h^{-1} Mpc. My guess is that it will be very difficult to get at clustering on scales of 100 Mpc with correlation studies of galaxies.

Tyson: Recently Kirshner, Oemler, and Schechter (Astron. J. 83, 1978) have obtained data on magnitudes, positions, and redshifts for a complete sample of galaxies to 16th magnitude and find no evidence for a drop-off at 9 Mpc. Would you please comment on this?

Groth: The KOS sample includes positions and redshifts. The redshifts were used to obtain positions along the line of sight. Thus, KOS had a sample of three-dimensional positions from which they could estimate ξ directly, rather than going through the calculation of the angular function. This has the advantage that ξ, having a much larger amplitude than w, is much easier to estimate from a small sample. On the other hand, this method requires that peculiar velocities be negligible. For example, if "field" galaxies, such as those in the KOS sample, have peculiar velocities of the order of a few hundred km s^{-1}, then the estimated ξ would be smeared out over several Mpc.

I believe that the proper way to treat a position-redshift sample is not to convert velocities to distance, but to try to estimate the correlation function in both position and velocity space, ξ (r, v). This will allow the study of dynamics of clusters as well as the shapes of clusters.

SPECTROSCOPY AND PHOTOMETRY OF FAINT GALAXIES:
HINTS AT THEIR EVOLUTION

Hyron Spinrad
Dept. of Astronomy
University of California, Berkeley

ABSTRACT

I discuss four optical methods to locate standard candles (giant
E galaxies), which should eventually lead us to reasonably large samples
at high z.   Redshift criteria and determinations are briefly discussed.
The observed faint galaxy colors fit well into a simple model of stellar
(galactic) evolution for systems with star-formation confined to a short
time interval. Finally a summary of the statistics of blue galaxies in
Coma-like clusters (Butcher-Oemler effect) is presented and interpreted.

I. A BEGINNING

The motivating force which drives some observers to the long and
tedious observation of faint galaxies, is the dual hope of obtaining
some understanding of galaxy evolution and, of course, reaching for the
"cosmological grail."

To begin a modern attack on the problem we need methods to select
our standard candles. We have to get a reasonably large sample at high
z. To push into the unknown world, at z > 0.7 is not easy, and time
spent deciding on selection criteria may be well-spent.

I can think of four ways to accomplish a selection of giant E ga-
laxies--the canonical standard candle. Some of the selective criteria
may be successful, and all are worth examination, as they lead to different biases. They are:

1) Use of the classic ensemble object--the brightest galaxy in a
moderately-populous cluster, discovered optically. Of course, at
z > 0.55 we begin to have discovery problems, as the clusters begin to
dip beyond the range of large Schmidt cameras with even IIIa-J/IIIa-F
emulsions. One may have to locate them at random on deep reflector
plates (<1/^{o} sq. per plate); unfortunately extra-optical techniques to
find these clusters have not yet paid off. Figure underline{1} illustrates a rich

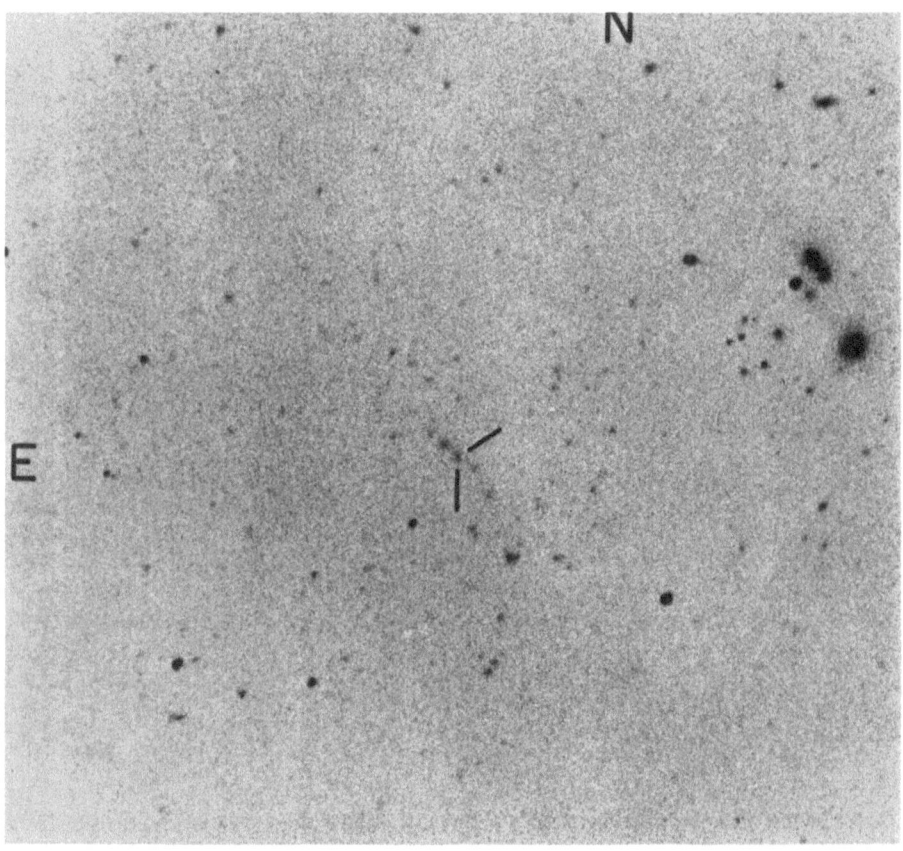

Figure 1. A print from a 4-m red photograph of the rich cluster
           0015.9 + 1610, at z = 0.541. The 1st-ranked E galaxy
           is marked.

galaxy cluster in SA 68 found by Richard Kron. The cluster redshift is z = 0.541 and the brightest galaxy has a red (F) magnitude of 20.4. Locating random clusters at z > 0.7 will be difficult from the ground, even with a CCD array detector. From ST we should do better on red clusters by observing near 1μ. The blue, very faint clusters are still a mystery!

2) We can also use radio galaxies as our probes to large distance. Hine and Longair (1979) show that over half the powerful 3CR classic double-lobed sources identified with galaxies have optical emission lines, which makes their redshift determination much easier. However, the non-thermal nuclear activity can make them optically untrustworthy in some cases, especially at emitted UV wavelengths (important observationally at large z) (c.f. Smith et al., 1979), so radio-galaxies are risky, albeit exciting candidates to include. Very few radio galaxies are underluminous (there is now one strange exception, 3C 258).

3) A new way to find standard candle ellipticals may be to look for very red galaxies, even in isolation. At V > 19 and color (J-F) $\geq 2^m0$ all galaxies should be giant E systems, $0.4 < z \leq 0.9$. This is because of a large K-correction in the J magnitude, the steeper ultraviolet continuum in the most luminous E's (Visvanathan and Sandage, 1977) and the modest color-evolution anticipated for them. These factors conspire to make luminous E galaxies in the mentioned red-shift range very red in (J-F) (Kron system). Of course, the high-z boundary is model-dependent.

A pilot program to obtain redshifts for several color-selected faint galaxies has begun at Lick Observatory. So far it seems promising, but again, somewhat demanding of spectroscopic observing time, as these red galaxies almost never show emission lines. Table 1 summarizes the start of this test; the magnitudes listed are all isophotal (to 26th in μ_r). The color-selected galaxies are almost, but not quite, up to the first-ranked cluster galaxy luminosity, which is about $M_V = -23.5$ (for $H_0 = 50$).

TABLE 1. Red-Color Galaxies - First Comparison

Galaxy Name	Associates?	F_{is}^{26}	V_{is}	z	ΔM_V^* (mag)
SA 57, R3	NO	18.9	19.9	0.330	+ 0.7
Herc, 1718+50,R1	A few	19.8	20.8	0.450	+ 0.6
3C 295, g1	Rich cluster	18.4	19.4	0.461	- 0.8

*Compared to ridge line of Kristian et al. (1978)

4) Finally, I suggest a new method using the reddest galaxies in clusters associated with quasars. Here we employ again the fact that E/S0 galaxies around $0.4 < z < 0.9$ have a special color signature, and

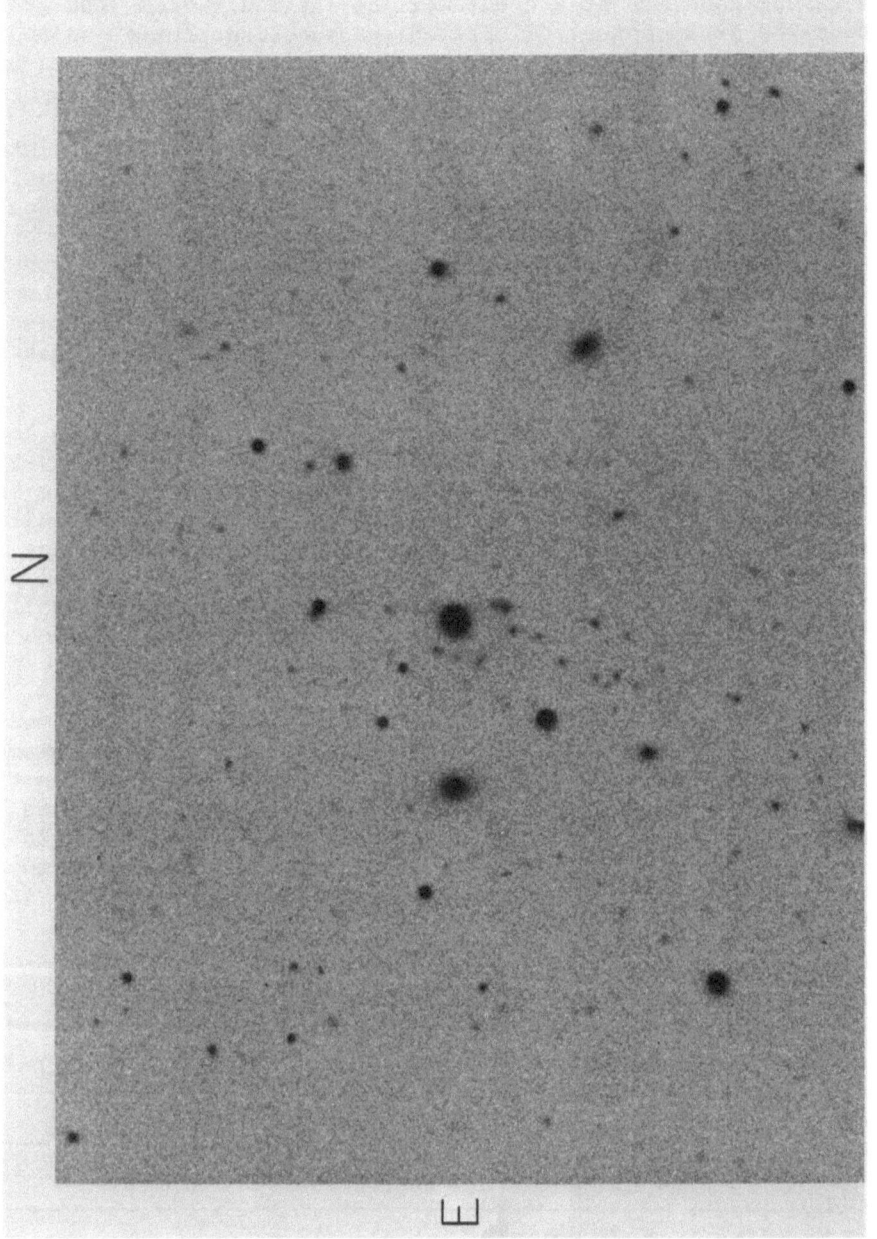

Figure 2.   A 4-m print of the vicinity of the quasar PKS 0405-123
(z = .57).   There is a cluster of faint, red galaxies extending off to the South – almost certainly physically associated with the QSO.

that very few random field objects are this red (Kron, 1978). Obviously a search for new quasar-clusters is in order, at z_e > 0.4. We know
from lower redshift quasars that normal galaxies are occasionally-to-
often in concert (Gunn 1971, Stockton 1978, Wehinger and Wyckoff, 1978).
Some more distant ones, for example PKS 0405-123 (z = 0.574) also have
apparent clusters and in this case (see Fig. 2) about half of the nearest
faint galaxies have the colors demanded for E-galaxy membership. A test
program of this selection procedure, with some controls, is planned for
Cerro Tololo in January. This particular test brings forth a doctri-
naire statement: we cannot let our former prejudices rob us of an opportunity to study the galaxies near quasars, from the ground or from
ST.

II. REDSHIFT CRITERIA AND DETERMINATIONS FOR LARGE-z GALAXIES

Besides the long integrations required to determine galaxy redshifts even with large reflectors, simply settling upon redshift criteria presents new problems at z > 1.0. The familiar λ_0 4000 break
moves far to the red, into the "OH-jungle."

We now can, with the helpf of the IUE satellite ultra-violet spectra
of stars and galaxies (c.f. Johnson 1979), spectrophotometry of two galaxies (z ∿ .5) to emitted λ_0 2500, and G. Bruzual's evolutionary predictions, make reasonable quantitative estimates about the UV spectra
of normal large galaxies. In the case of a gas-poor E or SO, seen
in its relative youth (∿ 8 billion years ago), we can anticipate the
G-star discontinuities at $\lambda\lambda_0$ 2900, 2640, and 2420 (in order of λ and
importance). For galaxies with hotter stars (spirals) the situation is
less clear, and I am not sure if any Si-edge near λ_0 1600 will survive
in the integral spectrum of an Sc galaxy. Probably the Lyman discontinuity at λ_0 912 will be a useful redshift criterion from ST (at
z ∿ 0.3), and conceivably even from the ground if primeval galaxies,
z ∿ 5, can be found.

A specific example of a redshift determined from the ultraviolet
is the case I make for 3C 427.1. This faint radio galaxy has no emission lines, but does show the $\lambda\lambda_0$ 2900, 2640 edges, yielding z = 1.172.

Emission-line galaxies at large-z could be more difficult than
anticipated, too. The emitted ultraviolet doesn't have strong emission
lines in HII region spectra; perhaps CIII] λ_0 1909 is the best feature.
It will not be an easily applied criterion with the inevitably poor
S/N ratios obtained for faint galaxy spectra. The situation is very
different from that which obtains in Seyfert nuclei and in QSO's which
do have strong ultraviolet emission lines. In the latter case(s) the
ionizing radiation is derived from a "hard" power-law spectrum extending
far to the UV.

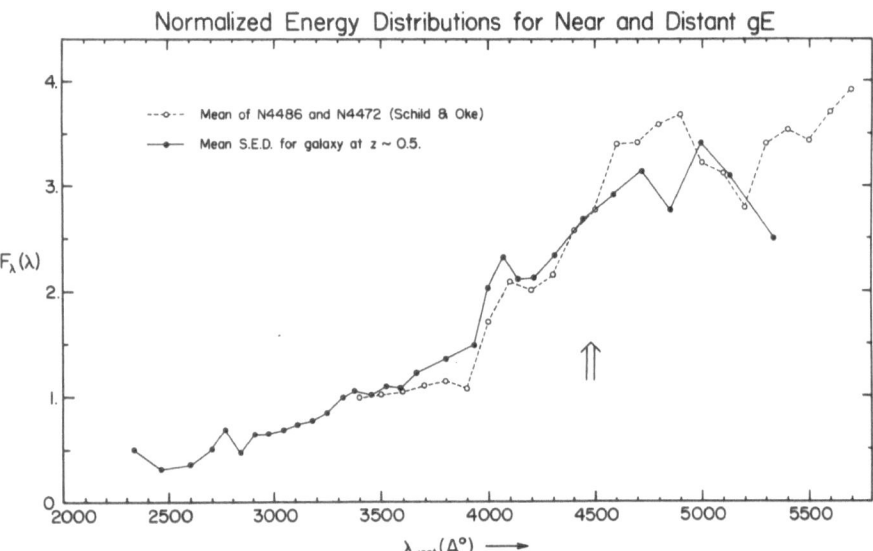

Figure 3. Normalized energy distribution for nearby (Virgo) and distant
(z = 0.5) galaxies. The distant E systems (an average of 3C
295, PKS 0400-64.3, and 3C 330, gal. 2) are somewhat bluer
in their rest-frame spectra.

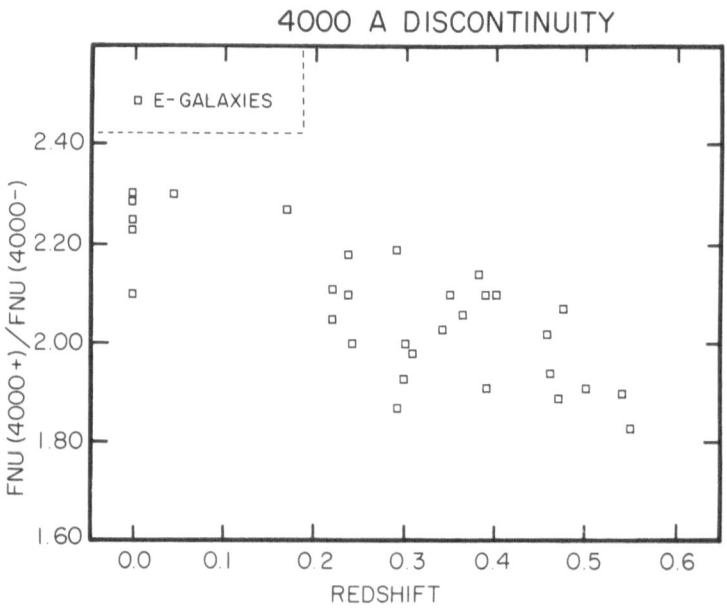

Figure 4. The decline of the amplitude of the $\lambda_0 4000$ break with redshift. Only "Thermal" E galaxy spectra have been used.

III. PREDICTIONS FOR EVOLVING MODEL GALAXIES

As part of his ongoing UCB thesis, Gustavo Bruzual has given me permission to mention a very few results pertaining to galaxies at high-z. The Bruzual models, patterned ater the assumptions made common by Tinsley, Larson, and others, make predictions of integral galaxy spectra and derive observables-magnitudes and colors, over the lookback times available to present and prospective future observation.

While these models have no claim to complete uniqueness in a domain where stellar abundance, age, mass-spectrum, and galaxy dynamics are all intertwined, a simple-minded interpretation can still be useful. With standard cosmological parameters of H_0 = 50 km^{-1} Mpc^{-1} and $q_0 \simeq 0.0$, galaxies are (arbitrarily) formed with a burst of star-formation 16 Gyrs ago. Another parameter describes the duration (exponentiation-time) of the star-forming-burst. A variety of observed E galaxies, $0 < z \leq 1.0$, fit the derived model colors over a wide wavelength interval, if the exponentiating decay time is roughly 1-2 Gyrs, a few galaxy collapse times. The last epoch of star-formation then occurred near z = 3. The observed $(J-F)_{max} < 2.3$ red color limit precludes a rapid single burst of stars at z > 4 with no consequent "dribbles" of young stars, in any type galaxy considered.

IV. OBSERVED SPECTRAL EVOLUTION OF GIANT ELLIPTICALS

An associated problem, detailed spectral evolution of elliptical galaxies, has a considerable literature. I reviewed it at Yale (Spinrad 1977); now the data is better, and we can claim that a positve change is detectable. Figure 3 shows the normalized (λ_0 4500) energy distributions of two nearby giant E's in Virgo, compared to the mean (rest-frame) s.e.d.'s for three galaxies at $\bar{z}$ = .5. The distant (younger) galaxies are bluer at proper $\lambda_0 < 4300$, but only by about 0^{m}15 in (B-V). Soon IUE data will extend the comparison to λ_0 2400.

A more telling illustration of elliptical galaxy spectral evolution comes from data on the amplitude of the λ_0 4000 discontinuity. Figure 4 shows the break plotted vs. redshift; the steep decline in its amplitude is simply due to an increasing proportion of the light arising in turnoff stars of types F8-G2V, as we look at progressively "younger" E galaxies. The scatter is partly observational and apparently partly cosmic. Bruzual's models, with some effort, can model this evolutionary change rather well. Some differential model tests are possible if any high S/N data on the λ_0 4000 break become available for cluster E's at z $\sim$ 1.0.

V. DYNAMICAL EVOLUTION OF GALAXIES IN CLUSTERS

Butcher and Oemler's (1978) discovery of substantial numbers of blue galaxies in concentrated, rich clusters has provided an exciting opportunity to observe the time-variation of the spiral content of Coma-like clusters--perhaps their residual dynamic collapse. Harvey

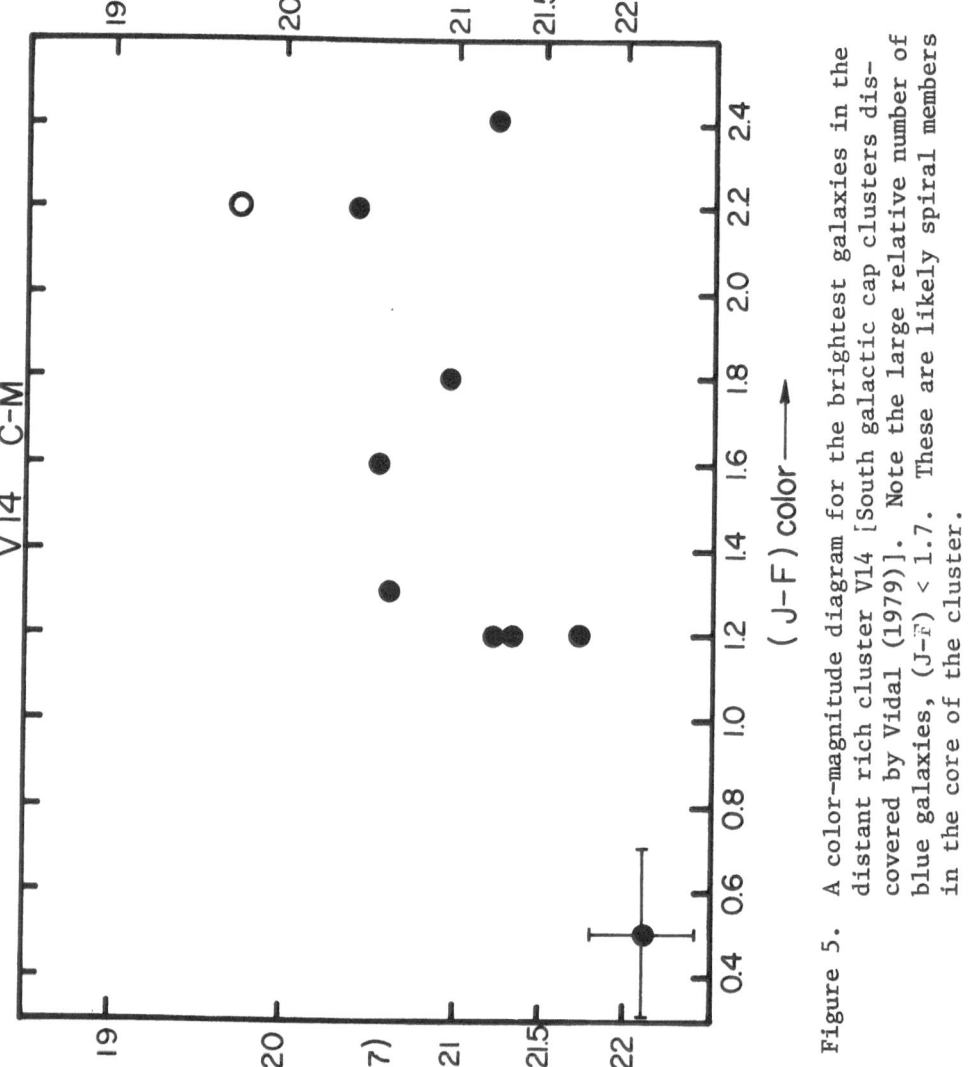

Figure 5.   A color-magnitude diagram for the brightest galaxies in the
distant rich cluster V14 [South galactic cap clusters discovered by Vidal (1979)].   Note the large relative number of
blue galaxies, (J−F) < 1.7.   These are likely spiral members
in the core of the cluster.

Butcher will discuss the KP work later in this volume, so my brief review will be encapsulated into a table which lists the fraction of (hopefully!) cluster members thought to be actively-star-forming (blue) spirals. There is a substantial trend with redshift.

Figure 5 shows a typical distant cluster C-M diagram from Berkeley PDS-photometry. V14 has a redshift of approximately 0.5.

TABLE 2.   Fraction of Luminous Cluster Galaxies at Least $0.^{m}4$ Bluer Than Giant Ellipticals (> Sb)

Cluster	z	% Blue[*]	Notes
Coma	0.02	5	Standard concentrated cluster.
II 2w 1305.4	0.24	13	Blue field galaxies removed.
Cor 0404	0.30	34	Slightly less regular morphol.
0949 + 44	0.38	33	Gunn-Oke cluster.
0024 + 16	0.39	27	BO cluster
3C 295	0.46	33	BO cluster, Berkeley photom.
V14	0.50±	55	Redshift from sizes & magns.
1305 + 2952	0.95:	55	Redshift not yet confirmed.

[*] In top 3^{m} of V luminosity function.

A simple interpretation of this trend comes from a recent preprint by Larson, Tinsley, and Caldwell (1979), who suggest stripping of a spiral's gas-rich envelope occurs during a cluster's first collapse, and the subsequent decay of the SFR then produces present-day SO's. The process takes roughly 4 Gyrs, so a systematic trend in the spiral galaxy content in condensed clusters (0.3 < z < 0.6) is a natural consequence. We should then never see clusters with a larger fraction of blue galaxies then the local field percentage of late-type spirals (∿60%).

Parenthetically, I should add, that at z > 1.5, all galaxies may appear blue, no matter what the destiny of their morphologies. Could Kron's blue clusters fit into the scheme in such a manner? If so, they could be very useful future probes of cosmology and evolution. I would like to acknowledge many useful discussions with Gustavo Bruzual, David Koo, Rich Kron, Ivan King, and Bea Tinsley. I again thank the NSF for their support.

REFERENCES

Butcher, H. and Oemler, A. 1978, Ap. J. 219, pp. 18–30.
Gunn, J. E. 1971, Ap. J. (Letters) 164, pp. L 113–118.
Hine, R. G., and Longair, M. S. 1979, M. N. 188, pp. 111–130.
Johnson, H. M. 1979, Ap. J. (Letters) 230, pp. L 131–136.
Kristian, J., Sandage, A. R. and Westphal, J. A. 1978, Ap. J.
      221, pp. 383–394.
Kron, R., Ph.D. Thesis, University of California, Berkeley.
Smith, H. E., Junkkarinen, V. T., Spinrad, H., Grueff, G., and
      Vigotti, M. 1979, Ap. J. 231, pp. 307–311.
Spinrad, H. 1977, Yale Conference on "The Evolution of Galaxies
      and Stellar Populations," pp. 301–338.
Stockton, A. 1978, Ap. J. 223, pp. 747–757.
Vidal, N. 1979, preprint.
Visvanathan, N., and Sandage, A. R. 1977, Ap. J. 216, pp. 214–226.
Wehinger, D., and Wyckoff, S. 1978, M. M. 184, pp. 335–340.

DISCUSSION

G. Burbidge: How sure are you of the redshifts of the three galaxies
                that you quoted in the beginning of your talk? In the
case of the galaxies around the QSO with a redshift of 0.57, will this
influence you when you attempt to measure the galaxy redshift?

Spinrad: 1. The 3 very red galaxies with $z = 0.33$, 0.48 and 0.53
                seem to have quite decent redshifts; the spectrum for the
2 brighter ones are respectable. The criteria for z were conventional,
the λ_0 4000 break and the inflexion at the G-band.

                2. I would think the QSO at $z = 0.57$ would influence the
                coice of grating tilt used for eventual spectroscopic
tests on potential cluster galaxies. I also expect to report the results
with honesty, no matter what the z-residuals are.

PHOTOMETRY OF REMOTE GALAXY CLUSTERS

H. Butcher†, A. Oemler*, and D. Wells†
†Kitt Peak National Observatory
*Yale University Observatory

ABSTRACT

   New data on the colors of galaxies in distant, rich clusters of
galaxies are presented. Of seven clusters examined having z>0.20, all
exhibit color distributions indicating that a significant proportion
of the member galaxies are undergoing star formation. The one cluster
studied with z<0.20 clearly shows a reduced fraction of such galaxies.

I. INTRODUCTION

   Several years ago, two-color photometry of the member galaxies
in two rich, centrally concentrated clusters at $z \sim 0.4$ indicated the
presence of an anomalously large population of blue objects (Butcher
and Oemler 1978a,b). The natural inference was that these blue objects
are galaxies in the process of forming stars, and that probably we are
observing <u>spiral</u> galaxies before they somehow lost their interstellar
matter and turned into SO's. If this hypothesis is correct, then we
are witnessing a rather great deal of evolution of galactic morphological types in clusters in the very recent cosmological past--so
recent, in fact, as to be entirely unexpected. If these two clusters
turn out to be representative of all concentrated, relaxed clusters at
such redshifts, then we will need to confront the fact that all or
nearly all of the giant clusters changed in an important way rather
abruptly some 2-3 billion years ago. For if cluster spirals stopped
forming stars any later, we would find today that SO's are distinguish-
able from ellipticals by their colors, which they are not (Sandage and
Visvanathan 1978), and if they did so much earlier, then these two
clusters must be quite unrepresentative for some reason.

   To address this question of how representative the first two
clusters really are, and at the same time to see if we might be able
to study the conversion process as it actually occurs, we have begun
a program of photometry of additional remote clusters. To date, we
have obtained two-color data on nearly two dozen distant ($z \gtrsim 0.2$), rich
(richness class $\gtrsim 3$) clusters for the analysis, and at present, we are

G.O. Abell and P. J. E. Peebles (eds.), Objects of High Redshift, 49–55.
Copyright © 1980 by the IAU.

in the process of examining this material. While not all the clusters, for one reason or another (but mostly because of background and fore-ground contamination problems), are found to be suitable for the study, we should soon be able to say something concrete about the galaxy cluster population between about 2 and 5 billion years ago. The purpose of the present report is to give some of the early, preliminary results of this study.

II. OBSERVATIONS AND REDUCTIONS

Recall that our technique is the very simple one of constructing the distribution of broad band colors for all galaxies in the core regions of the clusters. We then identify the peak in this distribu-tion due to the cluster members with classical elliptical galaxy-type colors, and attempt to use the size of this peak to estimate the frac-tion of spiral galaxies in the cluster. Note that we are unabashedly injecting into the discussion our biases concerning the nature of the blue objects. It is not yet certain that they are spirals, but we are supposing that they are, and at least for the moment, we will continue to refer to them as such.

Specifically, the data mostly comprise either ISIT vidicon obser-vations in the V and R passbands, or IIIaJ+GG385 and IIIaF+RG610 4-m prime focus plates. The vidicon data have been reduced by the methods described in Butcher and Oemler (1978a), and the photographic data by a new, automatic computer code which detects and measures all objects within a specified area.

We have tried to pay close attention to several effects which might compromise our final results. First, we have used metric dia-meters for both our measurement apertures and the radii in the clusters for inclusion in the final color distribution determination. Specifi-cally, we have assumed $q_0 = 0.05$, and have used an aperture of 6 arcsec diameter and cluster radius of 1.24 arcmin at z = 0.40, and then scaled these quantities to the redshift of each cluster before measurement. Similarly, we have assumed that our completeness limit occurs with R = constant at z = 0.40, and then varied the completeness limit as a function of color at each redshift in such a manner as to account approximately for the known effects of differing K-corrections for the different types and colors of galaxies. The depth in the luminosity function to which we go has not been maintained precisely constant, but in nearly all cases it is close to 4 magnitudes below the brightest cluster member. Finally, in our estimate of the fraction of classical E and S0 type galaxies in each cluster (and, hence, also the fraction of spirals), we have tried to account approximately for the effects of the color vs. magnitude relation for nearby E/So's, as given by Visvanathan and Sandage (1977).

III.   RESULTS AND DISCUSSION

As for the clusters discussed in Butcher and Oemler (1978b), our final results for each cluster consist of an estimate of the percentage of spirals in the cluster, and a concentration index to characterize the dynamical state of the cluster.   The latter quantity is defined as the logarithm of the ratio of two projected radii: that containing 60% of all cluster members, and that containing 20%.   Its definition is more or less arbitrary, but it does seem to quantify the well-known relation between cluster concentration and spiral galaxy content in the nearby clusters.   For reference, a uniformly filled sphere has an index of 0.26.

The results available at this time are given in the table and in Figure 1.   The filled circles in the Figure are for the sample of nearby clusters discussed by Butcher and Oemler (1978b), and the open circles containing redshift values are for the sample in the table.

TABLE 1.   Results for 8 Remote Clusters

Cluster	Redshift	Concentration	% Spirals
A2218	0.18	0.53	30
A2111	0.23	0.38	53
A2645	0.25	0.32	58
A1758	0.28	0.48	61
A370	0.37	0.44	59
Cℓ1446+26	0.38	0.30	64
Cℓ0024+16	0.39	0.50	50
3C295	0.46	0.58	59

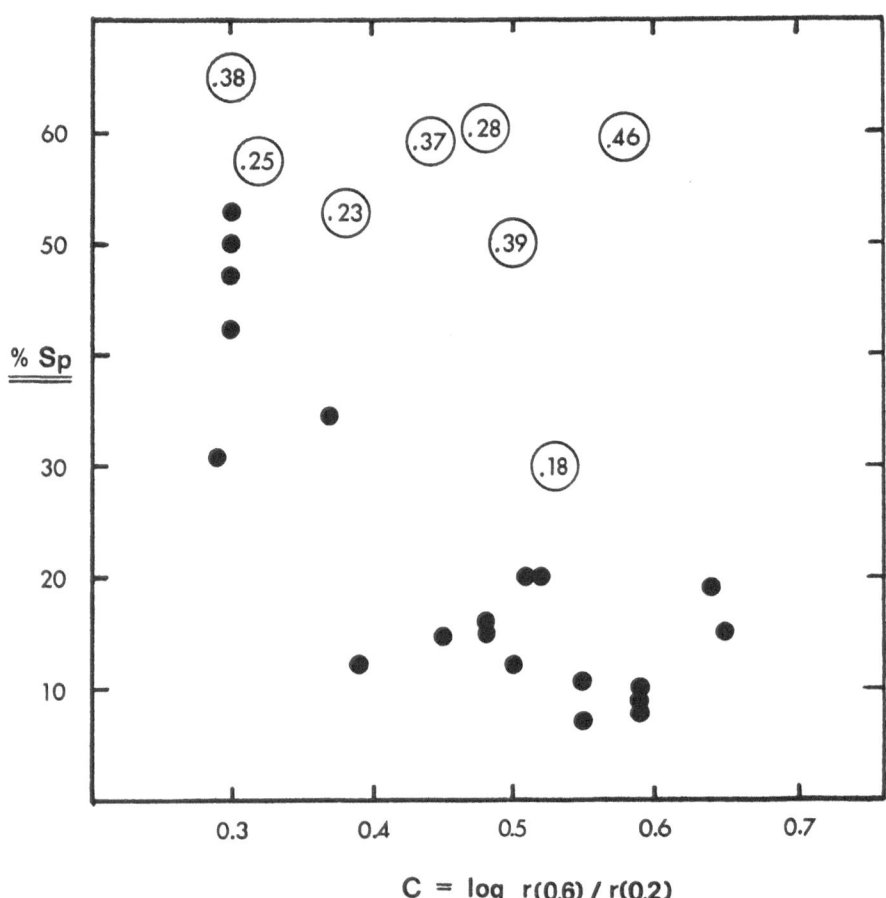

Figure 1.   Spiral content vs. cluster concentration, for
            nearby clusters (filled circles) and the present
            sample of remote clusters (open circles with
            redshifts indicated).

It is immediately evident that seven of the distant clusters have
essentially the same spiral content, some 50-60%, regardless of their
concentration indices. The sole exception, Abell 2218, has the lowest
redshift in the sample, and would seem to lie closer to the sample of
nearby, concentrated clusters than to the set of distant ones.

Apparently, then, it is not uncommon in giant clusters with
$z \gtrsim 0.2$ for there to be a significant population of blue cluster members
independent of cluster morphology. There also already seems to be
some suggestion that the observed quenching of star formation may be
a phenomenon related not only to the dynamical age of the clusters,
but also to the particular cosmological epoch near $z \approx 0.2$.

These new results also provide some hope that the detailed shapes of the various color distributions can provide us with clues to the progress of the transformation in each cluster at the time of observation. We find, for example, that the color distributions in Cℓ0024+16, A2645, and A370 are characterized by an elliptical color peak and a blueward tail stretching all the way to colors typical of the bluest known Sc galaxies. Such a distribution is similar to that in the field near zero redshift. On the other hand, A2111 and possibly also A2218 show only short, stubby blue tails, barely reaching into Sc type colors at all. Finally, Cℓ1446+26 and probably A1758 seem to exhibit color distributions at the opposite extreme, with perhaps half of the spirals showing the colors expected from typical Sc galaxies.

If this last sort of color distribution turns out to be real and not too infrequent, then we may want to begin thinking in terms of galaxy-cluster interactions having a significant effect on the evolution of cluster spirals. One might imagine, for example, that the act of cluster collapse could enhance star formation in member spirals for a brief period, perhaps even driving star formation close enough to completion for one or another of the various gas removal mechanisms to become important.

While such speculation is probably somewhat premature, we do get the impression from these early results that the study of the giant galaxy clusters as a function of redshift is likely to become a rich and rewarding field, one which may help us understand both the details of cluster evolution and certain aspects of the processes governing star formation in individual galaxies.

We wish to thank Jack Jewell for his help in shepherding our data through the KPNO CDC 6400 computer.

REFERENCES

Butcher, H. and Oemler, A. 1978a, Ap.J., 219, pp. 18-30.
Butcher, H. and Oemler, A. 1978b, Ap.J., 226, pp. 559-565.
Sandage, A. and Visvanathan, N. 1978, Ap.J., 225, pp. 742-750.
Visvanathan, N. and Sandage, A. 1977, Ap.J., 216, pp. 214-226.

DISCUSSION

Silk: Would you comment on the spatial distribution of the blue
               galaxies in the distant clusters that you have observed?

Butcher: Generally, the clusters do show a gradient in the ratio
               of red-to-blue objects with radius, in the sense of an
increasing fraction of blue galaxies with increasing radius. We try to
minimize the effects of such gradients by scaling the apparent size of
the region we include in our color distributions with redshift.

Segal: There appears to be a very simple explanation for the
               large proportion of blue galaxies that are apparently
members of clusters at redshift ~ 0.5. In fact, the chronometric cosmology preducts that for a cluster defined by apparent magnitude limits
in part, as is practice, a relatively large proportion of blue galaxies
is likely to be cluster members. This arises from the fact that a magnitude dimming of ~ 0.4 mag corresponds in the chronometric theory to a
displacement from $z = 0.5$ to $z = 1$, although in a Friedmann model, the
dimming is ~ 1.5 mag. Consequently, a large proportion of the objects
in a cluster defined (partially) by conventional magnitude limits would
be likely to be at redshifts > 0.5, quite possibly ~ 1, with consequent
shifting of the spectrum towards the blue. Would you not agree that
the hypothesis that the redshifts of your cluster members are not ~ 0.5,
but considerably greater, could explain your observation of an otherwise anomalously large proportion of blue galaxies?

Butcher: It certainly is a possibility we have not considered
               carefully.

M. Burbidge: Is anything known about X-ray emission from any of these
               clusters?

Butcher: Yes, the Einstein Observatory satellite has detected
               X-rays from the 3C295 cluster, as well as other clusters
near z ~ 0.4. Hence, one can probably infer the presence of a substantial intracluster medium in such clusters, present at the same time as
these blue galaxies are seen.

N. Bahcall: From one of the z ~ 0.4 clusters for which you find
               evidence for a large fraction of blue (spiral?) galaxies,
3C295, the Einstein observations have recently detected a rather strong
source of extended X-ray emission, similar to those found in the compact

nearby clusters. Since most of the hot intracluster gas is believed, in general, to have originated in the galaxies (due to the solar iron abundance found in the X-ray emitting gas), not too many normal spiral galaxies would be expected to be present in this cluster. Further observations, both in optical and X-rays, are needed in order to clarify this apparent "conflict," and better understand the relation between the X-rays, intracluster medium, and the galaxy type.

Koo: The color distribution of field galaxies resembles that of the fainter clusters. Would you comment on the possible contamination by the background?

Butcher: All of our clusters have been chosen to have richness classes of 3 or greater. Hence, they are at least a factor of 5, and often a factor of 10, above background. For each cluster we have determined the local background and subtracted it, but in most cases this step is of academic interest only and of little practical consequence.

SEARCHES FOR PRIMEVAL GALAXIES — PAST, PRESENT, AND FUTURE

Marc Davis*
Harvard-Smithsonian Center for Astrophysics

ABSTRACT. Most models of galactic evolution include a high luminosity epoch of bright star formation at a redshift 2 < Z < 100. Past searches have failed to detect galaxies in their "primeval" state, but recent advances in detector technology will soon enable much improved searches that will very seriously constrain the evolutionary models. Using the CCD detectors on the Space Telescope offers the opportunity for a thousandfold improvement over present search limits. We review here several models of primeval galaxies and comment on their observability in present and future experiments.

I. WHAT IS A PRIMEVAL GALAXY?

The evolution of galaxies is a subject of central importance in extragalactic astronomy and cosmology. The work of Butcher and Oemler (1976) suggests that in the recent past (Z ∿ .5) galaxies in clusters tended to be bluer, presumably because they contained more gas to form blue stars. Apart from this direct evidence for rich cluster members, there is little present constraint on the evolutionary history of normal galaxies. Number counts of galaxies versus magnitude may eventually provide constraints on galactic evolution for Z < 1, but at the present time, problems of uncertain luminosity functions and K corrections confuse the picture, as discussed by Ellis at this conference. What indirect evidence there is suggests the halo of our galaxy formed within one free fall time (Eggen, Lynden-Bell, and Sandage, 1962) and that the initial star formation rate was considerably higher in the past. Hence primeval galaxies (PG's) should have been quite luminous and may be detectable today, even if their formation epoch is Z > 5.

Extensive model calculations by Tinsley (1977, 1978), Tinsley and Larson (1978), Larson (1976) and others satisfactorily describe the colors and luminosities of galaxies of all types observed locally. Evolving these systems backwards in time provides evolutionary predictions that could be directly tested. The detection of primeval

*Alfred P. Sloan Fellow

G.O. Abell and P. J. E. Peebles (eds.), Objects of High Redshift, 57–64.

galaxies or lack of detection at a suitable limit, would serve to
seriously constrain these models and thereby improve our understanding
of galactic history.

In this talk I shall consider a primeval galaxy to be the progenitors of galaxies such as giant ellipticals or large spirals like
our Milky Way. There is no doubt that there exist "young" galaxies in
our local region, objects such as the dwarf peculiar galaxy II Zw 40
which is totally dominated by young blue stars. Presumably this object
was an intergalactic cloud recently perturbed and gone unstable (Searle
and Sargent, 1972). The progenitors of large galaxies may also have
resembled gigantic H II regions at one time (Sunyaev, Tinsley, and
Meier, 1978); it is these objects we designate as "primeval" galaxies,
because they belong to the earliest age of galactic evolution.

II. HOW BRIGHT AND HOW MANY?

Details of the early evolution of galaxies, and therefore of the
detailed appearance of primeval galaxies is of course highly specula-
tive. Nevertheless there are a number of relatively model independent
predictions that are of considerable interest.

The standard scenario of a primeval galaxy is a collapsing
protogalaxy, with halo stars forming in a time short compared to one
free gall time, and the disk stars forming after the inelastic collapse
of the gaseous disk.

The characteristic time scale of the primeval galaxy is its
collapse time which will be 10^8-10^9 years, and depends on the maximum
expansion size of the protogalactic cloud. In the standard scenario,
the bright phase begins when the clouds reach their maximum extent
after one collapse time, corresponding to $Z \sim 20 - 200$ in an $\Omega = 0$,
$H_O = 50$ km s^{-1} Mpc^{-1} universe, or to $Z \sim 3 - 16$ in an $\Omega = 1$,
$H_O = 100$ km s^{-1} Mpc^{-1} universe. The length of the bright phase depends
on the star formation rate, and can reasonably be expected to last one
collapse time.

Old disk stars are not particularly metal poor, so it is reasonable to suppose that the first generation halo stars Pop II.5-III
converted $\Delta X \sim 0.01$ of their hydrogen to metals. The energy conversion
efficiency for this process is .007, and presumably the bulk of this
energy will be radiated. If the present density of the remnants is
ρ_L, then the integrated sky brightness B from all the PG's will be
given as (Peebles, 1971)

$$B = \frac{0.007 \, \rho_L \, c^3 \, \Delta x}{4\pi \, (1+z)} \text{ ergs s}^{-1} \text{ cm}^{-2} \text{ ster}^{-1}$$

$$= \frac{8 \times 10^{-5}}{1+Z} \left(\frac{\Omega_L}{0.1}\right) \left(\frac{\Delta x}{.01}\right) \text{ ergs s}^{-1} \text{ cm}^{-2} \text{ ster}^{-1}$$

where Ω_L is the fractional closure density of this once luminous material and Z is the redshift at which the bright phase occurs. We are here using $H_0 = 50$ km sec^{-1} Mpc^{-1}. If we identify the once luminous remnant with the now dark massive halo around spiral galaxies and providing the virial mass of clusters, then $\Omega_L \sim .1$ is a not unreasonable estimate.

The best experimental limits are given by the ingenious experiment of Dube, Wilkinson, and Wickes (1977) which sets a 3σ upper limit of $B \leq 3 \times 10^{-5}$ ergs s^{-1} cm^{-2} ster^{-1} at 5100 Å. Because the emission spectrum is expected to be reasonably flat to the Lyman cutoff, this result implies that galactic halos, if they are comprised of burned out stars, must have formed prior to Z = 2.

If primeval galaxies are to be seen individually, how abundant are they expected to be? If we assume all large galaxies went through the bright phase and that their present density is $n_0 \sim .002$ Mpc^{-3}, then the number per square degree expected in an $\Omega = 1$ universe is

$$N \sim 2400 \ (1+Z) \left(\frac{\Delta t}{10^8 \text{ yrs}}\right) \text{deg}^{-2}$$

where again Z is the epoch of high luminosity and Δt is the timescale of the bright phase, which should roughly match the dynamical timescale of the galaxies. Thus it is not necessary to search large regions of the sky; it is much preferable to search a small region very deep.

III. PAST FAILURES, FUTURE HOPES

There have been several past attempts to detect individual primeval galaxies, all of which failed (Davis and Wilkinson, 1974; Partridge, 1974). These searches were based on the models of Partridge and Peebles, (1967a, 1967b) which predicted large fuzzy images ($\sim 10"$) not too centrally concentrated but with very low surface brightness. The resulting upper limits on flux (at ~ 7000 Å) and number density are plotted in Figure 1. Because the upper limits were large, the null results set no serious constraint on evolutionary models.

These earlier attempts were based on use of point detectors or red sensitive photographic plates. The ideal detector for such a program, and indeed for almost all astronomical programs, is a panoramic detector with high efficiency and negligible internal noise. Such devices are now becoming a reality; CCD's offer the chance for

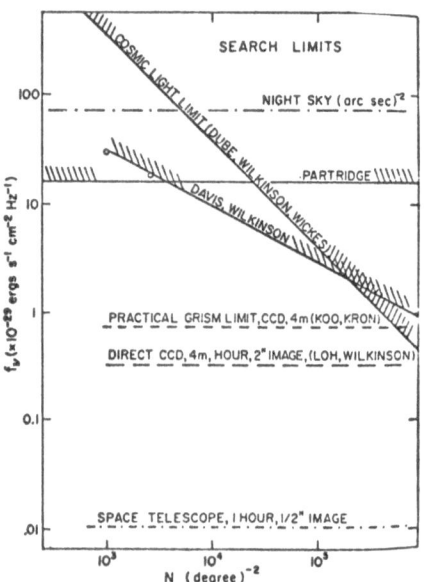

Figure 1. Present and expected flux and number density limits for searches of Primeval Galaxies.

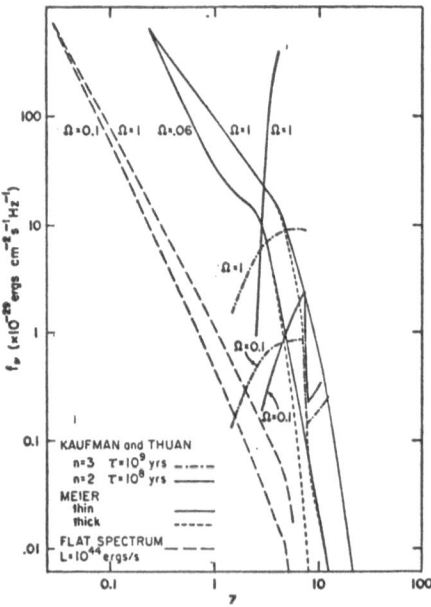

Figure 2. Expected flux at 7000 Å for several model primeval galaxies as a function of redshift. Each model is presented in the case of high and low density universe.

dramatic improvements in the search for primeval galaxies. Indeed at least two groups are currently trying to do just that. Shown in Figure 1 is the expected limit for detection of 2" images with signal to noise = 3, 1000 Å bandwidth, in one hour using the JPL CCD on the KPNO 4 meter (Loh and Wilkinson, 1979). Koo and Kron (1979) have used this same CCD for deep grism photographs. Their goal is to search for either Ly α emission or Lyman continuum break, and they estimate a practical detection limit of 1% of the night sky for a Ly α line. These limits are drawn in Figure 1. The Space Telescope facility offers further opportunities for detection of primeval galaxies because the sky background will drop four magnitudes in the red, and the seeing will improve to subarcsecond limits. Shown in figure 1 is the expected detection limit for 1000 Å bandwidth, signal to noise = 3, in one hour and 1/2 arcsecond images (Wilkinson, 1979).

IV. WILL PRIMEVAL GALAXIES BE DETECTED?

The limits above are impressive, but will they be good enough? I believe primeval galaxies will be detected at these limits, but they may not be recognized. Shown in Figure 2 are expected flux levels versus Z for several different, but typical, models. First consider a modest galaxy like our own with enough blue stars to increase its bolometric luminosity to 10^{44} erg/sec and to give an approximately flat spectrum to the Lyman break. The incident flux versus redshift (at 7000 Å) is shown for $\Omega = 1$ and $\Omega = 0.1$. Such an object would be readily detectable on the Space Telescope to redshifts ∿ 6.

Meier (1976a) has examined several evolutionary models of Tinsley and Larson; shown in Figure 2 is the incident flux at 7000 Å for a 10^{11} $M_\odot$ model collapsing from an initial radius of 30 kpc and undergoing rapid star formation (Larson, 1974). Such a model might result in an elliptical galaxy today. Presented are the cases $\Omega = 1$ and $\Omega = 0.06$, both for an optically thin and optically thick interstellar medium. These models use a Salpeter initial mass function and follow the evolutionary tracks of the stellar population. What is plotted is the apparent flux versus redshift of formation. Increasing the initial radius of collapse would increase the free fall time and decrease the initial star formation rate, but because the epoch is later, at smaller Z, the apparent luminosity is not dramatically decreased. All of these models are centrally concentrated with 50% of the emission occuring within 8 kpc, so these primeval galaxies would appear almost stellar in ground based observations.

Models of massive primeval galactic halos have been explored by Kaufman and Thuan (1977). They characterize the stellar birth rate with the form $B(m,Z) \propto m^{-n} e^{-t/\tau}$ where m is the stellar mass of the star and t is time since formation. Again they follow a galactic evolution code, and have considered the interstellar medium to be optically thin. The typical free fall time for these models is 10^9 years, as they collapse from a radius of 100 kpc. Their mass is of

order 3×10^{11} $M_\odot$. Plotted in figure 2 are evolutionary trajectories of
several of these models starting at their expected formation epoch with
n, τ, and Ω varied within their reasonable range. This type of primeval
galaxy will again be centrally concentrated, but with a 1/r brightness
profile out to 100 kpc, so they will appear distinctly fuzzy.

V.   WILL WE RECOGNIZE A PRIMEVAL GALAXY IF WE SEE ONE?

     It is apparent from comparison of Figures 1 and 2 that these
models of PG's should be readily detectable on the Space Telescope,
and very likely detectable from the ground today. Many alternative
models have been proposed, but the models of Fig. 2 give a general
picture of the expected range. It is entirely possible that galactic
disks formed slowly by infall and never underwent a bright phase.
However, to prevent dissipation, galactic halos and elliptical galaxies
almost surely underwent rapid star formation in a time scale of one
collapse time or less. If these objects are not detected by the Space
Telescope, the epoch of the bright phase will be pushed beyond $Z = 10$.

     How will we recognize a faint image as a PG? A primeval galaxy
detected at 27^m may be nearly stellar and without spectral information
it will be very difficult to distinguish it from a relatively nearby
dwarf elliptical galaxy. The signature of a PG is that it should have
the spectrum of an H II region, with weak metal lines and strong Ly α
emission, plus a likely cutoff at the Lyman continuum. The width of
the emission lines should not exceed the velocity dispersions in
galaxies, ∿ 300 km/sec. Thus spectroscopic searches such as being
undertaken by Koo and Kron will be very significant.

     Candidates for primeval galaxies should exhibit no polarization or
variability. They may be weak non-inverted radio sources, and should
be weak X-ray sources. If they are extended, the spectral features
should be the same in the core as in the outer fuzz.

     It is possible that some QSO's are really PG's. Indeed Meier
(1976b) has suggested that the spectra of OH471 and 4C05.34 strongly
resemble his models of PG's. QSO's with suitable narrow emission lines
are good candidates for primeval galaxies, and if so they should
appear fuzzy in good seeing. Mackay (1979) has taken Ly α filter
photographs of several high redshift QSO's in good seeing, which have
revealed nothing extending larger than 0.2". In the fields of
these QSOs in broadband (500 Å bandpass) photographs no objects with
unusual colors were detected down to m_B ∿ 23.

     Perhaps the past searches for primeval galaxies were premature.
Soon, however, it will be quite realistic to expect to detect galaxies
at these large redshifts. The detection and positive identification
of primeval galaxies would provide invaluable data for understanding
the dynamical and evolutionary processes of stellar systems.

REFERENCES

Butcher, H., and Oemler, A.E. 1978, Ap. J., 219, 18.
Davis, M., and Wilkinson, D.T. 1974, Ap. J., 192, 251.
Dube, R., Wilkinson, D.T., and Wickes, W.C. 1977, Ap. J. Lett., 215, L51
Eggen, O.J., Lynden-Bell, D., and Sandage, H.R. 1962, Ap. J., 136,
     748.
Kaufman, M., and Thuan, T.X. 1977, Ap. J., 215, 11.
Koo, D., and Kron, R. 1979, private communication.
Larson, R.B. 1974, M.N.R.A.S., 166, 585.
Larson, R.B., in Dynamics of Stellar Systems, I.A.U. Symposium No. 69
     (Ed. A. Hayli) (D. Reidel, Dordrecht, 1975) p. 247.
Larson, R.B. 1976, Comments Astrophys., 6, 139.
Loh, E., and Wilkinson, D.T. 1979, private communication.
MacKay, C. 1979, private communication.
Meier, D.L. 1976a, Ap. J., 207, 343.
Meier, D.L. 1976b, Ap. J. Lett., 203, L103.
Partridge, R.B., 1974, Ap. J., 192, 241.
Partridge, R.B., and Peebles, P.J.E. 1967a, Ap. J., 147, 868.
Partridge, R.B., and Peebles, P.J.E. 1976b, Ap. J., 148, 377.
Peebles, P.J.E. 1971, Physical Cosmology
Searle, L., and Sargent, W. L. W. 1972, Ap. J., 173, 25.
Sunyaev, R.A., Tinsley, B.M., and Meier, D.L. 1978, Comments
     Astrophys., 7, 183.
Tinsley, B.M. 1972, Ap. J., 178, 319.
Tinsley, B.M. 1977, Ap. J., 211, 621.
Tinsley, B.M. 1978, Ap. J., 220, 816.
Tinsley, B.M., and Larson, R.B. 1978, Ap. J., 221, 554.

DISCUSSION

Spinrad: The UV spectra of HII regions are, apparently, very different
        from that of QSOs; they have weak emission lines, not strong
Lyα, CIV and CIII of QSOs and Seyferts. So primeval galaxy spectra, if
like HII regions, may not be very spectacular.

Davis: Perhaps their non-spectacular nature will be the distinguish-
        ing element. That doesn't make it easy for the spectrosco-
pist, however.

Wolfe: There are some QSOs (3C 48, 3C 279, etc.) which have extended
        images with spectra resembling HII regions: the lines are
narrower than typical QSO emission lines. Thus, these look like the
objects you are talking about. But their redshifts are small, z ~ 0.5.
How do they fit into your theory?

Davis: Presumably, these are too rare to be the progenitors of
        bright nearby galaxies. They may be "young" galaxies which
for some reason have delayed the onset of star formation.

Boldt: Would there be significant X-ray emission associated with
              these protogalaxies?

Davis: They will be weak X-ray sources because of their O stars and
              supernovae remnant shells, but they should not be strong
X-ray sources like QSOs.

Rees: You showed a diagram which indicated that the space telescope
              would offer a great improvement in sensitivity compared with
a CCD or a ground-based 4-m telescope. Would this improvement really
be so significant if the intrinsic angular size of a photogalaxy exceeded
0."5?

Davis: The increased sensitivity of the space telescope is partly due
              to decreased sky background, which amounts to 4 magnitudes in
the red.

              This accounts for most of the difference between ground-based
              and space telescope CCD limits.

THE DYNAMICS OF SUPERCLUSTERS

Holland Ford[1], Richard Harms[2], Frank Bartko[3], Robin Ciardullo[1], and Erik Eason[1]
1. University of California, Los Angeles, USA
2. University of California, San Diego, USA
3. Martin Marietta Corp., Denver, USA

Relatively little is known about superclusters. Observations of superclusters should increase our understanding of these grandest aggregates of matter and may tell us a great deal about the large scale distribution of matter in the Universe. We report here observations of two superclusters chosen from the list by Murray et al. (1978, Ap. J. [Letters], 219, L89).

We used the Shane 3-m telescope to obtain Image Tube Scanner (ITS) spectra of the three brightest central galaxies in each of the Abell cluster candidate members. The redshifts of the galaxies were measured by reducing the ITS spectra to linear intensity, plotting the linear intensity versus the logarithm of the wavelength, and then registering the plot against a high signal-to-noise ratio reference spectrum. The average internal standard error of the redshift is 120 k/s.

The mean radial velocities of the Abell clusters within 2° of the center of MFJG # 18 (1451 + 22 = ABELL # 11) show that the five clusters A1976, A1980, A1986, A1988, and A2001 form a tight group and thus are a supercluster. The remaining clusters A1997, A2008, A2009, and a cluster which projects behind A2001 are 30% or more distant than the five supercluster members.

The supercluster MFJG # 19 is similar to MFJG # 18. Five clusters (A2158, A2172, A2179, A2183, and A2196) form a tight group. The remaining four clusters (A2187, A2192, A2190, and A2198) are 36% or more distant than the supercluster.

We conclude that superclusters are real, but, because of projection effects, often may be less rich than their appearances.

The observed dispersions of the cluster velocities about the supercluster means are surprisingly small. The weighted means and weighted velocity dispersions are:

G.O. Abell and P. J. E. Peebles (eds.), Objects of High Redshift, 65–67.
Copyright © 1980 by the IAU.

Supercluster	$<\bar{z}>$	$\sigma_{<\bar{z}>}$ (k/s)	Avg. Std. Error per Cluster (k/s)
MFJG # 18	0.1166	363	364
MFJG # 19	0.1364	360	500

The observed dispersions include the uncertainty in the cluster means, and are consistent with true dispersions as small as 100 k/s to 200 k/s.

As a bench mark, we consider a simple model of an empty (pure Hubble flow) supercluster. Its angular diameter is 4°, the clusters are distributed randomly, but uniformly, throughout a cubical volume (ℓ^3), and the mean redshift is 0.12. The observed velocity dispersion is then:

$$\sigma_o = \frac{H_o \ell}{\sqrt{12}} = \frac{\Theta \cdot c(z + z^2/2)}{(1 + z)^2} = 600 \text{ k/s}.$$

Inclusion of the uncertainty in the cluster means gives σ_{obs} = 700 k/s to 800 k/s.

The data lead us to two alternative conclusions:

1. Superclusters are roughly spherical, deacceleration is important, and superclusters may be bound.

2. Superclusters are flattened systems, or "cell walls" as proposed by Jôeveer and Einasto (1978, The Large Scale Structure of the Universe, ed. Longair and Einasto), seen face on.

If the latter conclusion is correct, we must radically revise our view of the distribution of matter in space. If the former is correct, we may be able to use superclusters to determine q_o. Observations of additional superclusters will be required to determine if either conclusion is correct.

DISCUSSION

Abell: The conclusion that superclusters can be flat ("pancakes") is
          borne out by recent studies of nearby superclusters (Coma,
Hercules, Perseus and the local supercluster) by Rood, Chincarini,
Thompson, Gregory, et al.

Tyson: If you were to include a few of your "background" clusters as
          members of your superclusters, the redshift dispersion would
easily reach 800 km/sec. Is there any evidence of a different number
density on the sky of your "background" clusters and the general field?

Ford: The density of background clusters does appear to be higher
          than average near the two superclusters that we observed. If
the nearest of the background clusters are associated with the superclusters, their geometrical shape becomes a cigar with a length-to-
diameter ratio ~ 6. Inclusion of additional background clusters would
make the superclusters even longer. We think that this geometry is
sufficient evidence to exclude the background clusters from supercluster
membership.

A CATALOGUE OF VERY FAINT CLUSTERS OF GALAXIES IN THE REGION OF THE SOUTH GALACTIC POLE

N.V. Vidal*
Royal Greenwich Observatory, Herstmonceux Castle, Hailsham,
East Suxxes BN27 1RP, England

In a search for very distant clusters of galaxies on a red-sensitive (6° x 6°) Schmidt plate taken at Siding Spring in the region of the south galactic pole, we found some 49 very faint and compact clusters. We tried to reduce the number of chance configurations, by limiting our survey to (a) very compact and isolated groups of galaxy-like images outside crowded regions; (b) some degree of concentration could be observed towards the brightest galaxy; (c) the size of the central object exceeded by several times the seeing disk diameter; (d) a minimum of 15 galaxies could be counted within a circle of ~ 30 arcsec in diameter. The red magnitudes of the brightest galaxies in each clusters are estimated to be fainter than 19.0 magnitudes. From the Hubble diagram for red magnitudes we estimate the redshifts to be about $z \gtrsim 0.5$. The integrated red magnitude of the first ranked galaxies in the faintest clusters are about 21.5. Taking into account seeing effects, a more realistic formula for the surface brightness at the centers of galaxies (rather than the $[1 + z]^{-4}$) is derived. It is shown that, assuming no evolution, the limit of our survey corresponding to the faintest clusters is about $z \gtrsim 1.0$. A list of clusters coordinates, magnitudes of the first ranked galaxies and finding charts can be obtained from the author.

*Present address: 14 Hofyen St., Ramat Aviv, Tel-Aviv, Israel.

G.O. Abell and P. J. E. Peebles (eds.), Objects of High Redshift, 69.
Copyright © 1980 by the IAU.

SIT SPECTRA OF FAINT FIELD GALAXIES

Edwin L. Turner
Princeton University Observatory

Digital SIT spectra have been obtained in collaboration with J.E. Gunn and W.L.W. Sargent for a sample of 49 field galaxies chosen from Kron's survey of selected area 68. Most have apparent visual magnitudes in the range 19 to 21. Redshifts have been measured for 48 of the 49 galaxies. The redshift (or, equivalently, luminosity) distribution for the sample is anomalous in that $\gtrsim 10\%$ of the sample has $z \gtrsim 0.3$ (and, thus, $L \gtrsim 6 L*$). These high z and L galaxies are among the bluest objects in the sample. These data require substantial evolution of at least some galaxies at small z. It is tempting to speculate that this luminous, blue field galaxy population is related to the Butcher and Oemler blue cluster galaxies found in the same redshift range.

DISCUSSION

G. Burbidge: You say that no galaxies with small redshifts have such large luminosities. Is it not the case that several galaxies in this category have been found?

Turner: It is true that a few nearby giant spiral galaxies are known to have such high lumonosities and blue colors. The point is that the relative frequency of such objects increases by rather more than an order of magnitude for galaxies at 20th magnitude as compared to samples at much brighter apparent magnitudes. These "extra" luminous blue objects have redshifts near 0.4.

Tyson: Assuming your proposed galaxies are standard candles of luminosity $L*$, our J-band galaxy number counts at 24th magnitude ($z \sim .5$) can set upper limits to the number of such galaxies.

Turner: In other words, if one wanted to explain these galaxies by simply increasing the luminosity of all galaxies at $z \gtrsim 0.4$ by some multiplicative factor, one would also have to predict an unobserved feature in the faint galaxy counts. I haven't yet tried to make any models, but that certainly sounds plausible.

G.O. Abell and P. J. E. Peebles (eds.), Objects of High Redshift, 71–72.
Copyright © 1980 by the IAU.

Hawkins: To what extent did your selection criteria avoid bias in favour of selecting more luminous galaxies?

Turner: The sample was selected such that each galaxy is intended to be a random galaxy of its particular apparent magnitude. I know of no bias in favor of giant, luminous galaxies. It seems more likely that a few dwarf galaxies of low surface brightness might have been omitted, but, obviously, this could not explain the bright galaxy anomaly.

DISTRIBUTION IN DEPTH OF QUASARS

Maarten Schmidt and Richard F. Green
Hale Observatories[1], California Institute of Technology

We discuss the distribution in depth of different kinds of quasars: quasi-stellar radio sources with steep radio spectrum, those with flat radio spectrum, and optically selected quasars. All exhibit an increase of space density with distance to a different degree. The optically selected quasars, in particular, show a steep increase of surface density with magnitude. The steepness of the increase is inconsistent with a uniform distribution of quasars in the local hypothesis. In the cosmological hypothesis the co-moving space density of optically selected quasars increases by a factor of 100,000 to a redshift of 2, and by factors of 1000 and 10 for steep-spectrum and flat-spectrum radio quasars, respectively.

The distribution in depth of quasi-stellar radio sources has been studied on the basis of samples from the 3CR, 4C, Parkes, and 6-cm NRAO catalogues; see Schmidt (1978), and Wills and Lynds (1978). Since samples are selected to optical and radio limits simultaneously, the V/V_{max} method (Schmidt 1968) is used in the analyses. Quasars with steep radio spectrum and flat spectrum show different $\langle V/V_{max} \rangle$ values, typically 0.67-0.70 and around 0.59, respectively. If we interpret the results in terms of a density law

$$\rho = \rho_0 \qquad K(t-t_0)/t_0$$

where t is the cosmic epoch and t_0 the age of the Universe (we assume a $q_0 = 0$ Friedmann model), then:
K = 10-12 for quasars with steep radio spectrum,
K = 4 for quasars with flat radio spectrum.

Samples of optically selected quasars are very scarce. In fact, the only published sample is that of Braccesi, Formiggini, and Gandolfi (1970). Spectroscopic work on the 175 objects is far from complete. A complete sample is constituted by 17 quasars with redshifts brighter than B = 18 over an area of 36 square degrees (Green and Schmidt 1978). We report on the preliminary results of the Palomar Bright Quasar Survey of optically selected quasars brighter than B ≈ 16. The survey is based on double (U, B)

G.O. Abell and P. J. E. Peebles (eds.), Objects of High Redshift, 73–76.
Copyright © 1980 by the IAU.

exposures obtained by Green with the 18-inch Schmidt telescope over an area
of approximately 10,000 square degrees. He selected as candidates some
3000 stellar objects with an ultraviolet excess. We have taken spectra of
essentially all these objects and have found around 105 quasars, of which
some 25 were known previously (half of them radio sources). Results are
still tentative, since the limiting magnitude in each of the many Schmidt fields
is yet to be determined precisely.

The counts of quasars in this survey, in the Braccesi sample to B = 18,
and the unpublished Sandage-Usher surface density to B = 18.5 show
$\log N (< B) = 0.93B + \text{const.}$, corresponding to an increase of the surface
density by a factor of 8.5 per magnitude.

Such a steep count slope is incompatible with any local hypothesis of
quasars in which the space distribution is uniform. In the latter case we
expect $\log N (< B) = 0.60B + \text{const.}$, regardless of the shape of the
luminosity function. The steep slope observed requires that the space density
of local quasars increase with distance, approximately as $r^{3/2}$.

In the cosmological hypothesis, the steep slope of the counts requires
$K = 18$ if the space density varies exponentially with cosmic time. The
corresponding space density (in co-moving coordinates) at redshift 2 is some
100,000 times the local density.

The density increase appears to depend on the intrinsic optical
luminosity of the quasars. Observed numbers of intrinsically luminous
quasars in our survey, in the Braccesi sample, and in objective prism
surveys (see next paper by P. Osmer) suggest a very steep increase, i.e.,
$K > 18$. Counts of faint stars reported by Tyson in this Symposium limit the
rate of evolution of intrinsically weak quasars to $K < 18$. With luminosity-
dependent evolution of quasars their luminosity function will change with
redshift.

Our criterion of ultraviolet excess for candidate objects discriminates
against redshifts larger than 2.3. We will soon be able to predict the number
of quasars expected for $B > 18$, $z > 3.5$, which is of interest in connection
with the continuing suspicion that there is a redshift cutoff or at least a sharp
decrease in evolution at a redshift around 3.5 (see next paper by P. Osmer).

We are planning to obtain radio observations of the new sample of
quasars. This should help in establishing the radio luminosity function of
quasars, which is needed to interpret the lesser evolution for radio quasars
referred to above.

Also planned are X-ray observations of the optically brighter quasars
with HEAO-2. This should allow improved estimates of the X-ray background contributed by faint quasars, discussed elsewhere in this Symposium.

REFERENCES

Braccesi, A., Formiggini, L., and Gandolfi, E.: 1970, Astron. Astrophys. 5, pp. 264-279.

Green, R. F., and Schmidt, M.: 1978, Astrophys. J. (Letters) 220, pp. L1-L4.

Schmidt, M.: 1968, Astrophys. J. 151, pp. 393-409.

Schmidt, M.: 1978, in IAU Symposium No. 79, The Large Scale Structure of the Universe, eds. M.S. Longair and J. Einesto (Boston: Reidel), pp. 289-293.

Wills, D., and Lynds, R.: 1978, Astrophys. J. Suppl. 36, pp. 317-358.

NOTES

[1] Operated jointly by the Carnegie Institution of Washington and the California Institute of Technology.

DISCUSSION

Segal: The chronometric cosmology is an inherently non-evolutionary cosmology that predicts an infinite $\partial \log N/\partial m$ for objects of spectrum $v^{-\alpha}$, $\alpha < 1$, just before it vanishes identically (in a single luminosity class), as appears consistent with the observed quasar counts and indicated constraints on the counts at fainter magnitudes based on X-ray observations with the Einstein satellite, in addition to which its other predictions have been shown consistent with all substantial published quasar samples. I wonder whether you are aware of any empirical basis within quasar observations for non-acceptance of the chronometric hypothesis (without prejudice to other possible hypotheses)?

Narlikar: I do not see why the steep slope (of 8.5 per magnitude) of the log N-apparent magnitude of quasars is "fatal" for the local hypothesis. In the local hypothesis the distance variation is not large since all observed quasars are claimed to be at distances of 30-100 Mpc. On the contrary, the variation in intrinsic luminosity is expected to be very large. The data on the number counts which you have presented today does not seem convincing enough to claim that the departure from the Euclidean value (of ~ 4 per magnitude) has any significance in the context of the local hypothesis. I think the numbers are still too small: the highest point at B = 18 is based on only 17 quasars.

Schmidt: If a slope of 4 per magnitude held for quasar counts between B = 16 and B = 18, then Braccesi's sample should have contained only 4 quasars with B < 18, rather than the 17 observed. Independent confirmation of the steep slope of 8.5 per magnitude is provided by the Sandage-Usher determination of a surface density of 1.3-1.6 per square degree for B < 18.5.

Davis: With the density and evolution of QSO's as determined by
              your survey, how does the QSO density at Z ~ 1 or 2 compare
to the galaxy density at that epoch?

Schmidt: The steeper evolution now found for optically selected
              quasars will leave space densities at z ≈ 1 fairly unchanged.
Densities will be lower at low redshift, higher at larger redshifts.

Koo: Since the optically selected QSOs were based upon ultravio-
              let excess, and since you have spectroscopic data for these
objects, would you please comment on the distribution of the spectral
index of the continuum and its possible effects on the problem of missing a large number of redder QSOs?

Schmidt: As you would expect on the basis of our criterion of ultra-
              violet excess, all quasars found had fairly flat optical
spectra. We would almost certainly have missed the class of very red
quasars (such as 3C 181). These represent about 6 per cent among 3CR
quasars (3 out of 50) and probably a smaller percentage of 6-cm NRAO
quasars. The bias against red quasars exists both in our survey and in
the Braccesi sample, so it is unlikely that the slope of the counts
has been affected.

SYSTEMATIC SURVEYS FOR QUASARS WITH THE SLITLESS SPECTRUM TECHNIQUE

Patrick S. Osmer
Cerro Tololo Inter-American Observatory
La Serena, Chile

ABSTRACT

Recent results with the slitless spectrum technique are reviewed and the total surface densities from five independent surveys are compared. The observational base for extending the technique to search for quasars with z > 3.5 is now large enough that either the apparent cutoff at z = 3.5 will be confirmed in new surveys or quasars with z > 3.5 will be found.

I. SCOPE

In the previous talk Schmidt has reviewed the distribution in depth of quasars as determined from radio-selected samples and from optical surveys with the ultraviolet excess technique. Here I discuss recent results from the slitless spectrum technique. This technique, which is best suited for redshifts larger than 1.8, is a natural complement to the ultraviolet excess method, which is best suited for redshifts between 0 and 2.3.

The excellent review by Smith (1978) describes in detail the slitless spectrum technique, its application to various problems in quasar research, and the results up to May, 1978. Therefore I shall limit myself to events occurring subsequent to Smith's article. In particular I shall discuss the question of where are the redshift four quasars. The statistical base for investigating the question is now solid enough for a feasible program either to find quasars with z > 3.5 or show that their space density is indeed low. Although the recent results have implications for the emission-line and absorption-line problems in quasars, time does not allow them to be discussed here.

II. DEFINITION OF THE SLITLESS SPECTRUM TECHNIQUE

The problem in carrying out optical surveys for quasars is how to distinguish them from the myriad star images. For example, there may be 100,000 stars on an exposure with the UK Schmidt that contains

G.O. Abell and P. J. E. Peebles (eds.), Objects of High Redshift, 77–82.

100-200 quasars (Smith 1978). The slitless spectrum technique is based
on the fact that quasar emission lines can be seen directly on objective
prism or grating-prism spectrograms. The strongest emission line in
quasar spectra, Lα, is the one normally detected, and the technique is
color independent. As used at Cerro Tololo, the technique is based on
spectra of about 1500 Å mm^{-1} dispersion, a value which I believe is
nearly optimum. At higher dispersion the limiting magnitude suffers;
at lower dispersion weaker-lined objects are missed. The UK Schmidt
spectra (2500 Å mm^{-1} at Hγ) mark the low end of the usable range (see
discussion by Smith 1978). Additional requirements are a large
telescope aperture and/or large field of view, good seeing conditions,
dark sky, and fine grain emulsions. The Curtis Schmidt telescope,
with a 60 cm aperture and 25 deg^2 field, turns up about 8 quasars per
25 deg^2 field; for most programs it represents the smallest telescope
one would want to use. The 4-m telescopes are suitable because of
their great light gathering power; about 4 quasars are found per
exposure in a 1/3 deg^2 field. The UK Schmidt, with medium aperture
and large field, is the most efficient instrument in terms of discovery
rate per exposure: 100-200 quasars (Smith 1978).

The detailed characteristics of the technique are described by Osmer
and Smith (1980) and Osmer (1980). In brief, 90% of the candidates
are confirmed to be quasars and 80% of the confirmed quasars have
$z \gtrsim 1.8$. The technique is indeed efficient for discovering large
redshift quasars.

Perhaps the main technical problem remaining with the survey technique
concerns the calibration of and correction for inhomogeneities in the
plate material. It is known that bad seeing degrades the resolution
of the slitless spectra and thereby reduces the detectability of the
quasar emission lines. Whether seeing effects are behind the broad
magnitude distribution in the technique remains to be seen.
Certainly the seeing effects must be considered in any analysis of the
plate-to-plate variation in the number of quasars and before any
statement can be made about their spatial uniformity. It is likely
that the application of automated measuring and identification techniques to the plate material will lead to progress in both areas. With
such techniques quantitative information on the detection threshold of
the emission lines and on the spectral resolution of the plates will be
available, so that well defined limits of completeness can be
established.

III. THE MAJOR SURVEYS

The principal surveys for new quasars with the slitless spectrum
technique have been carried out at Cerro Tololo with the Curtis Schmidt
and 4-m telescopes, at Kitt Peak with the 4-m telescope, and with the
UK Schmidt telescope. A brief bibliography of the surveys is given in
Table 1.

TABLE 1

Major Surveys[1]

CTIO

Curtis Schmidt	4-m
MacAlpine and Lewis (1978)	Hoag and Smith (1977)
Osmer and Smith (1980)	Bohuski and Weedman (1979)

KPNO 4-m	UK Schmidt
Hoag, Burbidge and Smith (1977)	Bolton and Savage (1978)
Sramek and Weedman (1978)	Smith (1978)

[1] By author according to last published reference, which in turn is a source for earlier work.

By now several hundred quasars have been discovered. Of these, more than 250 have follow up observations with slit spectroscopy. Lewis, MacAlpine and Weedman (1979) describe observations of 76 candidates from the Michigan-Tololo survey. Osmer and Smith (1980) present complete data, including a spectral atlas, for a sample of 120 candidates from the Curtis Schmidt survey and Osmer (1980) has carried out an analysis in similar format of the 71 independently discovered quasars in the Hoag-Smith (1977) 4 m sample.

With these recent results it is possible to intercompare the different surveys and check their consistency. The best measure for this purpose is the total surface density of quasars in each survey. Not all surveys have complete photometry and the magnitude scales are not homogeneous, so a more detailed comparison is not yet possible. It can be seen from Table 2 that the results are entirely consistent. Both surveys with the Curtis Schmidt give practically the same surface densities, as do the 4-m surveys, albeit at much larger values. The UK Schmidt result is intermediate between the other two.

Thus the internal agreement of the surveys is excellent. The main questions (Osmer 1980) concern the steepness of the increase from the Curtis Schmidt value, 0.34 deg^{-2} on the average, to the 4 m value of 13 deg^{-2} and the lack of agreement between the two surveys at magnitude 18, where they might be expected to give similar results. Until the systematic effects involved in different telescopes are understood better, considerable caution is required for any attempt to combine the results.

TABLE   2

Total Surface Densities
Number deg^{-2}

CTIO Curtis Schmidt	
Osmer and Smith (1980)	Lewis, MacAlpine, and Weedman (1979)
0.32	0.36

UK Schmidt	CTIO and KPNO 4-m	
Smith (1978)	Hoag and Smith (1977)	Hoag (1979)
~5	13	~13

IV.   WHERE ARE THE QUASARS WITH z > 3.5?

The Lα emission line in quasars should be visible on slitless spectrum
plates taken on IIIa-F emulsion to redshifts as large as 4.7.
Nonetheless, the largest redshift yet found in the slitless spectrum
surveys is 3.45, still second to the z = 3.53 value for OQ 172.
Carswell and Smith (1978) have nicely shown how the absence of quasars
with z > 3.5 in the Hoag-Smith 4-m sample can be accounted for by the
properties of the IIIa-F emulsion, the ultraviolet blaze of the grating,
and the steep luminosity function of quasars. According to their
calculations the data are not inconsistent with a constant space
density for 2.1 < z < 4.7 or a e$^{10}\tau$ form for the evolution, although a
$(1 + z)^6$ form is ruled out.

However, the absence of presence is not the same as the presence of
absence; that is, the definitive test for the existence or absence
of the z > 3.5 quasars with the slitless spectrum technique still needs
to be done. Both Osmer (1977) and Carswell and Smith (1978) realized
that in the case of the grating-prism surveys at the 4-m telescopes,
the use of a red-blazed grating should be significantly more sensitive
for the detection of quasars with 3.5 < z < 4.7. Now that the spectroscopy for the Hoag-Smith sample is complete (Osmer 1980), it is possible
to make a refined estimate of the expected numbers of quasars with
3.5 < z < 4.7 that should be detectable on 4-m plates optimized for the
red.

Although there are several ways to make the estimate, let me describe
the simplest one here. Slit spectra (Osmer 1980) confirm that the
Hoag-Smith sample contains a total of 7 quasars with 2.5 < z < 3.5
in an area of 5.1 deg^2. Carswell and Smith (1978) have made a

numerical estimate that the Hoag-Smith sample (IIIa-F emulsion with a grating blazed at $\lambda 3550$) should detect the same number of quasars with $2.5 < z < 3.3$ as a new survey with IIIa-F emulsion and a grating blazed at $\lambda 5700$ would for $3.3 < z < 4.7$. This estimate assumes constant density; if there is evolution of the type described by Schmidt in the previous article, there would be more quasars with $3.3 < z < 4.7$. If we scale this estimate to the redshift bins $2.5 < z < 3.5$ and $3.3 < z < 4.7$, we find that if an area of 5.1 $\deg^2$ is surveyed with the red-blazed grating, it should contain 2/3 as many quasars with $3.5 < z < 4.7$ as the Hoag-Smith sample did for $2.5 < z < 3.5$, namely 4.7 as opposed to 7. If there is density evolution of the form $c^{10}\tau$, 8 quasars are expected, and if the form is $e^{15}\tau$, 10.5 quasars are expected. I believe it is possible to do even better than the Carswell-Smith estimates by filtering out the sky shortward of 5700 Å, for a fainter limiting magnitude can be obtained.

The point is that a feasible survey optimized for the $\lambda\lambda 5700$–6700 region should contain at least several quasars with $z > 3.5$. This prediction is based on the simplest possible extrapolation of a known number of quasars in the redshift range immediately below 3.5. If such a survey fails to find quasars with $z > 3.5$, it will provide convincing evidence that the space density of luminous quasars turns down near $z = 3.5$. The motto for the survey might be "Redshift four or bust".

REFERENCES

Bohuski, T.J., and Weedman, D.W. 1979, Ap. J., 231, 653.
Bolton, J.G., and Savage, A. 1978, I.A.U. Sym. 79, 295.
Carswell, R.F., and Smith, M.G. 1978, M.N.R.A.S., 185, 381.
Hoag, A.A. 1979, private communication.
Hoag, A.A., Burbidge, E.M., and Smith, H.E. 1977, I.A.U. Coll. 37, 521.
Hoag, A.A.. and Smith, M.G. 1977, Ap. J., 217, 362.
Lewis, D.W., MacAlpine, G.M., and Weedman, D.W. 1979, Ap. J. in press.
MacAlpine, G.M., and Lewis, D.W. 1978, Ap. J. Suppl., 36, 587.
Osmer, P.S. 1980, Ap. J. Suppl., in press.
Osmer, P.S., and Smith, M.G. 1980, Ap. J. Suppl., in press.
Smith, M.G. 1978, Vistas in Astronomy, 22, 321.
Sramek, R.A., and Weedman, D.W. 1978, Ap. J., 221, 468.

DISCUSSION

G. Burbidge: Is it not the case that Arp has looked at the distribution of bright galaxies and your clumps of QSOs and has found that they are correlated?

Osmer: Arp has looked at the correlation between the brightest galaxies and the quasars with z > 2.5 in the Curtis Schmidt survey. I have been looking athe distribution of quasars in the much smaller areas of the Hoag-Smith 4-m survey. I have done a simple check of Arp's results which shows the excess of quasars near the galaxies is about 1.5 σ above the expected value. However, I do not pretend that the check is definitive and I am unable to comment further on the statistical significance of his result.

Turner: Why is there such a small difference in the predicted number of z ≈ 4 quasars between your "no evolution" and strong density evolution cases?

Osmer: The difference is small because the extrapolation is from z = 3.0 to z = 4.1. The evolutionary effects are relatively small over such an interval in z.

Abell: Clumping of quasars on a scale of 0.5° to 1° or so would be expected, would it not, if quasars, like everything else, are within superclusters of size ~ 100 Mpc?

Osmer: Yes, exactly. In fact, I have been looking for them. There certainly are groups of quasars with these dimensions in the Hoag-Smith sample, but it is not yet clear if the grouping exceeds the expectations for random fluctuations.

A SPECTROSCOPIC SEARCH FOR GALAXIES ASSOCIATED WITH QUASI-STELLAR OBJECTS*

Joseph S. Miller, Howard B. French and Steven A. Hawley
Lick Observatory, Board of Studies in Astronomy and
Astrophysics, University of California, Santa Cruz

It is generally assumed that quasi-stellar objects represent phenomena taking place in galaxies. There are at least three lines of reasoning that lead to this viewpoint. First, there appears to be a continuity of properties extending from Seyfert galaxies and N systems to QSOs with the differences between the various groups principally a question of the contrast between the luminous central object and its surrounding galaxy. Since Seyfert galaxies and N systems are definitely in galaxies, it is concluded that QSOs must also be in galaxies which are not directly visible because of the high luminosity of the central object. A second argument is based on the result that QSOs do not appear to differ dramatically in abundances of elements from those which are typical of normal galaxies. With the prevailing view that virtually all the elements heavier than helium originated as a result of stellar processes, it would be concluded that QSOs must be associated with galaxies of stars that produced the heavier elements. A third, rather indirect argument is based on the results of Stockton (see his paper in this volume) and others that QSOs are often found in groups of galaxies and therefore are likely to be located in galaxies themselves (guilt by association). But the fact is that direct evidence that QSOs are in galaxies is sadly lacking.

A few QSOs show some kind of very faint nebulosity associated with them. However careful studies of objects such as 3C 48 (Wampler et al. 1975), 3C 249.1 (Richstone and Oke 1977) as well observations of the jet in 3C 273 indicate that the extended emission in these objects results from gaseous emission or other nonstellar processes. It should also be pointed out that there is little spectral data available in the literature pertaining to the stellar component of N galaxies.

*Contributions from the Lick Observatory, No. 422.

G.O. Abell and P. J. E. Peebles (eds.), Objects of High Redshift, 83–87.

To pursue these matters further, we have been carying out an extended series of spectroscopic observations of active extragalactic objects. The goal has been to detect the spectra of the stars of associated galaxies, and where detected, to estimate the brightness of the galaxies. Portions of the results, those concerning BL Lacertae objects, have already been published (Miller, et al. 1978a; and earlier references given in that paper), and a thorough review of all results covering BL Lac objects, N systems and QSOs is in preparation (Miller 1980; based on review given at June 1979 meeting of Astronomical Society of the Pacific), so this account will be brief and limited principally to the results on QSOs.

All observations were made with the Lick Observatory Shane 3 m reflector equipped with the image-tube scanner (Miller, et al. 1976); details of the observations will be given elsewhere (Miller 1980). The general approach was to obtain very high signal-to-noise data by observing each QSO on several nights and at different grating tilts to minimize the possibility of system response irregularities appearing in the reduced data even at a very low level. In addition, the independent spectra obtained from the two entrance apertures were kept separate in the averaging of results from different nights, so that two independent spectra of equal observing time were ultimately obtained for each object (see Miller and Hawley 1977 for an illustration of the approach used). All observations were made with a rectangular entrance aperture 2".4 x 4".0, since the annular aperture used in another study (Miller, French, and Hawley 1978b) is of no use for essentially stellar appearing objects. As a supplement and check on the approach, observations identical in approach were made of a number of N galaxies, where detection of the stellar features of a galaxy was to be expected.

The type of galaxy to be expected in association with QSOs is of course an open question. Seyfert galaxies tend to be found in spirals, while N systems are often in ellipticals. If the galaxies associated with QSO were dominated in stellar light by stars of A type or earlier, the stellar component would be very difficult to detect spectroscopically because of the complete correspondence of strong emission features of the QSO with strong absorption features of the stellar component. Since galaxies with strong radio sources are generally ellipticals, it might be expected that QSOs which are radio-loud would also be located in E galaxies. The late-type stars of an E galaxy produce an absorption spectrum that has features which occur in regions free of strong QSO emission. In particular, the Mg I feature near 5175 Å is one of the strongest features in the spectra of galaxies. Strong absorption is also observed at Ca II H and K and Na D, but since these features can be produced by gas as well as stars, they are not as good an indicator of the stellar component as is Mg I. The continuum of E galaxies is also heavily inflected, with a sharp break in the continuum shortward of 4000 Å. On an F_λ scale the continuum is approximately level from 4000 to 4300 Å, beyond which the

continuum rises considerably to about 4800 Å, where it levels out
once again because of Mg H absorption. Just beyond Mg I, it abruptly
rises up again. For a more detailed discussion of these characteristics,
see Miller (1980).

Given the above considerations, this study was limited to
detecting E galaxies, and the majority of the objects selected were
radio-loud QSOs. The QSOs were selected for low luminosity, that is,
for being within 2 mag. of the Hubble line for first-ranked giant E's.
In all, 9 QSOs and 9 N systems comprise the sample for this study.
The overall results are straightforward to summarize: the diluted
stellar spectral features of a galaxy were detected for 7 of the N
systems, but for none of the QSOs. If galaxies comparable in type
and brightness to those detected for the N systems were associated
with QSOs, they would have also been detected given the quality of
the data. Furthermore, the N galaxies detected have magnitudes
inferred from the strengths of the spectral features observed that
places them generally 1-2 mag. fainter than a first-ranked giant E.
This implies that whatever may be the type of galaxies associated
with the QSOs in this study, they are certainly not first-ranked
giant E's; if they are E galaxies at all, they are at least 1-2 mag.
fainter then first-ranked ones. The table below summarizes the
results for the QSOs. Limits given are based on using the Sandage
(1972) curve of growth for E galaxies to extrapolate the value
derived for the aperture used to his standard metric diameter. Also
we adopted $H_0 = 75$ km s^{-1} Mpc^{-1} and $q_0 = 1$; in this system a first-
ranked giant E is -22.4 V mag. For 3C 215 and 1635+119, an extremely
weak feature was seen in the data near Mg I, but it is quite likely
that feature was not real. For more details on these results and
those on N systems, see Miller (1980).

QSOs

Object	3C 47	PHL 1093	0736+01	3C 215	1217+02
Z	0.42	0.26	0.19	0.41	0.24
M_v(galaxy)	>-22.9	>-21.1	>-21.0	$\gtrsim$-21.1	>-20.8

Object	3C 277.1	B264	3C 323.1	1635+119	
Z	0.32	0.10	0.26	0.15	
M_v(galaxy)	>-19.9	>-19.1	>-21.7	$\gtrsim$-20.6	

The principal result of this study, that QSOs in this sample are
not located in first-ranked giant E galaxies, or even in E galaxies
comparable in luminosity to those detected in N systems, raises a

number of questions. Are these QSOs in galaxies at all? They could
be spirals, but then one would have to acknowledge spiral galaxies
can produce strong radio sources. Since BL Lac objects and N systems,
active extragalactic objects that are radio-loud, are generally
found to be in E galaxies when the galaxies can be detected, it is
reasonable to assume that at least the radio-loud QSOs in this study
are in E galaxies also. It is important to consider whether these
results could be a result of the definition of QSOs. Since the QSOs
were selected for low luminosity and for not being called N systems,
this selection process would directly eliminate those QSOs with very
luminous galaxies: the galaxies would be visible and they would have
been classified as N galaxies. This would imply that the principal
difference between the N galaxies and the QSOs in this study would be
the brightness and hence visibility of the galaxy. However, preliminary analyses of the spectra of the two groups suggests that, for
the most part, N galaxies and QSOs can be distinguished by spectroscopic characteristics alone. For N systems the lines appear stronger
relative to the continuum, and the rise in the continuum shortward of
4000 Å, commonly seen in QSOs, is rarely seen in N systems; a more
detailed and quantitative discussion of these matters will be
presented elsewhere (Miller 1980).

To conclude, it is clear what kind of galaxies these QSOs are
not in, but whether they are in galaxies at all and what type remains
for future work.

This research was supported in part by NSF Grant AST 76-20843.

REFERENCES

Miller, J. S. 1980, Pub. A. S. P., in preparation.
Miller, J. S., French, H. B., and Hawley, S. A. 1978a, in Pittsburgh
    Conf. BL Lac Ob., p. 176.
Miller, J. S., French, H. F., and Hawley, S. A. 1978b, Ap. J. (Letters)
    219, L85.
Miller, J. S., and Hawley, S. A. 1977, Ap. J. (Letters), 212, L47.
Miller, J. S., Robinson, L. B., and Wampler, E. J. 1976, in Advances
    in Electronics and Electron Physics (New York: Academic Press)
    Vol. 408, p. 693.
Richstone, D. O., and Oke, J. B. 1977, Ap. J., 213, 8.
Sandage, A. 1972, Ap. J., 173, 485.
Wampler, E. J., Robinson, L. B., and Burbidge, E. M. and Baldwin, J. A.
    1975, Ap. J. (Letters), 198, L49.

DISCUSSION

P. Veron: What are the main differences between the spectra of the N-galaxies and of the QSOs?

Miller: The analysis of the line spectra of the two groups of objects is still in a preliminary stage. It appears that forbidden lines of [OII] and [OIII] have much larger equivalent widths in the N systems than in the QSOs in our sample. Also, only one N galaxy shows the increase in "continuum flux" toward the Balmer limit which is observed in nearly all the QSOs in the sample.

Arp: You said you compared BL Lac to a giant E galaxy, but you actually showed it compared to M32. Does that mean the spectrum is a better match to a dwarf elliptical?

Miller: The strengths of absorption features seen in normal elliptical galaxies are functions of both the absolute magnitudes of the galaxies and the radial distance from the center of the observed regions. The BL Lac observations were made with an annular aperture which corresponds to a distance interval of 6 to 10 Kpc when projected on BL Lac. Data in the literature indicate that the spectrum of the nucleus of M32 should provide a reasonable match for the much more luminous BL Lac galaxy in the outer regions which we observed, and this was borne out by the observations.

Wehinger: What size apertures did you use? Were the apertures annular like the ones used to observe the BL Lac objects?

Deep ESO 3.6-m prime-focus IIIa-F plates show extended structure (in the form of log intensity profiles) in the range ~ 2 to 5 arc sec from the QSO, at surface brightness of ~ 10 to 20% night sky (at best). Extended asymmetric structures have integrated apparent magnitudes of $\gtrsim$ 22 to 23 for quasars in the range z = 0.2 to 0.5. This work was done in collaboration with Sue Wyckoff (ASU) and Tom Gehren (MPIA).

Miller: The QSOs were typically observed with apertures of rectangular sizes of about 2".5 x 4".0. The annular apertures had inner and outer diameters of 6" and 10", respectively. The faint features you report near QSOs are too faint to be observable with our equipment on Mt. Hamilton.

Grandi: I would like to emphasize that Seyfert 1 spectra do show the continuum rise to the ultraviolet that is seen in QSOs but not, in general, in N galaxies. Thus, there exist objects inside galaxies with spectra equivalent to QSOs.

Miller: The spectra of Seyfert 1 galaxies is indeed very similar in this respect to those observed in QSOs.

ASSOCIATIONS BETWEEN QSOs AND GROUPS OF GALAXIES

Alan Stockton
Institute for Astronomy
University of Hawaii
Honolulu, Hawaii 96822

ABSTRACT

Evidence that *most* presently known low-redshift QSOs are associated with groups of galaxies has led to a program aimed at determining the typical properties of these groups. For the 3C273 field,
observations of essentially all galaxies with $m_B \lesssim 21.5$ over a region
10.1' (~1.5 Mpc) in diameter yield four galaxies having redshifts
close to that of the QSO. Consideration of the present evidence concerning the richness of groups associated with QSOs leads to the conclusion that, for at least those QSOs that are radio sources, their
galactic environment is similar to that of radio galaxies, except that
they seldom, if ever, occur in the centers of rich clusters of
galaxies.

I. INTRODUCTION

If QSOs are events occurring in the nuclei of galaxies, and if
their redshifts are cosmological in origin, the known clustering
properties of galaxies make it virtually inevitable that at least
some QSOs should be found in association with other galaxies. The
first positive evidence for such associations came with Gunn's (1971)
discovery of a galaxy near PKS 2251+113 with a similar redshift.
Other fairly bright QSOs found to be apparently associated with galaxies were 3C323.1 (Oemler *et al.* 1972) and 4C37.43 (Stockton 1973).
A few years ago, I undertook a systematic survey of galaxies near
low-redshift QSOs, the results of which were published last year
(Stockton 1978, henceforth LRQ survey); since I shall be referring to
this study in what follows, let me quickly recapitulate its main
features.

The QSO sample for the LRQ survey comprised all those known to me
as of mid-1976 with: (1) $z \leq 0.45$, (2) V (as listed by Burbidge *et al.*
1977) < 19.12 + 5 log z, and (3) $-15° < \delta < 55°$. Around each of the

G.O. Abell and P. J. E. Peebles (eds.), Objects of High Redshift, 89–97.

27 QSOs meeting these criteria, all galaxies brighter than the red
Sky Survey limit within 45" of the QSO were selected for the spectroscopic program: a total of 29 galaxies were found in 17 fields, the
other 10 fields having no qualifying galaxies.

At the time the results of this survey were submitted for
publication, definite redshifts had been obtained for 25 of the 29
galaxies along with a tentative redshift for one more.   Of these, 13
galaxies in eight fields had redshifts within 1000 km s^{-1} of their
respective QSOs.   Two of the three galaxies for which no redshift
information was available are of relatively minor interest as far as
the identification of groups is concerned, since they occur in fields
which already have at least one galaxy agreeing in redshift with the
QSO.   Further observations of the one remaining galaxy, Q1048−090(1),
show it to have a redshift of 0.3456, where that of the QSO is 0.344.
Thus nine of the original 27 fields have at least one galaxy meeting
the specified selection criteria and agreeing in redshift with the QSO.

It is important to recognize the limitations of this survey.   It
was designed, not to find all the associations between QSOs and galaxies that might exist for this particular sample of QSOs, nor to
determine the detailed properties of those that were found, but simply
to test the cosmological hypothesis for the redshifts of QSOs.   In
order to cover as many fields as possible in a tractable observing program, both the field size and the galaxy magnitude limit were severely
restricted.   At the low-redshift end of the QSO sample, a 45" radius
samples only a few percent of the total volume over which a typical
group of galaxies is spread.   At the high-redshift end, this radius
samples something like half the volume of such a group; but in this
case only the very brightest galaxies will be members of the galaxy
sample.   The importance of this magnitude cutoff in limiting the
detectability of associated groups is demonstrated by Table I, which
shows the number of fields in the sample and the number of detected
associated groups for different intervals in QSO redshift.   Note that
13 of the QSOs in the sample have redshifts above 0.35, but for only
two of these have associated galaxies been found, whereas below this
redshift fully half of the QSOs have detected associated galaxies.   It
is clear that, once the limitations and selection effects inherent in
the survey are taken into account, the evidence is that *most* QSOs selected in a manner similar to the LRQ sample will be located in groups
of galaxies.

II.   THE 3C273 FIELD

As a first attempt to explore further the properties of groups
around QSOs, the fields of five low-redshift QSOs have been selected
for more detailed study.   The intention is to obtain redshifts and
magnitudes for all galaxies brighter than $m_B \sim 21.5$ within a radius of
$0.8/z$ arcminutes (~750 kpc for $H_0 = 75$) of each QSO.   The only QSO
field for which the observations are substantially complete is that of

3C273. This field is of particular interest not only because 3C273 is by far the brightest of the very low redshift QSOs, but because, among the six lowest-redshift QSOs in the LRQ survey, only for 3C273 and PG 0026+129 were no associated galaxies found.

TABLE I

STATISTICS OF DETECTION OF QSO-GALAXY ASSOCIATIONS
AS A FUNCTION OF QSO REDSHIFT

z_{QSO}	Number of Fields	Detected Associations
< 0.20	2	0
0.20--0.25	3	3
0.25--0.30	5	2
0.30--0.35	4	2
0.35--0.40	7	1
0.40--0.45	6	1

The field of 3C273 is shown in Figure 1, and the galaxies for which spectroscopic observations were undertaken are indicated. The observed magnitudes and redshifts for these galaxies are given in Table II. Four of the galaxies - nos. 2, 3, 4, 5, - have redshifts agreeing with that of 3C273 to within reasonable limits (the association of no. 2 with 3C273 has already been reported [Stockton 1978b]). Figure 2 shows the spectra of these four galaxies, and Table III lists their absolute blue magnitude, M_B, their projected separation from 3C273, R_p, and their radial velocity difference with respect to 3C273, ΔV_r.

TABLE II

DATA ON GALAXIES IN 3C273 FIELD

No.	m_B	z	No.	m_B	z
1	20.5	0.248	7	20.6	0.21:
2	19.9	0.1577	8	20.4	0.1698
3	20.6	0.1592	9	20.0	0.180
4	19.6	0.1601	10	20.5	0.322
5	20.1	0.1600	11	18.5:	0.0895
6	21.6:	--	12	19.2	0.1778

This group around 3C273 is an example of one that probably would not have been detected at any redshift by the LRQ survey. The galaxy with the smallest angular separation, no. 2, would only come

within the 45" radius field for redshifts > 0.26, at which point it
would be over a magnitude fainter and would probably have been eliminated by the magnitude limit of the galaxy sample.

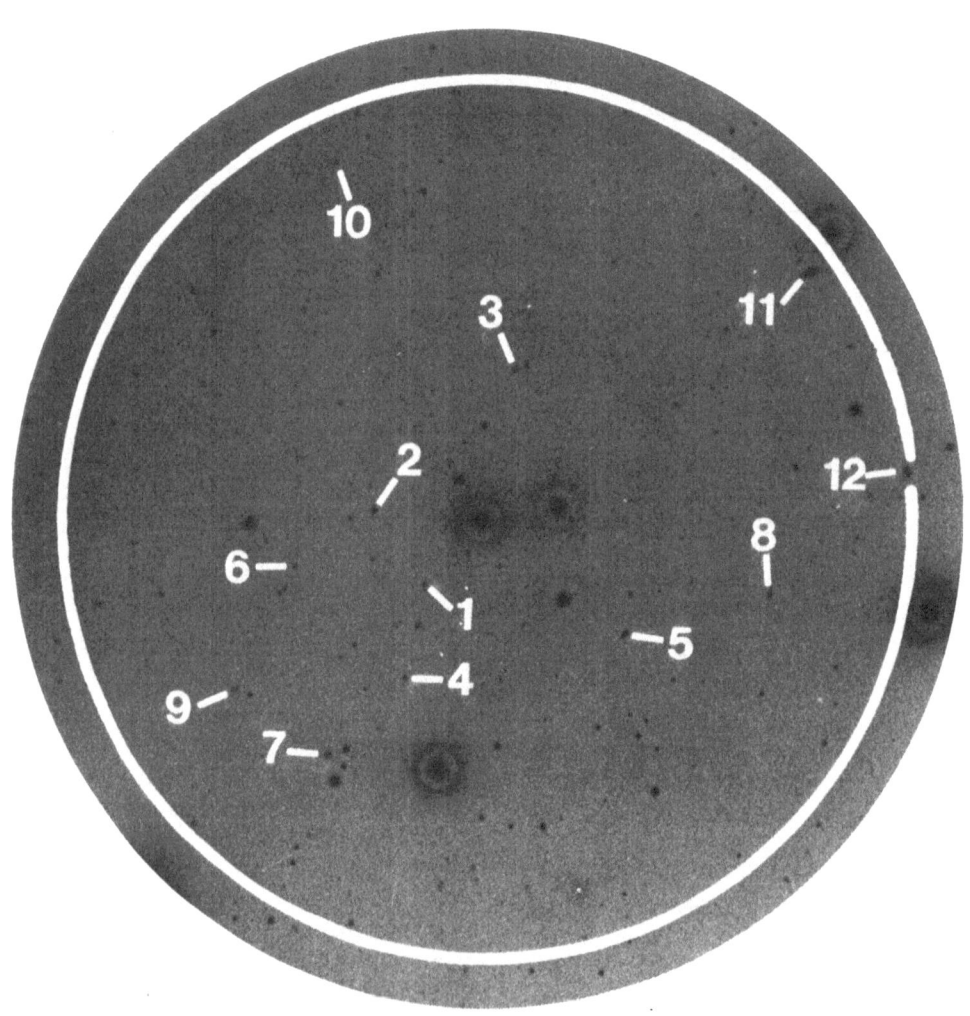

Figure 1. The field of 3C273 (centered), taken from an 098 plate
exposed through an OG570 filter. The white circle has a radius of
5.06', and all galaxies within that radius with $m_B \leq 21.5$ are marked.

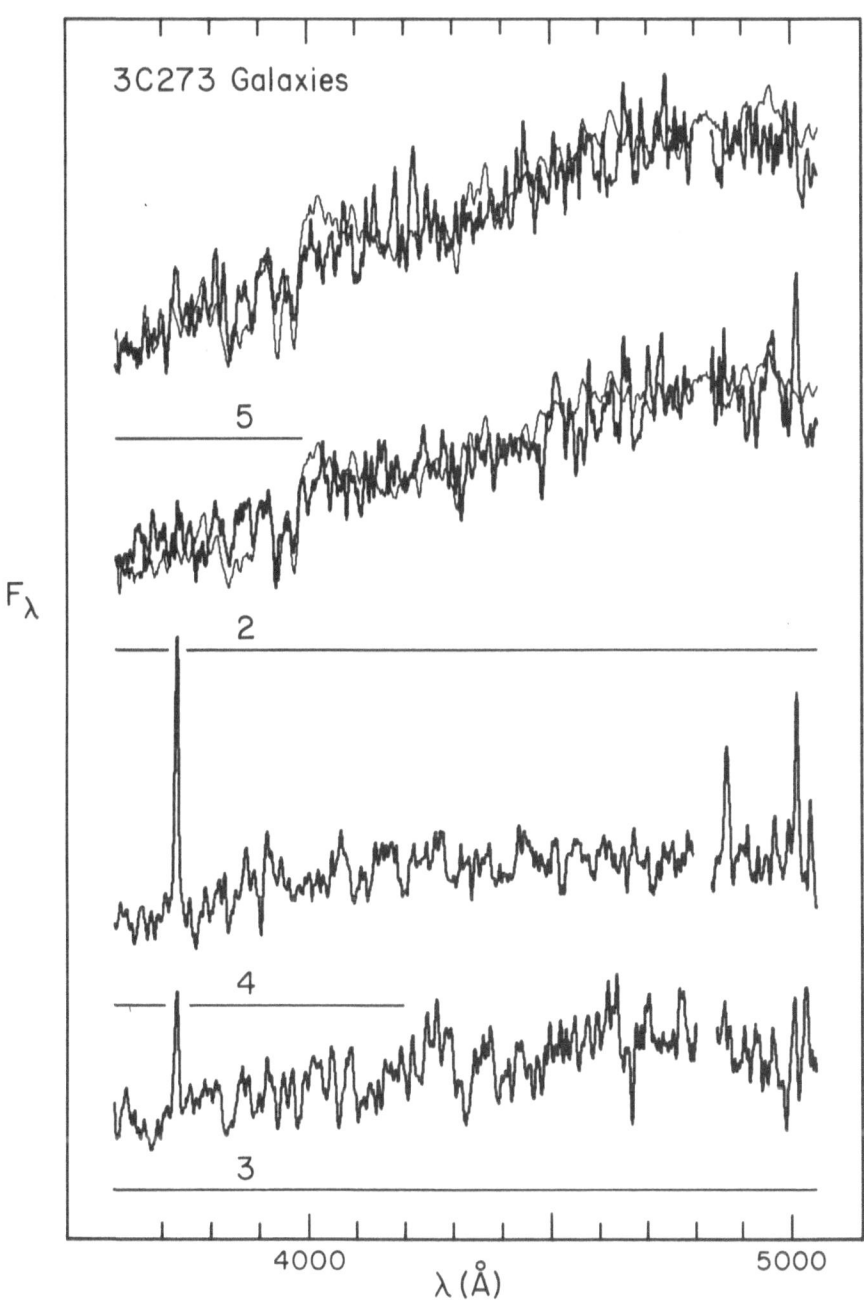

Figure 2. Spectra of the four galaxies with redshifts near that of 3C273, reduced to the rest frame. The light trace on the two upper spectra is Spinrad's spectrum of the inner region of M31, shown for comparison.

TABLE III

PROPERTIES OF GALAXIES ASSOCIATED WITH 3C273

$(H_o = 75, q_o = 0)$

No.	M_B	R_p (kpc)	V_r (km/s)
2	-19.8	185	-80
3	-18.8	260	300
4	-19.8	295	530
5	-19.5	310	510

III. PROPERTIES OF GROUPS ASSOCIATED WITH QSOs

Only after several groups around QSOs have been surveyed in detail will we begin to realize fully the possibility of learning,
through the study of their galactic environment, something about the
nature of QSOs and their relation to other forms of energetic phenomena occurring in galaxies. But even with the sketchy information
available now, it is possible to arrive at some interesting tentative
conclusions and to isolate some of the research areas that seem most
worth following up over the near term.

A.   Richness of Associated Groups

No QSO has ever been found in a rich cluster of galaxies
(although Butcher *et al.* (1976) have found a distant cluster centered
on the BL Lac object 3C66A), and it has therefore long been suspected
that QSOs do not occur in rich clusters with any great frequency.
From QSOs and 3CR galaxies covering the same range of redshift and
distribution on the sky as Abell clusters, Roberts *et al.* concluded
at a 95% confidence level that QSOs are associated less frequently
with clusters of Abell richness class 1 or greater than are strong
radio galaxies. This conclusion is strengthened by the fact that none
of the QSOs in the LRQ sample, which is largely disjunct from the
Roberts *et al.* sample, occurs in a rich cluster.

On the other hand, Weymann *et al.* (1978) have shown that if the
galaxy-QSO correlation for the LRQ sample of QSOs were the same as
the average galaxy-galaxy correlation, the survey of galaxies that
was carried out would have been expected to find only 2.4 correlated
galaxies rather than the 14 actually found. They therefore concluded
that galaxies are correlated more strongly with this sample of QSOs
than they are, on the average, with each other.

The LRQ sample of QSOs is biased strongly towards radio sources, and it is well-known (see, e.g., Seldner and Peebles 1978) that radio galaxies show a similar tendency to be galaxies with which other galaxies are strongly correlated. The present evidence, then, is that, *while both radio galaxies and radio QSOs preferentially occur in regions where the density of galaxies is significantly higher than average, radio galaxies often are located at the centers of rich clusters but radio QSOs seldom, if ever, are.*

It is important to emphasize, however, that any conclusions reached from a sample of low-redshift QSOs may not have universal applicability. For example, even an unequivocal demonstration that QSOs occurring near our own epoch *never* are found in the centers of rich clusters would not prove that those at a redshift z = 2 are not found in rich clusters: with the evolution of conditions in clusters, there might very well have been a corresponding evolution of the characteristic locus of QSO activity with respect to cluster richness.

B.   Effects of the Intracluster Medium

For radio galaxies there is a correlation between the richness of the cluster in which the galaxy lies and its radio spectral index, in the sense that the steeper index sources are in the richer clusters; this correlation is interpreted as being due to the influence of the density of the intracluster medium on the development of the radio source. Guthrie (1977) had suggested that radio spectral indices of QSOs might be useful as a means of detecting QSOs in clusters, but he was unable to find any QSOs with the very steep indices characteristic of radio galaxies in rich clusters. However, there is some indication that QSOs that are flat-spectrum radio sources are systematically in poorer groups that are steep-spectrum QSOs: considering QSOs in the LQR sample with z < 0.35, four of the five with steep spectra but only one of the five with flat spectra were found to have associated galaxies in the original survey. If this effect is real, it might also indicate an influence of the ambient gas density in these groups.

Another effect involving the intracluster gas may have some promise as a means of detecting QSOs in rich groups or in clusters, even at redshifts beyond those for which the cluster galaxies can be seen. Hintzen and Scott (1978) have proposed that distortions in the morphology of double radio sources associated with QSOs be used as an indicator of the presence of surrounding ambient gas and thus, presumably, a cluster. This technique does have limitations: it cannot detect the presence of gas that has both zero relative velocity and a symmetrical density gradient with respect to the source, so it would miss any QSOs that are analogous to the central radio galaxies in rich clusters. In addition, any attempt to compare, say, the relative frequency of QSOs in groups above a certain richness for two different epochs using this approach would have to deal not only with selection effects (which probably are manageable) but a number of presently unknown evolutionary effects.

C.   Future Observations

It will be important to continue observations of groups associated
with QSOs with a coverage comparable to that obtained for the 3C273
field in order to determine typical properties for such groups. Otherwise, the greatest need is for similar observations for a sample of
radio-quiet QSOs. If they should be found to be distributed (as the
radio QSOs are) in regions of higher-than-average galaxy density, then
arguments that they are closely related to the classical Seyfert galaxies will have to be re-examined, since it appears that Seyferts tend
to be, if anything, *more* isolated than the average galaxy (van den
Bergh 1975).

Finally, the fact that radio QSOs tend to be found in environments
similar to those of radio galaxies, except the centers of rich clusters,
indicates that it would be worthwhile to attempt to delineate as carefully as possible the differences between radio galaxies in the centers
of rich clusters and radio galaxies found elsewhere. The center of a
rich cluster is a unique environment, and at least the cD galaxies
found in cluster centers have apparently had a unique evolutionary
history. It is therefore at least conceivable that radio galaxies in
and outside of cluster centers have arrived at similar physical states
along different paths, only one of which can sometimes lead to the QSO
phenomenon.

This work has been supported in part by NSF Grant AST 78-21946.

REFERENCES

Burbidge, G. R., Crowne, A. H., and Smith, H. E.: 1977, Astrophys. J.
    Suppl. 33, pp. 113–188.
Butcher, H. R., Oemler, A., Tapia, S., and Tarenghi, M.: 1976,
    Astrophys. J. Letters 209, pp. L11–L15.
Gunn, J. E.: 1971, Astrophys. J. Letters 164, pp. L113–L118.
Guthrie, B.: 1977, Astrophys. Space Sci. 46, pp. 429–441.
Hintzen, P., and Scott, J. S.: 1978, Astrophys. J. Letters 224,
    pp. L47–L50.
Oemler, Jr., A., Gunn, J. E., and Oke, J. B.: 1972, Astrophys. J.
    Letters 176, pp. L47–L50.
Roberts, D. H., O'Dell, S. L., and Burbidge, G. R.: 1977, Astrophys. J.
    216, pp. 227–236.
Seldner, M., and Peebles, P. J. E.: 1978, Astrophys. J. 225, pp. 7–20.
Stockton, A.: 1973, Nature Phys. Sci. 246, p. 25.
Stockton, A.: 1978a, Astrophys. J. 223, pp. 747–757.
Stockton, A.: 1978b, Nature 274, pp. 342–343.
van den Bergh, S.: 1975, Astrophys. J. Letters 198, pp. L1–L2.
Weymann, R. J., Boroson, T. A., Peterson, B. M., and Butcher, H. R.:
    1978, Astrophys. J. 226, pp. 603–608.

DISCUSSION

Longair: M. Seldner and I have studied the clustering of galaxies about strong extragalactic radio sources using cross-correlation functions. We study the cross-correlation of the positions of 3CR radio galaxies in a complete statistical sample with the Lick counts of galaxies. We set up a scale of clustering in terms of the amplitudes of the spatial cross-correlation function. For galaxies selected at random, the scale of clustering is 1. For Abell clusters, the scale has value 13. Intermediate scales of clustering are defined on a linear scale between these values. For 3CR radio galaxies in the redshift range z < 0.1 to which this technique is sensitive, the average clustering lies intermediate between the Abell clusters and galaxies selected at random. However, when one splits the sample into classical double radio sources and more complex radio morphologies, the complex sources lie in much higher regions of galaxy clustering than the classical doubles. In fact, the clustering about the classical doubles is not significantly different from galaxies selected at random in the universe.

The significance of this result for the quasars is that the quasars found most frequently in low-frequency radio surveys are classical double sources and, consequently, we would not expect them to lie in regions of strong clustering of galaxies. This is consistent with the lack of prominent clustering of galaxies about low redshift quasars. We discuss the implications of our results in a paper soon to be published in Monthly Notices of the R.A.S.

It will be most interesting to express the strength of the clustering of galaxies about quasars found by Dr. Stockton in terms of cross-correlation functions to find if our very different approaches are in agreement.

Stockton: I think that this sort of investigation is very important, and I am glad to hear of what you have done. I understand that you also have found a relationship between the optical spectra of radio galaxies and the strength of the clustering around them, which I believe is an important result.

The result of Weymann et al. for the LRQ sample of QSOs would, I believe, indicate a galaxy density enhancement of about a factor of 5 above that found around a typical galaxy, but I am not sure how this value would translate into your clustering scale. It will certainly be ironic of galaxies should turn out to be more strongly correlated with radio QSOs than with radio galaxies having similar radio structures, after we have supposed the contrary for so many years!

Jaffe: A number of studies have shown that the probability of a radio galaxy occurring in an optical galaxy of given magnitude does not depend on whether that galaxy is in a rich Abell-type cluster or not.

EVIDENCE FOR NON-COSMOLOGICAL REDSHIFTS - QSOs NEAR BRIGHT GALAXIES AND OTHER PHENOMENA

G. Burbidge
Kitt Peak National Observatory*, Tucson, Arizona U.S.A.

All of the speakers at this meeting except me have assumed, or will assume, without question that redshifts are measures of distance. This reflects a point of view which is not borne out by all of the evidence. I shall attempt to give you some of the flavor of the evidence which so many wish to ignore. It comes from individuals who perhaps not surprisingly, have not been invited to speak.

We would all agree that patterns of redshifts, or very different redshifts associated with objects which appear to be physically associated provide direct evidence for non-cosmological redshifts. We start briefly with recent investigations by Tifft on galaxies. For some years he has shown that there appears to be correlation between the nuclear magnitudes of galaxies and their differential redshifts in clusters. He has found that there appears to be a "quantized" value of $c\Delta z$ = 72.5 km sec^{-1}.

In a recent paper Tifft[1] has found a similar effect by plotting the redshift differences between pairs of galaxies in close binary systems. In Fig. 1, I reproduce his results from one sample--pairs of galaxies whose redshifts have been accurately measured using the 21 cm line. When the distribution of redshifts is divided into boxes which are multiples of 72.5 km sec^{-1}, the number deduced from his earlier and very different investigations, a strong effect is seen. I see no way of explaining this away by selection effects. The implications are profound.

I now turn to associations between QSOs and bright galaxies. The first investigation of this effect using properly chosen samples was that involving the QSOs in the 3CR catalogue and the galaxies in the Shapley Ames Catalogue[2,3]. A highly significant effect was found, with five out of 50 QSOs lying very close to bright galaxies, and even more remarkable, a plot of log θ for these five QSOs (θ is the angular separation between

---

*Operated by the Association of Universities for Research in Astronomy, Inc., under contract with the National Science Foundation.

G. O. Abell and P. J. E. Peebles (eds.), Objects of High Redshift, 99–105.
Copyright © 1980 by the IAU.

the galaxy and the QSO) against log z galaxy showed a slope of -1 suggesting that the QSOs are all at approximately the same projected
distance from their (parent?) galaxies.

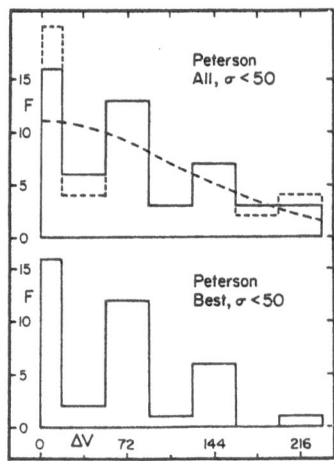

Fig. 1   The distribution of redshift differences for double
galaxies with redshift differences less than 250 km s^{-1} and
$\sigma < 50$ km s^{-1} according to Peterson. The upper histogram
includes all pairs, the lower histogram includes corrections
for optical pairing and overlapping signals in close pairs.
From Tifft (1979).

Even earlier than this Arp[4] had argued that many radio sources,
some identified with QSOs, were aligned across bright galaxies and were
thus physically associated with them.

In the last decade many more QSOs have been found to lie close to
bright galaxies. Arp has found a large number and others have been
found by a variety of individual studies. Recently, Hewitt and Burbidge[5]
have compiled a list of all of the cases known. They comprise at present
(September, 1979) 94 QSOs close to 65 galaxies. Of these 68 QSOs lie
within 10' of 54 galaxies.

I have attempted to look at the statistical significance of this
latter sample. Following the review by Wills[6] I have taken as the surface density of QSOs derived from several surveys 10 per square degree
$\leqslant 20^m$, $3 \leqslant 19^m$, $1 \leqslant 18^m$, $0.3 \leqslant 17^m$. These are very conservative values
especially at the bright end. On the assumption that at most 200 fields
near bright galaxies have now been surveyed, we can calculate the number
expected by chance $\langle n \rangle$ and compare it with the total number found, N_0,
and the total number found by Arp designated by N_{OA}. The results are
shown in Table 1. It can be seen that for very small separations the
result is highly significant, while for very large separations the total
number seen is less than we would expect to find by chance.

This latter result may be due to the fact that Arp and others have not searched extensively enough at distances $\geqslant 5'$ from the galaxies. Alternatively it may be that QSOs are rarer than we have assumed in using the values of the surface densities given above. In this case the significance of the close associations is enhanced.

Table 1. A comparison between the number (N_O and N_{OA}) of QSOs found near bright galaxies and the number $\langle n \rangle$ expected by chance.

Apparent Magnitude	$\theta \leq 60"$			$61" \leq \theta \leq 120"$			$121" \leq \theta \leq 180"$			$181" \leq \theta \leq 300"$			$301" \leq \theta \leq 600"$		
	N_O	$\langle n \rangle$	N_{OA}	N_O	$\langle n \rangle$	N_{OA}	N_O	$\langle n \rangle$	N_{OA}	N_O	$\langle n \rangle$	N_{OA}	N_O	$\langle n \rangle$	N_{OA}
≤ 17	1	0.05	0	1	0.16	0	1	0.24	1	1	0.82	0	4	3.9	1
≤ 18	5	0.17	0	6	0.52	4	1	0.86	1	2	2.8	0	9	13.0	2
≤ 19	9	0.52	1	12	1.55	9	4	2.4	2	4	8.2	0	15	39.1	8
≤ 20	12	1.73	3	16	5.2	12	6	8.6	4	7	27.6	2	18	129.6	10

Apparent Magnitude	$\theta \leq 180"$			$\theta \leq 600"$		
	N_O	$\langle n \rangle$	N_{OA}	N_O	$\langle n \rangle$	N_{OA}
≤ 17	3	0.46	1	8	5.2	2
≤ 18	12	1.56	5	23	17.3	7
≤ 19	25	4.62	12	42	51.9	20
≤ 20	34	15.5	19	59	173	31

No attempt has been made to assess the statistical significance of the fact that in cases recently found by Arp[7][8] QSOs with similar redshifts are aligned across bright galaxies. One example of this kind (Arp, 1979) is shown in Fig. 2. An even more striking case not involving a bright galaxy has recently been found by Arp and Hazard[9] where six QSOs with three similar pairs of redshifts close to 0.5, 1.6 and 2.1 are aligned in two triple systems. If lines joining the pairs with similar redshifts are drawn they appear to have a common origin (Fig. 3).

It seems to me that examples of this kind make an almost overwhelming case for the reality of large non-cosmological redshifts.

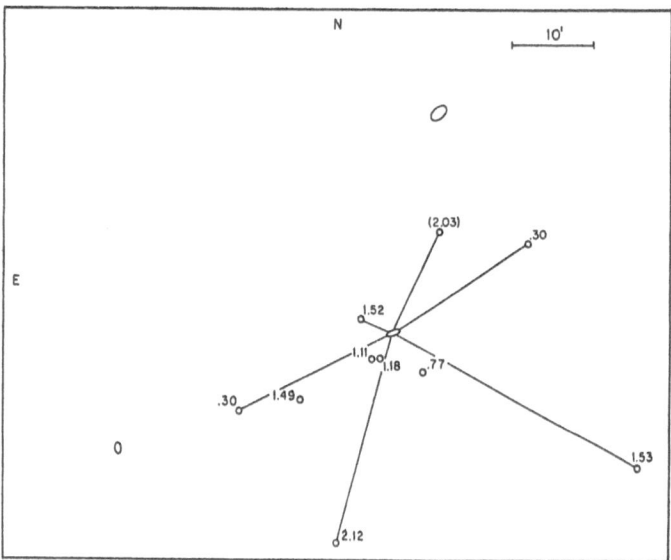

Fig. 2

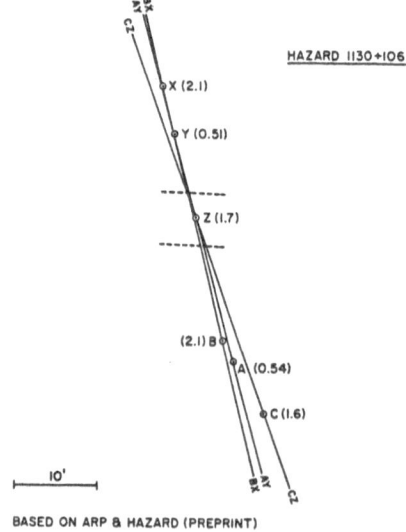

Fig. 3

Finally, there is another test that we can make for physical association. In Fig. 4 we show a plot of log θ against log z galaxy for all of the 94 QSOs close to 65 galaxies listed by Hewitt and Burbidge[5]. This is an update of the original plot of Burbidge, O'Dell and Strittmatter[3]. The equation of the line shown is

log θ = -1.17 log z_{galaxy} + constant.

The range of slope for this line at the 99% confidence level is between -0.93 and -1.56. The correlation between log θ and log z is significant with a correlation coefficient of 0.68.

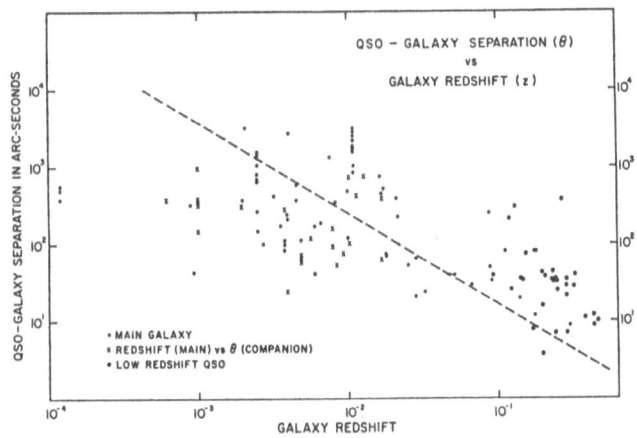

Fig. 4

I have left out many other pieces of evidence. For example, the existence of 3 QSOs within 2' of the center of NGC 1073[10] with "magic" redshifts of 0.599, 1.411 and 1.945[11] is most significant.

I believe that however much many astronomers wish to disregard the evidence by insisting that the statistical arguments are not very good, or by taking the approach that absence of understanding is an argument against the existence of the effect, it is there and many basic ideas have to be revised.

A revolution is upon us whether or not we like it.

I am indebted to H. C. Arp, A. Hewitt, J. V. Narlikar and W. Tifft for giving me access to material not yet published and for much help in preparing this paper.

REFERENCES

1. Tifft, W. 1980, Ap.J. in press.
2. Burbidge, E.M., Burbidge, G.R., Solomon, P.M. & Strittmatter, P.A.
      1971, Ap.J. 170, p. 230.

3. Burbidge, G., O'Dell, S.L. and Strittmatter, P.A. 1972, Ap.J.,
       175, p. 601.
4. Arp, H.C. 1966, Science 151, p. 1214.
5. Hewitt, A. and Burbidge, G. 1980, Ap.J.Suppl. in press.
6. Wills, D. 1978, Physica Scripta 17, p. 333.
7. Arp, H. 1979, Proc. 9th Texas Symposium on Relativistic Astro-
       physics, Munich (Ann. NY Acad. Sci.) in press.
8. Arp, H. 1979 preprint.
9. Arp, H. and Hazard, C. 1979 preprint.
10. Arp, H. and Sulentic, J.W. 1979, Ap.J. 229, p. 496.
11. Burbidge, E.M., Junkkarinen, V., and Koski, A.T. 1979 preprint.

DISCUSSION

Tyson: You have given many examples of associations involving
                several combinations of angular distance, redshift,
lines, and other patterns. I think I counted eight separate "games."
I am reminded of the origin of constellations. Although we probably
ought to remain open-minded when searching through the "noise" of
stars, galaxies, and redshift, the statistical significance of any discovery of some single kind of association is considerably reduced if
we allow the rules of the game to change during the search.

G. Burbidge: There is really only one "game," namely the collection
                of evidence for physical association of QSOs with large
redshifts with galaxies with small redshifts. We can test by looking
for luminous bridges by finding QSOs very close to galaxies using complete samples and working out probabilities of chance effects or by
finding alignments and pairing of reshifts.

                I agree that the game is not being played always accord-
                ing to a list of rules generally agreed on, but in prac-
tice, while statistical arguments are important, every scientist will
evaluate evidence in his own way. All I ask is that people do have
open minds, that they look at the data and are not too thoroughly
influenced by current beliefs and convictions to ignore their data or
unfairly criticize the methods of those who present them.

Arp: It is so rare to hear a really unprejudiced speaker in
                this subject that I listened with great interest. I
would only like to remark that in the preprint of the paper by Arp and
Hazard, the group of quasars that form a small cluster is a little
more than 2 degrees away from the two triplets that are in the same
paper. Saslaw and Hazard described in this cluster from the preprint
two more triplet alignments which had not been seen before. As before,
both those triplets had exact alignment, within the 2 arc sec image
sizes on the photograph. One of the triplets had extremely similar
redshift to redshifts in the first two triplets discovered.

Rees: It's still, of course, difficult to assess the quantitative significance of the various alignments and associations involving quasars. (In fact, the argument is reminiscent of
another long-running controversy -- the astronomical significance of
stone alignments in megalithic sites.) However, even if one accepts
these effects as real, it's important to emphasize that this doesn't
necessarily force us to invoke "new physics." The irregular collapse
of a massive system in a galactic nucleus could, quite plausibly, lead
to "slingshot" ejection of compact objects at high speed. Also, there
is now increasing evidence for production of beams and jets in galactic nuclei, and the analogous galactic object SS433 shows that jets
can consist of <u>cool</u> material coasting along at well-defined relativistic speed. It is by no means a wild extrapolation from this evidence
to suppose that the material in such jets can sometimes condense into
compact objects.

G. Burbidge: I agree. The first step is to gain acceptance of the
evidence that there is a generic connection between QSOs
with high redshifts and galaxies with low redshifts, for astronomers
to realize that large (and small) non-cosmological redshifts exist and
may be commonplace. At that point we would look for an explanation
and exhaust all other possibilities contained within conventional
physics before we consider "new physics."

THE ORIGIN OF ABSORPTION SPECTRA IN QUASI-STELLAR OBJECTS

Ray J. Weymann
Steward Observatory, University of Arizona

ABSTRACT

A classification scheme for QSO absorption line spectra is
described which ascribes the origin of the lines to at least four
mechanisms: (A) Explosive ejection of material at speeds up to 0.1 c.
(B) Absorption by highly ionized material moving in a rich cluster in
which the QSO is embedded. (C-1) Cosmologically distant intervening
material with 'normal' abundances, probably associated with large
galactic halos. (C-2) Cosmologically distant intervening material
consisting of primordial uncondensed gas. Examples of each type of
spectra are given and their ionization and other spectral characteristics discussed. The similarity between the development of novae
spectra and a possible evolutionary sequence of the explosive ejecta
of type A is striking and suggestive. Several difficulties and unsolved problems involving this scheme are noted. Finally, we speculate
on the interpretation of two interesting objects (PKS 0237-23 and the
'twin quasars' 0957+56A,B) in the context of this scheme.

1. INTRODUCTION

Several years ago, Bahcall (1971) proposed a two-component classification scheme for absorption spectra in Quasi-Stellar Objects. On
the basis of more recent work, briefly summarized below, it now seems
possible to recognize at least four distinct origins for QSO absorption lines. Although a reasonable case can be made for such a scheme,
we shall spend more time discussing the various weaknesses and unsolved
problems this scheme presents than we shall justifying it.

2. BRIEF DESCRIPTION OF QSO ABSORPTION LINE CLASSIFICATION SCHEME

The scheme (modified slightly from the discussion of Weymann et.
al. 1979, hereafter WWPT) is summarized in Table 1.

G.O. Abell and P. J. E. Peebles (eds.), Objects of High Redshift, 107–117.

TABLE 1

QSO ABSORPTION LINE SYSTEM CLASSIFICATION

TYPE	EXAMPLES	IONIZATION AND OTHER CHARACTERISTICS	CLOUDS ON HORIZON AND MAIN UNSOLVED PROBLEMS
A EXPLOSIVE EJECTION	PHL 5200 MCS 275 1303+308	High to Very High N V often >> Ly α Not ≡ Emission Line Region; Outside ELR	1. Why hardly ever strong radio source? 2. Total energetics? Total mass? 3. Ejection mechanism? 4. Evolutionary Sequence? 5. Is 0237-23 a member?
B INTRINSIC B-1 Intrinsic Galaxy	3C191	Broad Range Si II F.S. Present	1. No evidence this class exists! 2. Acceleration of disc material?
B-2 Intrinsic Cluster	0736-06 PKS 0119-04	High; Mg II << C IV NV often present	1. Velocity Dispersion ≈ rich cluster, but no low z QSOs in rich clusters. 2. Some examples of extreme velocity? 3. Parameters consistent with class C-1?
C INTERVENING C-1 Metal Enriched	Q1101-264 PHL 938	Moderate, but definitely higher than disk. CII F.S. usually, but not always << less than disk of our galaxy	1. OVV/Mg II absorption correlation? 2. Meaning of occasional F.S. lines? Disk of a galaxy? 3. Why galaxy ⟶ absorption but not absorption ⟶ galaxy so far?
C-2 Primordial	All high z QSOs shortward of Ly α	Unknown	1. Why lack of clustering? 2. Clear evidence of extreme metal deficiency? 3. Behavior at small z? 4. Helium abundance? (need knowledge of ionization).

The prototype of the first class, type A, PHL5200, was first discovered by Lynds (1967) during the early phase of spectroscopic studies
of QSOs and for many years was considered to be almost unique. However
it now appears that type A objects are fairly common and may comprise
up to 10% of the optically selected QSOs with redshifts greater than
about 1.6. The type is characterized by very broad absorption troughs
with inferred ejection velocities up to 0.1 c. Although there is no
direct proof that they in fact involve ejected material (it has been
suggested that they could arise from intervening supernova remnants)
the fact that (a) The absorption very frequently sets in just at the
emission line redshift and (b) There are instances of multiple troughs
well separated in velocity argues very strongly for ejection by the
QSO. (See also discussion remarks by Wampler.) Two examples of this
type of absorption are shown in Figure 1.

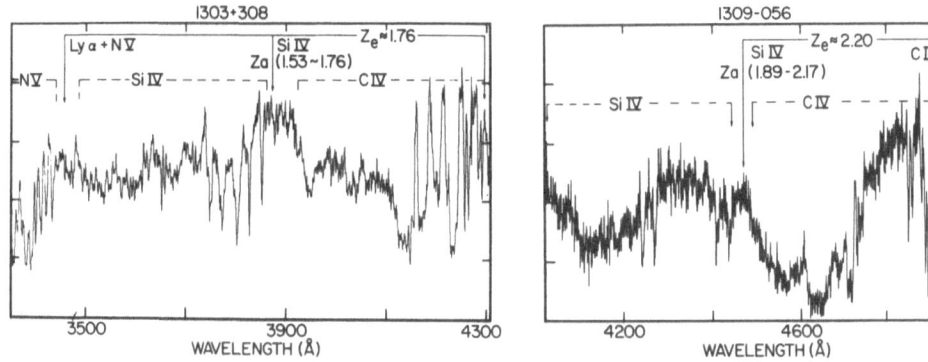

Fig. 1. Two broad-line QSOs in early (1309-056) and late (1303+308)
stages of development. (1309 illustration courtesy Carswell and Smith)

When a histogram is made of the distribution of the (apparent)
ejection velocities of the CIV doublet (Figure 1 of WWPT) one finds
a pronounced maximum at zero velocity, but this peak extends to
infalling material and has a characteristic dispersion of about 1000
km/sec. Such infalling material cannot be due to 'cosmologically distant' material, although it could be attributed to the gravitational
potential of the QSO itself. However, arguments involving the CII fine
structure line and applied to a single object (PHL1222, Williams and
Weymann 1976) strongly suggest that the absorbing material is at a
distance of several hundred kpc from the QSO. This scale and the
velocity dispersion are characteristic of rich clusters of galaxies
and this suggests that the zero velocity peak arises from gas moving
within a rich cluster in which the QSO is itself moving.

Very sharp absorption lines of the metals (especially CIV and MgII) in which the absorption redshift can be much less than the emission redshift are frequent in QSO spectra. In addition, shortward of Lyman α emission line in high redshift QSOs there are a large number of sharp absorption lines which are now recognized to be Lyman α. The prevailing opinion is that most of both these types of lines (class C-1 and C-2) are due to cosmologically distant intervening material, with the metal lines being associated with extended halos of galaxies. There are two lines of argument supporting this view. First, there has been direct detection of absorption lines in a QSO having the same redshift as a nearby galaxy, the best known example being 3C232, where 21 cm absorption was found by Haschick and Burke (1976) and the corresponding Ca II absorption lines subsequently found by Boksenberg and Sargent (1978). Second, there are indirect astrophysical arguments which place the absorbing material at such large distances from the QSO that the ejected masses and energies associated with the shell become exceedingly hard to explain. These arguments involve the failure to detect scattered Ly α radiation around certain QSOs (e.g. Sargent and Boroson 1979) together with the failure to detect the 1335 CII line arising from the fine structure level (e.g. Turnshek, Weymann and Williams 1979).

Until recently it has not been clear whether the large number of Ly α absorption lines shortward of the Ly α emission lines represented merely low column density counterparts to the metal line systems just described or in fact represented a different component of intervening material, perhaps uncondensed primordial gas. Sargent et. al. (1979) have now presented convincing evidence that the latter interpretation is correct. They find that the CIV lines and Ly α lines have different statistical properties, the former tending to clump in velocity space at a characteristic splitting of about 150 km/sec while the latter show no tendency to clump. These authors proceed to draw several important conclusions and inferences about the properties of these clouds and an inferred confining intergalactic medium.

3. PROBLEMS AND DIFFICULTIES

With this very brief outline of the proposed origins of various types of QSO absorption lines we now consider some of the major unsolved problems posed by this scheme.

3.1 Ejected Systems

Although we have described PHL5200 as the prototype of the ejection systems, in fact its very broad smooth troughs with relatively little structure differ significantly from those of other members of this class, some of which show more than one distinct trough, sometimes well separated from the emission line (MCS 275) while others show very complex and occasionally quite sharp components spread over several thousand km/sec (1303+308). All seem characterized by a very high

level of ionization. In Turnshek et. al. (1979) it is speculated that
this range in appearance represents an evolutionary sequence, with
objects like PHL 5200 and 1309-056 representing a phase early in the
ejection episode and objects like 1303+308 representing a much later
phase when the ejecta have spread out and the column density has
decreased. Then, only points in the profile where the material remains
highly clumped in velocity space (or where such clumps develop due to
instabilities) have high enough opacity to produce observable absorption.
Very indirect support for this speculation is provided by the development of the absorption spectrum in some novae. This is illustrated in
Figure 2 in a sequence of spectra of Nova Delphini obtained by Dr. L.
Kuhi. One observes a stage in which there is more than one broad
trough and the development of sharp components reminiscent of the
sequence PHL 5200, MCS 275 and 1303+308.

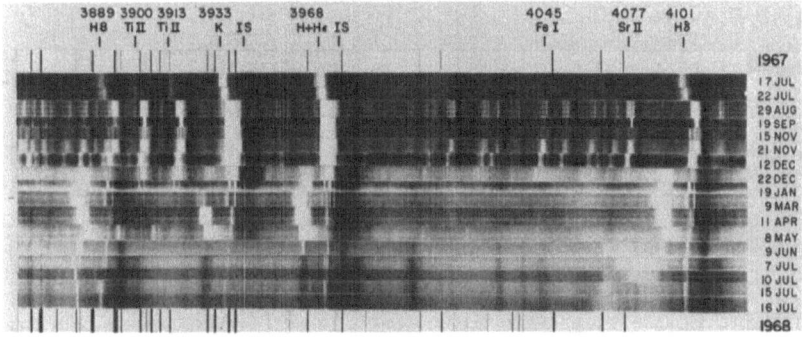

Fig. 2. Nova Delphini (Courtesy L. Kuhi).

The widespread belief that both pre-novae and QSO nuclei involve
accretion discs adds interest to the analogy but before it can be
seriously pursued some fundamental questions about these objects must
be answered. What are the masses and energies involved in the ejecta?
What is the duration of the ejection? What is the true incidence of
this phenomonon among all QSOs? PHL 5200 has been observed for over
a decade and no clear evidence for a change in the absorption structure
has been observed. This suggests that a quasi-steady wind lasting
~ 10 years is involved or that the scale of the flow is larger than a
few tenths of a parsec. Photoionization models generally require
material in the emission line region to be at distances of 1 to 10
parsecs from the continuum source. The residual intensity and shape
of the profiles in some of these objects, together with an apparent
difference in the ionization level between the emitting and absorbing
regions (e.g. the MgII/CIV ratio) seems to require that the absorbing
material be separate from, outside of, and cover most of the emission
line region. In a recently studied sample of QSOs (Turnshek et. al.
1979) discovered with objective prism techniques by the University of
Michigan group (McAlpine and Lewis, 1978) 3 out of 15 objects were
members of this class. Though this summer is very small and selection

effects may be important it appears that the covering factor for absorbing material in this sample must be at least 0.1. These considerations suggest ejecta masses of 10 solar masses or more and energies of at least 10^{53} ergs. Some violent outbursts in QSOs (Eachus and Liller 1975) have released about this amount of energy in visible radiation. Photons, relativistic gas or hot thermal gas may all play an important role in the acceleration process.

An important clue to the nature of these outbursts is surely contained in the fact that with one or possibly two borderline exceptions these objects are not strong radio sources, but we have no idea which is cause and which is effect in this correlation. Perhaps strong radio sources establish a clean channel through which relativistic plasma and hot gas flows before radiative cooling occurs in the thermal gas. Perhaps geometrical orientation is crucial.

3.2 Intrinsic Intervening Systems

Many examples are known in which absorbing material is apparently falling in towards the QSO with speeds up to about 3000 km/sec. (Note that such infalling material is not confined to QSOs: NGC 1275 also shows the same phenomenon.) As mentioned previously, largely on the strength of the lack of the $\lambda1335$ CII line in a single object, PHL 1222, it has been argued (Weymann et. al. 1977; WWPT) that these systems arise in rich clusters where both the extremes in velocity differences and the mean dispersion in velocity are adequate to explain the phenomenon. However, as emphasized by Perry, Burbidge and Burbidge (1978), this model runs into the difficulty that QSOs of low enough redshift such that rich clusters in which the QSO is embedded could be detected avoid such clusters. Instead, such QSOs seem to be frequently associated with small groups having very much smaller velocity dispersions. (Stockton 1978; note that Stockton's data also seem to rule out both systematic and random differences between the measured emission redshift and the true redshift of the QSO large enough to avoid the difficulty.) Since the redshift required to detect such CIV systems (comparable Mg II lines are rarely seen) from the ground is at least 1.2, it is still conceivable that there are evolutionary effects giving rise to differences in rich cluster-QSO associations between low and high redshift QSOs. Observations with Space Telescope would settle this. If this alternative is eliminated we may have to explore the possibility that the presence of a QSO in a large volume of gas having the dispersion of rich clusters suppresses the formation of normal galaxies. If it is true that QSOs are frequently found in the nuclei of otherwise normal spiral galaxies, then we should occasionally see absorption lines (possibly shocked and mildly accelerated) arising from this gas. The high densities and moderate distances in 3C191 (Williams et. al. 1975) may be an example of such a situation. The small outflow velocity may be the result of a QSO wind as discussed recently by Dyson, Falle and Perry (1979).

3.3 Cosmologically Distant Intervening Material

The hypothesis that essentially all of the sharp, highly displaced metal lines are due to cosmologically distant material implies that associated with many galaxies are clouds with approximately solar abundances extending to many Holmberg radii. 'Effective cross sections' implying radii for halos of about 100 kpc for galaxies like our own are required, but this number depends upon assumptions concerning the Hubble constant, galaxy luminosity function normalization, and whether all classes of galaxies contribute to such absorption. Moreover, since the material is certainly patchy, the actual radii out to which absorbing clouds are found must be substantially larger than this. The clouds responsible for most of these displaced systems are not typical of those responsible for interstellar lines in the disc of our galaxy. In the QSO lines the ionization is in general higher than it is in the disc. Recent observations of stars in the Magellanic Clouds by Savage and deBoer (1979) in which the line of sight passes through our halo suggest there is a substantial column density of CIV well above the disc. Occasionally, however CII fine structure lines are apparently seen in highly displaced QSO absorption systems. Does this represent a chance passage of the line of sight through a galaxy disc? The fact that the absorbing material is usually associated with highly ionized gas may explain two recent results of Wolfe (1979). Wolfe found that observable 21 cm absorption was detected in only 1 of 16 QSOs showing FeII or MgII absorption and, that the results of a search for 21 cm emission were incompatible with the incidence of MgII absorption by halos of intervening galaxies unless the properties of the galaxies in the 21 cm survey were different from the hypothetical intervening ones or that the gas in the halos of the galaxies was more than 90% ionized.

In any event some basic problems concerning the generation and/or confinement of such metal-containing clouds at distances of at least 100 kpc from the galaxy nucleus remain to be understood.

At least one potential observational inconsistancy remains in this picture: The incidence of MgII absorption in optically violent variables (Miller 1979) and in some QSOs which are compact radio sources (Burbidge 1979) seems much higher than expected on the basis of the survey of WWPT. This point needs to be settled by examination of a control group with data of exactly the same quality and coverage as used in the OVVs and compact sources.

As noted earlier, the discovery by Sargent et. al. (1979) that the Ly α lines are distributed uniformly in redshift with no tendency to bunch in velocity space, and with properties which are independent of the particular QSOs examined argues strongly for cosmologically distant intervening material of a character different from the metal-containing clouds described above. These authors discuss the properties of the clouds and infer that they must be confined by a warm intergalactic medium which ought to be detectable by a general depression in the

continuum of all QSOs shortward of HeII Ly α. It will be of great
interest to see if the approximately constant density of Ly α lines
with redshift holds right down to the present epoch. In addition,
if these clouds really represent primordial material they should be
nearly pure hydrogen and helium. With space telescope one ought to
be able to observe oxygen in its most abundant ionization state and
detect lines in some of the larger clouds even if the abundance is
down by a factor of 1000 from solar abundance.

4. TWO SPECIAL OBJECTS

4.1 The Twin Quasars

There are two objects which pose especially interesting problems
for the scheme described above. One is the pair of nearly identical
quasars 0957+561A,B (Walsh et. al. 1979). At present, the various
radio, spectroscopic, X-ray and direct photographic evidence does not
conclusively rule out the gravitational lens hypothesis, although substantial difficulties are posed by it. Suppose the lens hypothesis is
eliminated. Then, as noted by Walsh et. al. and Weymann et. al. (1979)
the explanation for the velocity difference of less than 15 km/sec
between the absorption systems in the two objects involves some
unlikely coincidences. If the velocity difference of about 2000 km/sec
between the emission line and absorption line redshifts is used to
classify the lines as of type B-2, then agreement to within 15 km/sec
of two clouds chosen at random out of a sample with a dispersion of
750 km/sec is unlikely, especially when coupled with the fact that
very low ionization systems such as the ones in 0957+561 are very rare
in type B-2 systems. Alternatively, we could consider the clouds to
be an intervening halo (type C-1). Since the characteristic velocity
dispersion within a halo is very much smaller, the agreement in velocity
between A and B is less unlikely, but this advantage is offset by the
a priori improbability that an intervening system would occur only
2000 km/sec to the blue of the emission line redshift.

4.2 PKS 0237-23

The absorption spectrum of this object has defied a convincing
interpretation despite many years effort and some excellent data.
Boroson et. al. (1978) proposed that the concentration of CIV doublets
represents an intervening supercluster. The discovery of several type
A objects with complex troughs extending to at least 0.1 c suggests
the possibility that the CIV complexes in this object represent a more
extreme and later phase of the same phenomenon. There is still some
dispute about the presence of the SiII fine structure lines and it is
important to resolve this point (cf. Roberts 1978). If the ejection
occuring in the type A objects is accompanied by some kind of collimat-
ing process than very late stages in this type may have rather small
covering factors and the energetics of an object like 0237-23 with
marginally present SiII fine structure lines may not be insuperable.

Even if this interpretation is correct, the extremely high velocities together with the fact that PKS 0237-23 is such a strong radio source makes it a unique object.

I wish to thank Drs. J. Bahcall, E.M. Burbidge, R.F. Carswell, W.L.W. Sargent, M.G. Smith, A. Wolfe, and A. Wright for the use of both published and unpublished material in preparing this review and for several helpful discussions. Much of the work summarized here was done in collaboration with colleagues at the Steward Observatory and I wish to acknowledge their contributions, especially H. Stockman, P. Strittmatter, D. Turnshek, B. Peterson, J. Scott and R. Williams.

REFERENCES

Bahcall, J. N. 1971, A. J., 76, p. 283.
Bokenberg, A., and Sargent, W. L. W. 1978, Ap. J. 220, p. 42.
Boroson, T. A., Sargent, W. L. W., Boksenberg, A., and Carswell, R. F. 1978, Ap. J. 220, p. 772.
Burbidge, E. M. 1979 (private communication).
Dyson, J. E., Falle, S. A. E. G., and Perry, J. J. 1979, Nature, 227, p. 118.
Eachus, L. J., and Liller, W. 1975, Ap. J. (Letters) 200, p. L61.
Haschick, A. D., and Burke, B. F. 1975, Ap. J. (Letters) 200, p. L137.
Lynds, C. R. 1967, Ap. J., 147, p. 396.
MacAlpine, G. M., and Lewis, D. W. 1978, Ap. J. Suppl. 36, p. 587 and references to lists I, II and III therein.
Miller, J. S. 1979 (private communication)
Perry, J. J., Burbidge, E. M., and Burbidge, G. R. 1978, Pub. Ast. Soc. Pac. 90, p. 337.
Roberts, D. H. 1979, Ap. J. 228, p. 1.
Sargent, W. L. W., and Boroson, T. A. 1979, Ap. J. 228, p. 712.
Sargent, W. L. W., Young, P. J., Boksenberg, A., and Tytler, D. 1979, Ap. J. Suppl. (in press).
Savage, B. D., and deBoer, K. S. 1979, Ap. J. (Letters), 230, L77.
Stockton, A. N. 1978, Ap. J., 223, p. 747.
Turnshek, D. A., Weymann, R. J., and Williams, R. E. 1979, Ap. J., 230, p. 330.
Turnshek, D. A., Weymann, R. J., Liebert, J. W., Williams, R. E., and Strittmatter, P. A. 1979, Ap. J. (in press).
Turnshek, D. A., Green, J., Liebert, J. W., Strittmatter, P. A., Weymann, R. J., and Williams, R. E. (in preparation).
Walsh, D., Carswell, R. F., and Weymann, R. J. 1979, Nature, 279, p. 381.
Weymann, R. J., Williams, R. E., Beaver, E. A., and Miller, J. S. 1977, Ap. J. 213, p. 619.
Weymann, R. J., Williams, R. E., Peterson, B. A., and Turnshek, D. A. 1979, Ap. J. (in press).
Weymann, R. J., Chaffee, F. H., Davis, M. and Carleton, N. P. 1979, Ap. J. (Letters) 233 (in press).

Williams, R.E., Strittmatter, P.A., Carswell, R.F., and Craine, E.R.
     1975, Ap. J. 202, p. 296.
Williams, R.E., and Weymann, R.J. 1976, Ap. J. (Letters) 207, L143.
Wolfe, A.M. 1979, paper presented at "The Universe at Large Redshift"
     Copenhagen, June 25-29, 1979.

DISCUSSION

Aller: You mentioned the metallic absorption lines attributed to
                    absorption by halos of intervening galaxies. Can we draw
any firm conclusions about gas kinetic temperatures, densities, and
levels of excitation in these halos?

Weymann: The level of ionization seems definitely higher in QSO
                    spectra than in our disc. Likewise, the ratio of strengths
of the fine structure CII line $\lambda 1335$ to the ground state line $\lambda 1334$ is
generally lower in QSO spectra than in interstellar lines, implying that
the electron density is lower in the halo (which would explain the
higher ionization). Note that the CIV at zero redshift seen in 3C 273
is consistent with this picture.

Wampler: I would like to comment about the suggestion that the
                    strong absorption seen in PHL 5200 is due to an interven-
ing supernova explosion. In the PHL 5200 system one does see a nearby
symmetric Lyα emission feature in the absorption trough of NV. Now,
as you pointed out, MgII $\lambda 2800$ does not show as strong absorption features. This can be understood if Mg^+ is in a cloud in front of the
region producing the strong absorption. In that case the Lyα is easily
explained as emission coming from the Mg^+ region, too. But if the
absorption is produced by an intervening supernova, the Lyα line in the
QSO should be absorbed by the NV feature in the SN and then the Lyα
line from the SN shell at $z \approx 2$ should be too weak to detect.

Weymann: No comment except to note that the suggestion has been
                    made by others, not by me.

Rees: The blanketing of the continuum by the "Lyman-α forest"
                    increases with redshift as $(1 + z)^2 (1 + \Omega z)^{-1/2}$ x (a
factor allowing for the z-dependence of cloud properties). Maybe the
effect could frustrate the attempts by Dr. Osmer to discover quasars
with $z > 4$.

Weymann: I doubt it. The forest shouldn't eat into the red wing of
                    Lyα, so Lyα emission detection techniques should work.

Chaffee: Do you have any evidence for the existence of H_2 absorp-
                    tion toward any QSO?

Weymann: H_2 has been looked for in the spectra of a number of high
                    redshift QSOs and has not been found. It would not be

expected in either highly ionized galactic halos or in the hydrogen
bubbles of Sargent et al.

G. Burbidge: What is your opinion concerning line locking? If the
                 ratio 1.11 is real, will this give you a problem in terms
of your various scenarios for absorption line production?

Weymann: If the 1.11 line locking ratio were to be demonstrated
                 to be a statistically significant phenomenon in a certain
subclass of absorption systems, we would certainly be forced to the
conclusion that they presented ejecta at 0.1c, possibly the remnants
of what I have called "class A" systems, which apparently eject material at $\leq$ 0.1c. My own opinion, however, is that the most persuasive
cases for line locking may be found in CIV doublet overlapping and
SiIV doublet overlapping. This still implies ejection, and is much
easier to understand theoretically.

STANDARD CANDLES IN QSO's?

E. Joseph Wampler
Lick Observatory, University of California, Santa Cruz 95064

We present here the current status of a continuing program to investigate the possibility that certain resonance lines in the spectra of QSO's can be used as luminosity calibrators for the spectra.

I.   INTRODUCTION

Baldwin (1977) discovered a strong negative correlation between the equivalent width of the Lyα and C IV λ1549 emission lines in QSO's and the luminosity of the underlying continuum. The effect was used by Davidsen, Hartig and Fastie (1977) to determine the luminosity of 3C 273. From a comparison between the derived luminosity of 3C 273 and the luminosity given by Baldwin (1977) for high redshift QSO's, Davidsen et al. (1977) found a formal value for $q_o \approx 1.0$.

Possible systematic effects which could produce the correlation found by Baldwin include: a) observational selection effects that were not properly taken into account in the initial "random" sample of QSO's, b) effects associated with the radio properties of the QSO's, c) evolutionary effects that would be a function of z, and d) observational errors associated with a sample of spectra that was taken for other purposes over a period of years with an instrument whose physical configuration was changing.

In order to reduce or remove errors associated with effects a, b, and d and to investigate possible evolutionary effects (c), a moderately large complete sample of flat radio spectra QSO's was chosen from the lists of Schmidt (1977) and Wills and Lynds (1978). It is important to note that the QSO's studied here were chosen solely on the basis of their radio properties and the positional coincidence between the radio and optical positions. The properties of their optical spectra were not a criterion for inclusion in the study; in fact, a number of BL-Lac objects were included in the survey. The number of QSO's chosen was sufficiently large that the correlation between line strengths and continum luminosity could be studied as a a function of z.

G.O. Abell and P. J. E. Peebles (eds.), Objects of High Redshift, 119–123.
Copyright © 1980 by the IAU.

The results given here are the product of the continuing observations by Baldwin, Burke, Gaskell and Wampler.

II.   CURRENT STATUS OF THE PROJECT

A.   The C IV Luminosity Indicator

C IV $\lambda 1549$ was originally selected by Baldwin (1977) as the most
useful of the lines that showed a strong luminosity correlation because
a) the line is not as strongly distorted by absorption as $Ly\alpha$ in high
redshift QSO's, and b) the line is observable in QSO's with a wide
range of redshifts ($1.1 \lesssim z \lesssim 3.5$).

In a preliminary status report on the observations of this flat
spectrum sample, Baldwin, Burke, Gaskell and Wampler (1978) found that
the correlation described earlier by Baldwin (1977) was present in this
sample of QSO's and that the correlation existed for subsets of the list
that were chosen to have comparatively narrow ranges in z.

Because only the very best observing conditions were suitable for
observing the faintest objects in the list, the paper by Baldwin et al.
(1978) contained a limited selection of the faintest objects in the
observing list. Additional observations have increased the sample of
faint QSO's and we find that the new data points lie along the line
defined by the earlier observations. We have now observed all but a few
of the objects for which we can detect C IV $\lambda 1549$ from the ground and
we find no reason to change our earlier conclusions.

The BL-Lac objects in our sample present a problem. Since we
cannot measure the equivalent width of C IV $\lambda 1549$, if, indeed, C IV $\lambda 154$
is present, we do not know where to plot the BL-Lac objects on our graph
In addition to the absence of emission lines, these objects have much
steeper spectra than the other QSO's in the survey. We have taken the
somewhat arbitrary point of view that they represent a class of objects
that can be ignored for the purpose of this survey.

Since our survey began, a few space observations of C IV $\lambda 1549$ in
individual sources have become available. The data for QSO's are
consistent with the Baldwin (1977) relationship if $q_o \approx 1$ (Gaskell,
private communication). The C IV $\lambda 1549$ lines in Seyfert galaxies are
too weak to fit the regression line defined by the QSO's. Clearly more
space observations are needed to investigate the relationship for low
redshift QSO's and to provide a wider range of redshifts to determine
the value of q_o.

B.   Mg II $\lambda 2800$ as a Luminosity Indicator

In our search for a luminosity indicator that would be useful for

low redshift QSO's, Baldwin et al. (1978) pointed out that the
equivalent width of the resonance line Mg II λ2800 also showed a
negative correlation with the continuum luminosity.

New data show that while there does seem to be a correlation
between the strength of Mg II λ2800 and continuum luminosity the
correlation is not so tight as for C IV λ1549 and the slope of the
correlation seems to be slightly steeper than that found for
C IV λ1549. In fact for Mg II λ2800, $\log I(Mg~II) \approx 2/3 \log L_{cont.}$
instead of the C IV dependence, $\log I(C~IV) \approx 1/3 \log L_{cont.}$

As has been noted by numerous authors, e.g., Davidsen et al. (1977)
the continuum level near Mg II λ2800 is not a simple interpolation of
the continuum slope in regions away from λ2800. The anomolous excess
radiation near λ2800 may be affecting the W Mg II-$L_{cont.}$ relationship.
More data will be needed to clarify the situation.

C.   The Value of q_o

The new Mg II λ2800 data have been tied to the C IV λ1549 data
by measuring both lines in a number of QSO's of intermediate redshift.
Despite the scatter in the Mg II λ2800 data noted above, the new data
strengthen the conclusion reached by Baldwin et al. (1978) and
Davidsen et al. (1977) that the value of q_o as determined by calibrated
QSO's is high enough to close the universe. The new UV satellite data
of C IV λ1549 in low redshift QSO's also support this conclusion
(Gaskell, private communication).

Of course it is possible that an undetected evolutionary effect
is giving an anomolous value for q_o, but such an effect, if present,
must leave unchanged the relationship between W C IV and $L_{cont.}$ as a
function of redshift. We have found no evidence to suggest that the
emission line spectra of high redshift QSO's are inherently different
from those of low redshift QSO's.

References

Baldwin, J. A. : 1977, Ap.J.,214, 679-684.
Baldwin, J. A., Burke, W. L., Gaskell, C. M., and Wampler, E. J.: 1978,
     Nature 273, pp. 431-435.
Davidsen, A.F., Hartig, G. F., and Fastie, W. G.: 1977, Nature 269,
     pp. 203-206.
Schmidt, M.: 1977, Ap.J.,217, pp. 358-361.
Wills, D., and Lynds, R.: 1978, Ap.J. Suppl. 36, pp. 317-356.

DISCUSSION

Murdoch: The upper three points around 21^m on the m_v vs log W(CIV)
               diagram in the paper by Baldwin et al. (Nature, 273, 431)
have all decreased in intensity by ~ 1^m5 compared to their discovery
values in the finding surveys. Is this also true of the new points in
this region of the diagram? If so, there remains only an ill-defined
cloud of points in the centre of the diagram apart from the two points
in the lower left of the diagram which are not part of the complete
sample.

Wampler: Our measured continuum intensity is less than the stated
               limiting magnitude of the surveys. By including the emission
lines part of the discrepancy is removed but we still find the objects
fainter than the limiting magnitude. This could either be because the
objects have decreased in brightness since their discovery or because
the photographic estimates were incorrect. We now have no way of
deciding between these possibilities. However, note that there is still
a correlation (although weaker) even if the faint objects are not used.

Murdoch: If the objects near 21^m appeared on the sky survey above their
               true continuum luminosities because of strong emission lines
rather than being variable, then in order to have a truly complete
sample one would need to consider other objects with faint continuum
magnitudes which may well have weak emission lines.

Wampler: We have not been able to find faint objects with weak CIV.
               For MgII we have a few examples of such objects. Perhaps
just as important, we have found no bright objects with strong lines.
We find that even objects such as OQ172, which are reported to have
strong lines, have relatively weak lines when compared to faint QSOs.
The important result is that the use of this spectroscopic callibration
procedure can reduce the scatter in the Hubble diagram by an order of
magnitude.

Penston: I think this is very interesting and looks quite encouraging
               but, in fact, the Seyfert galaxies and BL Lac objects do not
fit this relationship. How does an active nucleus know it is a "quasar"
rather than a Seyfert galaxy or a BL Lac object?

Wampler: I don't know. We commonly assume that QSOs, Seyfert galaxies
               and BL Lac objects are different examples of the same phenom-
ena, but we must remember that this is an assumption.

Osmer: Isn't the point of the discussion on variable quasars that if
               the continuum luminosity varies and the line luminosity stays
constant, then you get a track nearly parallel to your luminosity
effect?

Wampler: Not quite. · For CIV we find $I_{(CIV)} \approx L^{1/3}$ and not $I_{(CIV)} \approx L^0$.
               I think that the data excludes the latter relation.

D. Roberts: You have included the same variable (the continuum flux) in both ordinate and abscissa. What does the plot look like if you just throw out the continuum flux and plot the number of CIV photons observed versus the cosmological factors?

Wampler: There is still a correlation, but in the case $I_{(CIV)} \sim L_{cont}^{1/3}$ or $I_{(MgII)} \sim L_{cont}^{2/3}$.

IUE OBSERVATIONS OF QUASARS

M. Schmidt, R. F. Green, J. R. Pier
Hale Observatories
California Institute of Technology
Carnegie Institution of Washington
                 and
F. B. Estabrook, A. L. Lane, H. D. Wahlquist
Jet Propulsion Laboratory

Spectra of six quasars have been obtained with the International Ultraviolet Explorer Satellite. Five of the six show no evidence for strong Lyα absorption between the redshifted and rest wavelengths, for $.23 \leq z_{em} \leq 1.72$. In addition, the quasar PG 1115+080 at z=1.72 shows no evidence for strong He I absorption from the resonance transition at $\lambda 584$ Å. These results confirm that the intergalactic medium must be both tenuous and hot enough to produce an optical depth <0.1 in neutral hydrogen and helium. In no case was the Lyman edge detected in absorption near z_{em}. Four of the objects produce an average Lyα/Hβ intensity ratio of 6.3, in disagreement with the theoretical prediction for Case B optically thick recombination of 30. Also, two of the objects show Lyγ in emission, a result unexpected from Case B line transfer assumptions. The Lyα emission line in 3C 351 shows the identical sharp core plus 20,000 km s^{-1} broad wings observed in Hβ and Mg II, implying a common origin in the same dynamical ensemble of emitting regions. These quasars show systematically steeper spectral indices when the energy distributions are fit from the ultraviolet through the visible than those derived from the visible spectra alone. PG 1115+080 shows a featureless continuum down to an observed $\lambda 1173$ Å. The ionizing spectrum, with $f_\nu \alpha \nu^{-2.0}$, therefore persists beyond 2 Rydbergs. The spectrum of PG 1247+268, with z=2.038, contains a strong absorption line at observed $\lambda 2697$ Å, with no net flux detected from $\lambda 2000$ Å down to the observed limit at $\lambda 1150$ Å. This result is interpreted as absorption in Lyα and the Ly edge at z=1.218. Low dispersion optical spectra show no evidence for Mg II or C IV absorption in the same system; the signal to noise ratio is too low in the IUE spectrum to confirm Lyβ. We conclude that the line of sight intersects a metal-poor cloud with $\tau \leq 1$ in the Ly continuum, at $(1+z_{em})/(1 + z_{abs}) = 1.37$.

G.O. Abell and P. J. E. Peebles (eds.), Objects of High Redshift, 125–126.

DISCUSSION

Aller: These observations of such faint objects are truly quite
             remarkable. I presume that you selected quasars that had
small interstellar extinction (which can put such a severe restraint
on measurements of many galactic objects).

Green: It is true that these quasars are generally at high galactic
             latitude. The worst case is 3C 351, at b = 36°. I examined
the effects of interstellar reddening on this object by assuming an
average cosecant reddening law, in the form proposed by Sandage, and by
applying the Code et al. extinction curve. There was no effect on the
fitted spectral index, and the Lyman to Balmer line ratios were increased
by 20%. Since the qualitative consequences are not changed and this is
the extreme case, the data as shown were not corrected for interstellar
reddening.

Longair: In the case of 3C 390.3, the line profiles can be decomposed
             into broad-line and narrow-line components, the narrow lines
having the standard recombination ratio of Lyα to Hα while the broad
lines have the anomalous ratio. Have you been able to perform this
decomposition for your quasars and what are the answers?

Green: The only quasar of the four with observed Lyman and Balmer
             emission lines for which a decomposition seems possible is
3C 351, but it has not yet been done. The total Lyα:Hβ ratio is around
3, but the sharp-line component contributes only ~ 10% to the total Hβ
flux according to Grandi and Phillips. It is therefore possible that
3C 351 resembles 3C 390.3 in line component ratios, and that will be
investigated. In the case of PKS 1302-102, Lyγ is seen in emission,
so the standard model cannot be strictly applicable.

ANOMALOUS REDSHIFTS OF QSOs

J. V. Narlikar
Tata Institute of Fundamental Research, Bombay 400005, India

P. K. Das
Indian Institute of Astrophysics, Bangalore 560034, India

The evidence for association of high redshift QSOs with low red-shift galaxies is subject to controversies. If it can be convincingly argued that the observed associations are accidental and could occur with reasonably high probabilities (say > 5%) then the cosmological interpretation of the redshifts of QSOs remains unaffected by the data. If, however, there is increasing evidence, either statistically or in direct physical terms, for these associations to be real, then the excess redshifts of the QSOs become anomalous. Taking the latter alternative seriously, a suggestion was made by Narlikar (Annals of Physics, 107, 325, 1977) that the anomalous redshift of the QSO in a typical QSO-galaxy pair could arise because the particle masses in the QSO were systematically smaller than those in the companion galaxy, as predicted by a theory of gravitation based on Mach's principle. One astrophysical consequence of this effect is that the QSO should appear younger than the galaxy.

If the QSO is ejected from the galaxy at some stage in its life, would it remain in the neighbourhood of the galaxy? The mass of the QSO is initially zero and grows with epoch. We have examined the dynamics of such a variable mass QSO ejected from the companion galaxy. In the specific case of the galaxy NGC 3067 and the QSO 3C 232, numerical integration of the dynamical equations shows that the effects of low mass and large velocity last for a very short time so that the QSO remains trapped within the gravitational field of the parent galaxy and its separation is in agreement with that observed. Other observable consequences of this idea are being investigated.

G.O. Abell and P. J. E. Peebles (eds.), Objects of High Redshift, 127–129.
Copyright © 1980 by the IAU.

DISCUSSION

Weymann: Do you not agree that the arguments that a <u>significant</u>
              <u>fraction</u> of QSOs are cosmological <u>are</u> compelling (i.e., the
Green-Schmidt 8/mag increase; Stockton's results)? Thus, if Arp's
objects are local, should they not show some quite different properties
(e.g., in different properties of the absorption spectra shortward of
Lyα) from the "real" cosmological QSOs? This is certainly testable!

Narlikar: The numbers in the Green-Schmidt survey are still too small
              to convince me that the steep slope has any significance.
Regarding Stockton's quasars, as far as I can make out some have z_Q =
z_G, whereas some have $z_Q \neq z_G$. If we concentrate only on the former,
we can assume that the redshifts of these quasars are cosmological,
although they are small. The Arp-quasars are of course considerably
closer than these Stockton quasars, and any effect or test which makes
use of the column density of the intergalactic medium should be able
to distinguish between the two.

Davis: It seems to me that your theory makes a very specific predic-
              tion of a strong correlation between the redshift of a QSO
and its proximity to a bright galaxy of low redshift.

Narlikar: Yes. It is possible to calculate the maximum angular separa-
              tion as a function of the redshifts of the galaxy and the QSO.

Green: In your theory, would a particle new to the Universe, for
              instance, created by pair production in the laboratory, start
out with 0 mass and then grow in mass as its own horizon expands?

Narlikar: There is a difference between pair creation in the laboratory
              and the type of creation envisaged here. In the former, e.g.,
in $e^- e^+$ creation, we have in the Feynman diagram a world line which first
goes backward in time and then turns round (at the creation event) and
goes forward in time. In the latter we have creation as in a big bang
(only delayed and on a smaller scale) so that the world lines have a
genuine beginning at the creation event. The rule of mass growth with
epoch applies from the epoch of such genuine beginnings.

Scheuer: How does the incidence of "delayed big bang" depend on the
              surrounding density in your theory? Would it only happen in
the centers of galaxies?

Narlikar: I expect it to occur more often near places of high density
              such as in the nuclei of galaxies.

Peebles: Does your theory not predict the number-magnitude relation
              $N \propto 10^{0.6m}$ for quasars?

Narlikar: If the galaxies are distributed uniformly out to distances of
              ~ 30-100 Mpc, I expect the QSO population to follow the same

rule. However, there are fluctuations in the luminosities of QSOs and one has to take into account the (Hz)^2 factor in the denominator for optical flux. In a large enough sample, these effects would be ironed out and one would get a $N \propto 10^{+0.6m}$ dependence. But in small samples such as the one discussed by M. Schmidt yesterday, these fluctuations as well as any local inhomogeneities will be important. For this reason I do not consider the observed steepness of the quasar number count to be significant.

A PROBABLE BL LAC OBJECT WITH ABSORPTION REDSHIFT 1.49

R.W. Hunstead, School of Physics, University of Sydney
J.C. Blades, Anglo-Australian Observatory
H.S. Murdoch, School of Physics, University of Sydney

The radio source 1309-216 is identified with an 18^m stellar object having a steep non-thermal continuum ($\alpha = 2.6$, $f_\nu \propto \nu^{-\alpha}$) with no emission features. High resolution spectra obtained with the AAT reveal strong absorption due to the CIV doublet at redshifts 1.361, 1.489 and 1.491. In the strongest absorption system ($z_a = 1.489$) the SiIV doublet and possibly SiII are also detected. No information on the optical polarization is available but spectrophotometry at several epochs shows evidence for optical variability and the object is therefore classified tentatively as a BL Lac object.

The species found in absorption in 1309-216 are consistent either with ejection from a central source or absorption in the haloes of intervening galaxies. The velocity dispersions and column densities derived for the absorbing regions are comparable to those for gas in the LMC and in the halo of the galaxy.

DISCUSSION

Gaskell: I would like to add another object to Rich's sample of one -- this is PKS 0215+015. It also shows two high ionization absorption redshifts of 1.5489 and 1.6480, and as well as these it also has a lower ionization (MgII, FeII, etc.) system at z = 1.3448 ($\pm$ 0.0005). This was found as part of the complete luminiosity criterion survey that Joe Wampler reported on yesterday. It also has a steep optical spectrum, is polarized, and has varied -- so it is definitely a BL Lac object.

Weymann: 1. It should be stressed that while CIV absorption may be anomalous in BL Lac objects compared to MgII, it is much more common in QSOs with $Z_e \gtrsim 1.2$, so I interpret your discovery as simply representing one of the few known high-redshift BL Lacs, but having normal absorption.

G.O. Abell and P. J. E. Peebles (eds.), Objects of High Redshift, 131–132.

2.   I believe the SiIV/CIV does not imply anything anomalous
about velocity structure, but can be interpreted in terms
of normal clouds, and abundances given the lower ionization potential
of SiIV. This same phenomenon is present in many QSOs, also.

DEEP IMAGING OF QUASAR FIELDS

P.A. Wehinger
  Max Planck Institute-Heidelberg
S. Wyckoff
  Arizona State University
T. Gehren
  Max Planck Institute-Heidelberg

Large scale (19 arc sec/mm), deeply exposed photographs (26.5 mag/sq arc sec) of a sample of 12 QSO's over the redshift range, z = 0.158 to 0.528 have been obtained. Baked IIIa-F plates were used at the prime focus of the ESO 3.6-m telescope. The band pass was 5700-6900A. Logarithmic intensity contour maps and logarithmic image profiles were constructed from PDS microdensitometer scans. Seven of the 12 QSO's show extended (5-10 arc sec), asymmetric structure on the contour maps; while 9 of 12 QSO's show broader image profiles than stars of the same magnitude. The intensity contour maps provide two essential bits of information for follow-up spectroscopy: 1) a guide for locating the slit aperture and 2) a means of estimating integration times.

DISCUSSION

Roberts: How much time would be required to obtain a spectrum of one of the extensions?

Wehinger: Using a 4-meter telescope with a digital detector and careful sky subtraction techniques one might obtain spectra (of continuum and possible absorption-line features) in ∿5-10 hours for extended structures with integrated magnitudes of ∿22-23.

Beichman: Are the qualitative characteristics of the extensions and asymmetries repeatable from night to night?

Wehinger: Yes. We have two plates of some of the QSOs and they show the same extensions within the limits of detection (∿1-3% night sky in the red).

Scheuer: Does the proportion of light in the wings of the image depend on the luminosity of the quasar?

Wehinger: We have looked into this question, but as yet have no definite correlation based on the present sample of (12) objects.

G.O. Abell and P. J. E. Peebles (eds.), Objects of High Redshift, 133–134.

M. Burbidge: Your point about accurate positioning of maximum intensity
                 and location of asymmetric low luminosity material is
going to be very important for Space Telescope observations.

Wehinger: Yes, exactly.

OPTICAL IDENTIFICATIONS OF EXTRAGALACTIC RADIO SOURCES

M.S. Longair,
Mullard Radio Astronomy Observatory,
Cavendish Laboratory, Cambridge, England.

1. THE STATE OF THE ART

The ability to identify the many tens of thousands of extragalactic radio sources now known depends upon the precision with which the radio positions and structures are known as well as the availability of high quality plate material. For the brightest radio sources, positions with accuracy better than 1 arcsec are now routinely available and radio structures with angular resolution 1-5 arcsec can be readily measured with instruments such as the Westerbork Synthesis Radio Telescope, the Cambridge 5-km Telescope and the VLA. The sensitivity of the VLA is such that these observations can be extended to sources in the flux density range $\sim$ 1-10 mJy.

The standard procedure for identifying extragalactic radio sources is first to use the surveys of the northern and southern skies made with Schmidt telescopes, the Palomar-National Geographic Society Sky survey, the U.K. Schmidt survey and the ESO Schmidt survey. For the northern survey, the limiting magnitude on red plates is about 20 and generally speaking identifications of bright radio sources with objects brighter than this limit are easy. The UK Schmidt survey has a somewhat fainter limiting magnitude because of the use of IIIaJ plates. To extend optical identifications to significantly fainter magnitudes requires the use of telescopes in the 4-5 metre class. Until recently, the faintest magnitudes attainable under good seeing conditions was $\simeq$ 23, either by direct prime-focus photography using plates or with an image intensifier. For this work, radio positions with accuracy $\gtrsim$ 1 arcsec and radio structures of comparable angular resolution are essential.

Most recently, CCD cameras have been used on large reflectors and their very high quantum efficiencies in the red and near infrared wavebands ($\sim$ 60%) make them ideal for searching for distant galaxies in the fields of the radio sources. Because of the linearity of the detector, the limiting magnitude attained depends only upon how long one is prepared to integrate. Typically, stellar objects having m $\sim$ 25 can be readily

G.O. Abell and P. J. E. Peebles (eds.), Objects of High Redshift, 135–144.
Copyright © 1980 by the IAU.

M. S. LONGAIR

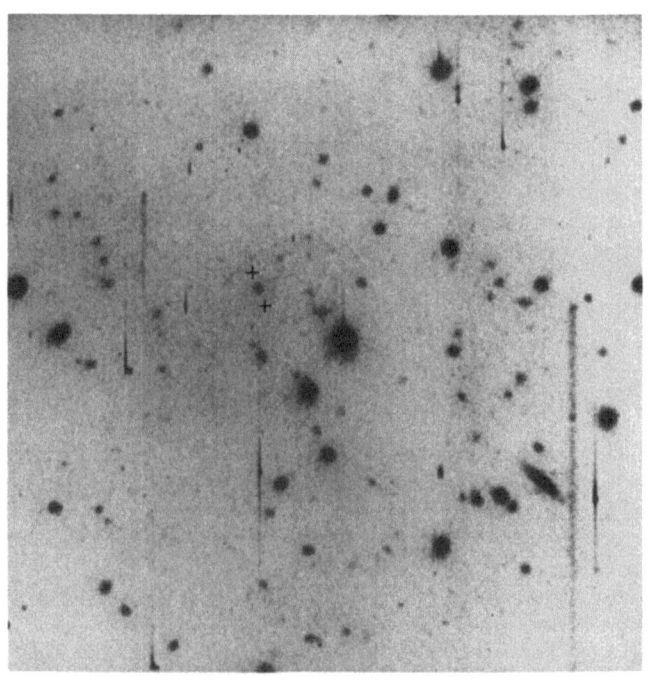

3C 289

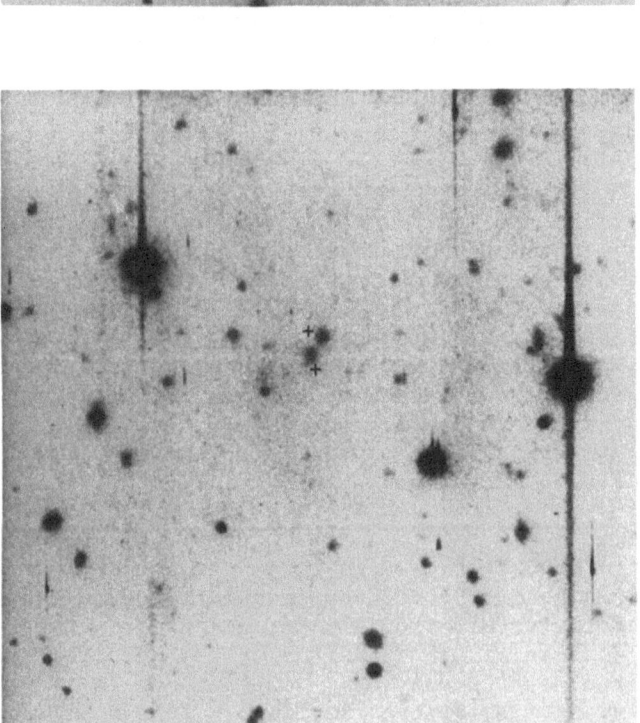

3C 280

Figure 1.   CCD images of the fields of the radio sources 3C 280 and 289.   Each field is 3 arcmin
in size.   The observations were made with the 500 x 500 pixel CCD camera at the prime focus of the
Hale 5-metre telescope.   The black crosses indicate the positions of the maxima in the radio brightness distributions of these double sources.

detected in a 1000-sec exposure. Our experience in identifying very faint radio galaxies and quasars shows that one can obtain about $1-1\frac{1}{2}$ magnitude improvement in identifying these objects with a 1000-sec exposure over what we have been able to achieve with an image intensifier and IIIaJ plates on the Hale 5-metre telescope (see e.g. Figure 1).

The astrophysical objectives of these studies are multifold. First, the optical identifications are an essential first step in trying to disentangle how the radio source population has evolved with cosmological epoch. Figure 2(a) shows a recent compilation of different counts of radio sources by Wall (1979). It is well known that the counts at all the frequencies shown in Figure 2(a) disagree with the predictions of uniform world models, as is illustrated in Figure 2(b) for the counts at 408 MHz. Second, the optical identification of distant radio sources is one of the most successful methods of discovering distant massive galaxies and quasars. These may be used to study how the properties of massive galaxies have evolved with cosmological epoch and may conceivably eventually lead to an estimate of q_0. The question of a redshift cut-off for quasars may be studied using unbiased radio selected samples. Third, the optical identifications are often objects exhibiting evidence of other high energy astrophysical activity such as strong emission-line spectra, strong continuum radiation and X-ray emission. Optical identifications have also led to unexpected discoveries such as quasars themselves and more recently the spectacular double quasar, 0757+561.

2. OPTICAL IDENTIFICATION SURVEYS

Low frequencies   Most effort has been devoted to the identification of the brightest radio sources at low frequencies because historically these sources were among the first to be detected and most subsequent work with high resolution aperture synthesis telescopes has concentrated on these samples because most of the sources are extended with resolvable radio structures. Until the beginning of this year, the optical identification content of a statistical sample of 166 3CR radio sources was about 85-90%, this work using prime focus photography, either direct or with an image intensifier (Longair and Gunn 1975, Smith, Spinrad and Smith 1976, Laing et al. 1978, Riley et al. 1980). This spring, Jim Gunn and I had access to the 500 x 500 pixel CCD camera built by JPL as part of the development programme of the Space Telescope. With this device, we were able to identify all the remaining unidentified sources in a complete sample of 60 sources which was a subset of the 166 3CR statistical sample (see Figure 1 for an example of the quality of this material). The apparent magnitude distribution of this sample is indicated in figure 3a. Most of the objects fainter than m = 20 are galaxies. As can be seen from the figure, a problem in interpreting these results is that there are very few redshifts available for objects fainter than about m = 20, essentially all of which are radio galaxies. The redshift distribution for the 41 objects with redshifts is shown in Figure 3(b). It is likely that the redshifts of the remaining radio

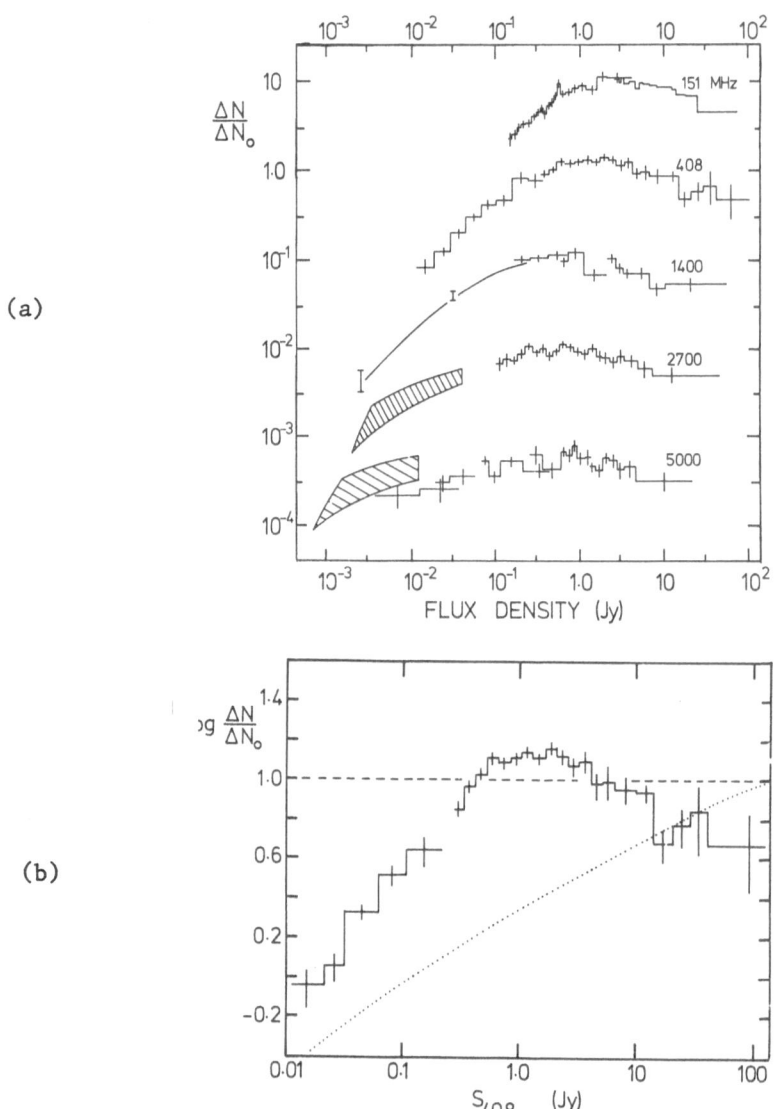

(a)

(b)

Figure 2(a)   Counts of radio sources at five frequencies in differen-
tial form. ΔN is the number of sources in the flux density interval
S to S+ΔS and ΔN_o is the expected number in a Universe in which
$N(\geq S) \propto S^{-1.5}$ (Wall 1979).

(b) The differential source counts at 408 MHz compared with the law
$N(\geq S) \propto S^{-1.5}$ (dashed line) and the expected relation for a Friedmann
world model having Ω= 1 assuming the sources are uniformly distributed
(dotted line).   (Wall 1979).

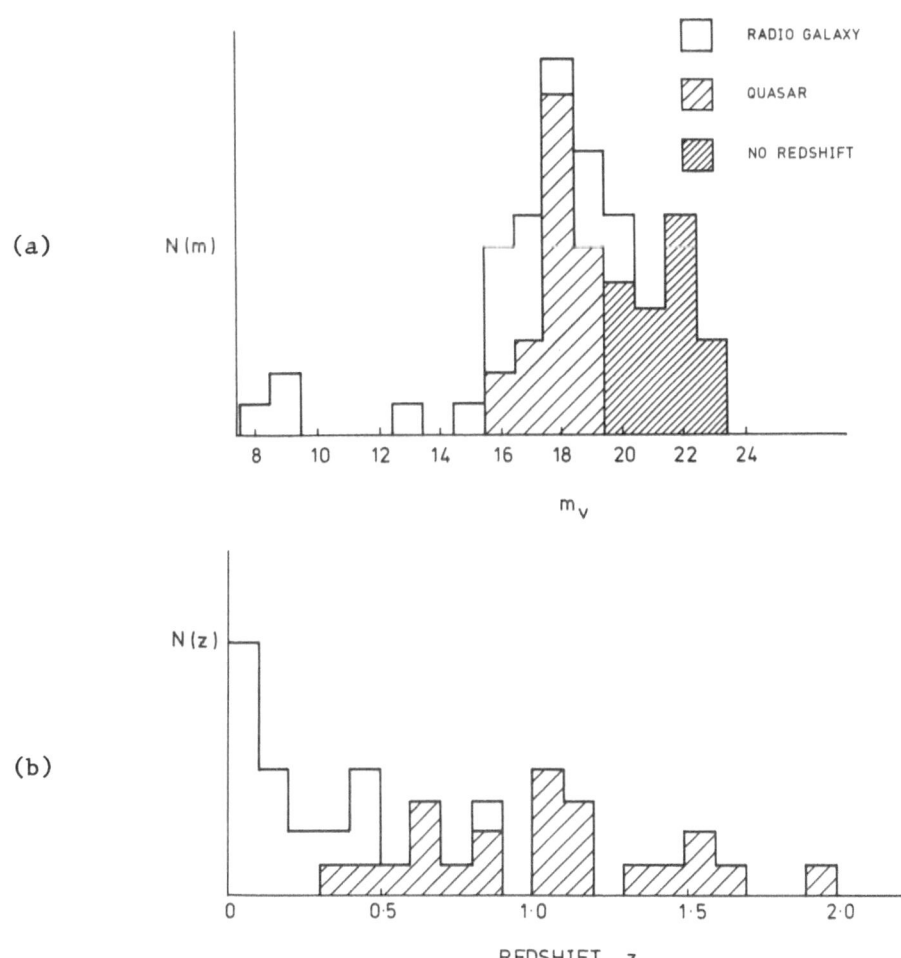

Figure 3(a)   The apparent magnitude distribution for a complete sample of 60 3CR radio sources all of which now have optical identifications (Gunn et al. 1980).

(b)   The redshift distribution for the sample of 60 3CR radio sources. Redshifts are only available for 41 of these, the apparent magnitudes of those without redshifts being indicated in Figure 3(a).

galaxies fall in the redshift range $0.5 < z \lesssim 1.5$.   In fact, for the
complete 166 sample of sources, there are only 4 sources for which
there is at present not even a tentative identification, 3C 65, 68.2,
294 and 437.   None of these fields has been studied using the CCD
camera.

It is the combination of optical identification and redshift
information at high flux densities combined with radio source counts
known with excellent statistical precision that enable models for the
cosmological evolution of the radio source population to be constructed
(e.g. Wall, Pearson and Longair, 1977).   It is instructive to present
these models in terms of "enhancement factors", $f(P,z)$ which indicate
the increase in comoving space density of sources as a function of radio
luminosity P and redshift z.   If the world model were uniform
$f(P, z) = 1$.   Figures 4(a) and (b) show two models which can account
for the source counts at 408 MHz (Figure 2(b)) and the V/V_{max} data for
quasars in the sample (Wall et al. 1977).   Some models have cut-offs
at redshifts $z \sim 3-4$ and others have different rates of evolution of
the strongest sources.   Notice that the strongest evolution is only
associated with the most luminous sources - intrinsically weak radio
sources show at best weak cosmological evolution.

The models differ in the predicted identification content and redshift distributions at low flux densities.   An example of these differences for the three successful models described by Wall et al. is shown
in figure 4(c) from which it can be seen that the differences between
the models are large.   Detailed optical identification searches have
been made by Grueff and Vigotti to $S_{408} \sim 1$ Jy and by Perryman (1979 a,
b), both of whom used IIIaJ plates taken with the Palomar 48-inch
Schmidt telescope.   Grueff and Vigotti (1977) achieved an overall
identification percentage of about 63% consisting of 40% galaxies and
23% quasars.   At 408 MHz, Perryman achieved an identification percentage of only about 20% at $S_{408} \geq 0.01$ Jy.   Even assuming that all the
latter objects are radio galaxies with $z \lesssim 0.6$, it is evident that these
statistics are as yet inadequate to discriminate between models.

Deep identifications at 1400 MHz   In contrast deep optical identification surveys at 1.4 GHz by de Ruiter et al. (1977) and Perryman
(1979a, b) have found significantly higher identification rates at the
lowest flux densities, $S_{1.4} \gtrsim 2$ mJy.   De Ruiter et al. find more than
40% of the faint sources to be identifiable on deep 4-metre plates while
Perryman achieved a figure $\sim 35-40\%$ for the 5C6 and 7 surveys using
very high quality IIIaJ 48-inch Schmidt plates taken by Dr. M.V. Penston.
The cause of this difference in optical identification rate as compared
with 408 MHz is attributed by Perryman to the fact that there is a
larger fraction of sources with spectral indices $0.4 < \alpha < 0.7$ in the 1.4
GHz sample which are of larger angular size than 5C sources in general.
These sources are more readily identifiable optically.   This result is
consistent with the known correlation between radio luminosity and
spectral index for radio galaxies recently re-analysed by Laing and
Peacock (1980).   Originally, Wall et al. (1977) hoped that this high

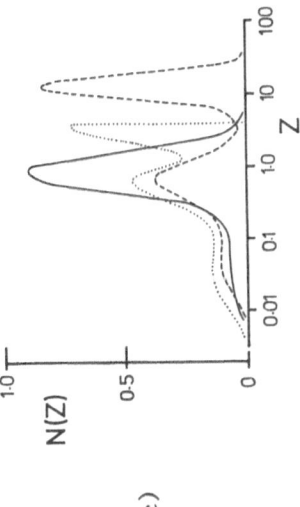

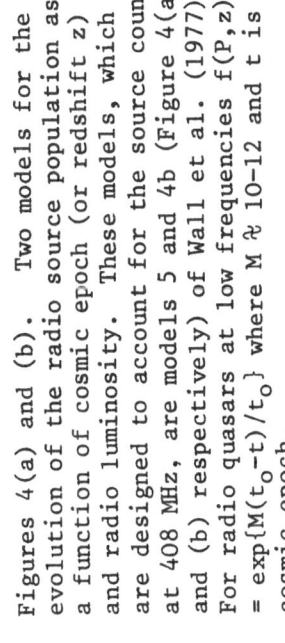

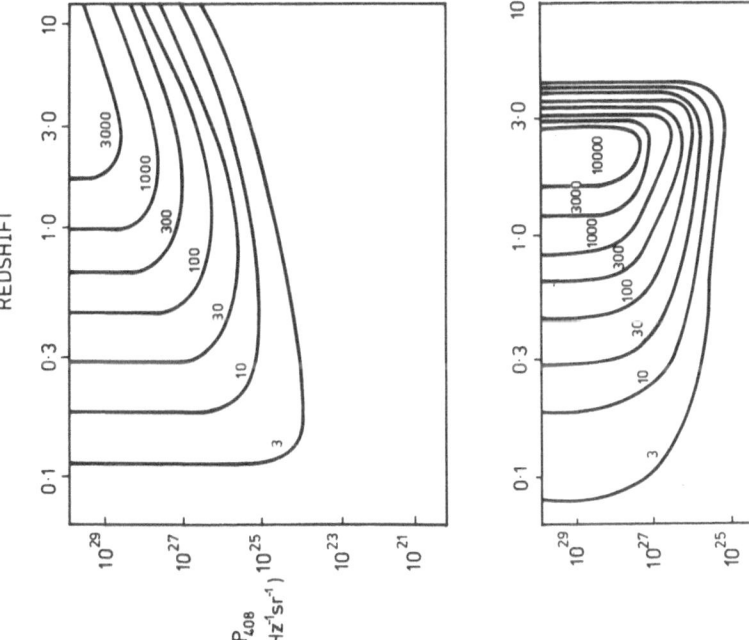

(a)

(b)

Figures 4(a) and (b). Two models for the evolution of the radio source population as a function of cosmic epoch (or redshift z) and radio luminosity. These models, which are designed to account for the source counts at 408 MHz, are models 5 and 4b (Figure 4(a) and (b) respectively) of Wall et al. (1977). For radio quasars at low frequencies $f(P,z) = \exp\{M(t_o-t)/t_o\}$ where $M \simeq 10$–12 and t is cosmic epoch.

(c) The predicted redshift distributions at $S_{408} \geq 10$ mJy for models 4a, 4b and 5 of Wall et al. (1977).

identification percentage would enable a tentative discrimination to be
made between the models illustrated in figure 4(c) but this new result
indicates that the sample of identified 5C sources at 408 MHz, the
frequency at which the models were constructed, is only 20%.   A note-
worthy feature of the recent Westerbork work (Katgert et al. 1979) is the
indication that the faintest identifications are significantly bluer
than those down to m = 20.   They interpret this as evidence for colour
evolution of massive galaxies at redshifts z $\sim$ 0.5-0.6.

Identifications at 2.7 GHz and higher   At high radio frequencies,
2.7 GHz and greater, the fraction of sources with flat radio spectra,
α < 0.5, is very much larger and continues to increase to the very
highest frequencies.   Recently, Peacock and Wall (1980) have defined
various statistical samples of the brightest sources at 2.7 GHz and the
optical identification content of a high flux density sample is shown
in Table 1.   The problems of interpretation of the source counts are
different at 2.7 GHz.

Table 1

Identification statistics at 2.7 GHz for a sample of 155 sources having
$S_{2.7}$ ≥ 1.5 Jy (Peacock and Wall 1980)

	Extended sources α > 0.5	Compact sources α < 0.5
Galaxies	66	19
Quasars	8	43
No identification	2	4
Neutral objects	–	2
New CCD survey		11*
	76	79

* 7 identifications roughly half of which are galaxies and 4 empty
fields.

The sources with steep radio spectra are similar to those found in
surveys at low radio frequencies and their source counts are also consistent with the low frequency counts.   However, the flat spectrum
sources have a much flatter count.   If the counts are interpreted in
terms of exponential evolution models, $f(z) = \exp\{M(t_o-t)/t_o\}$ where t is
cosmic time and t_o the present epoch, M $\approx$ 10-12 for sources with steep
spectra but only $\approx$ 3-5 for flat spectrum sources.   Wall et al. (1980)
show that there are further complications in understanding the 2.7 GHz
counts at lower flux densities at which it may be necessary to invoke
the presence of an evolving low-luminosity source population.

    Thus, the need for further identifications and redshifts for high
as well as low frequency samples is pressing at all flux densities
since it is clear that the evolution of the overall radio source population cannot be wholly understood from observations at a single frequency.

3. PROSPECTS FOR FUTURE OPTICAL WORK

I believe the prospects for future optical identification work and for measuring the redshifts of distant radio sources are good. First of all, studies of samples of bright radio sources associated with galaxies and quasars show a number of encouraging trends (Hine and Longair 1979). For example, the fraction of sources with strong non-thermal optical continua from their nuclear regions increases with increasing radio luminosity and correspondingly, even for radio galaxies alone the fraction of galaxies with strong emission line spectra increases to about 70% at the highest radio luminosities (for a brief survey of these and other optical properties of radio sources, see Longair 1979). Inspection of figures 4(a) and (b) shows that the models predict that it is sources of these properties which are the main cause of the "excess" of faint sources in the counts. Second, there is increasing evidence that at redshifts z $\gtrsim$ 0.5, galaxies and, in particular the brightest galaxies in clusters, are significantly bluer than comparable nearby objects. Thus, Butcher and Oemler (1978) find distant clusters to be bluer, Tinsley (1979) reports an excess of blue galaxies in the most recent galaxy counts, Oke (1979) has reported that the fraction of brightest galaxies in clusters with a blue excess increases rapidly at redshifts z $\gtrsim$ 0.55 and Katgert et al. (1979) find directly that the faintest radio galaxies are excessively blue. These trends make the problem of identifying the most distant radio galaxies significantly easier since the K-correction which is mostly responsible for the faintness of distant galaxies in, say, the V-band, no longer increases so rapidly with increasing redshift.

These programmes are prime candidates for study with the new generation of CCD cameras on ground-based telescopes and with the Space Telescope.

REFERENCES

Butcher, H. and Oemler, A., 1978. Astrophys. J., 219, 18.
De Ruiter, H.R., Willis, A.G. and Arp, H.C., 1977. Astron. Astrophys. Suppl. 28, 211.
Grueff, G. and Vigotti, M., 1977. Astron. Astrophys., 54, 475.
Gunn, J.E., Hoessel, J.G., Westphal, J.A., Perryman, M.A.C. and Longair, M.S., 1980. Mon. Not. R. astr. Soc., (in preparation).
Hine, R.G. and Longair, M.S., 1979. Mon. Not. R. astr. Soc., 188, 111.
Katgert, P., de Ruiter, H.R. and van der Laan, H., 1979. Nature, 280, 20.
Laing, R.A., Longair, M.S., Riley, J.M., Kibblewhite, E.J. and Gunn, J.E., 1978. Mon. Not. R. astr. Soc., 183, 547.
Laing, R.A. and Peacock, J.A., 1980. Mon. Not. R. astr. Soc., (in press).
Longair, M.S., 1979. Review article in "Scientific Research with the Space Telescope", IAU Colloquium No.54, (eds. M.S. Longair and J. Warner), MSFC publication (in press).
Longair, M.S. and Gunn, J.E., 1975. Mon. Not. R. astr. Soc., 170, 121.

Oke, J.B., 1979.   Comments in "Scientific Research with the Space
    Telescope", op. cit.
Peacock, J.A. and Wall, J.V., 1980.   Mon. Not. R. astr. Soc., (in
    preparation).
Perryman, M.A.C., 1979a.   Mon. Not. R. astr. Soc., $\underline{187}$, 223.
Perryman, M.A.C., 1979b.   Mon. Not. R. astr. Soc., $\underline{187}$, 683.
Riley, J.M., Longair, M.S. and Gunn, J.E., 1980.   Mon. Not. R. astr.
    Soc., (in preparation).
Smith, H.E., Spinrad, H. and Smith, E.O., 1976.   Publ. astr. Soc.
    Pacific, $\underline{88}$, 621.
Tinsley, B.M., 1979.   Review article in "Scientific Research with the
    Space Telescope", op. cit.
Wall, J.V., 1979.   Proc. Roy. Soc., (in press).
Wall, J.V., Pearson, T.J. and Longair, M.S., 1977.   "Radio Astronomy
    and Cosmology", IAU Symposium No.74 (ed. D.L. Jauncey), 269,
    Reidel and Co.
Wall, J.V., Pearson, T.J. and Longair, M.S., 1980.   Mon. Not. R. astr.
    Soc., (in preparation).

DISCUSSION

Rees: You presented a correlation between optical line width and the
             fraction of the radio flux coming from a central component.
Do you get a better or worse correlation between line width and the
radio luminosity of the compact component?

Longair: There is definitely a trend for the sources with greater
             intrinsic central component luminosities to have greater line
widths. We have not quantified our correlations because as yet the data
are very sparse. We work in terms of ratios of luminosities because
these tend to be less sensitive to distance-dependent selection effects
in a sample selected by flux density.

Wills: So far as you can tell, do the new faint optical identifica-
             tions of 3C sources continue to be exclusively galaxies?

Longair: For objects well above the limits of detectability, the new
             3CR identifications are all with extended objects. Close to
the plate limits, it is obviously difficult to distinguish stars and
galaxies.

ISOTROPY OF FAINT SOURCES

Carla Fanti
Laboratorio di Radioastronomia, Bologna, Italy

How matter distributes in the Universe is still an open question in spite of the many efforts people have done since early 1924 when Hubble first studied the distribution of bright galaxies.

Today a point which can be considered firmly established is that galaxies do cluster. Their two-point correlation function is given to good accuracy by the simple power law model:

$$\xi(r) = (r_0/r)^{1.77}, \quad r_0 \sim 5 \ h^{-1} \ \text{Mpc}, \quad r \lesssim 10 \ h^{-1} \ \text{Mpc}, \quad h = H/100$$

(Peebles, 1978). Clusters of galaxies are clustered as well on linear scales of a few tens of Mpcs (Peebles, 1978).

But this is how the near universe, that is for redshifts smaller than ~ 0.5 looks.

Information on the very far universe ($z \sim 1000$) is given to us by the microwave background radiation, which appears to be highly isotropic ($\Delta T/T < 0.001$ to 0.0001) on very large angular scales.

Intermediate epochs can be investigated through radiosources, most of which are believed to be at redshifts of the order of 2 or 3. And this will be the subject of this talk, mainly from the experimental point of view.

Since the limit one puts on the fluctuations in the source counts in inversely proportional to the square root of the number of sources, only recently have surveys become available which are sufficiently extensive for this type of analysis.

A comprehensive discussion of the problem has been given in 1976 in Cambridge at the IAU Symposium No. 74 on "Radioastronomy and Cosmology" (D.L. Jauncey, 1977) and the situation has not changed much since then.

As an example, I shall briefly illustrate the work done on the B2 survey by Fanti et al., 1978 (paper I). For a more detailed treatment

?.O. Abell and P. J. E. Peebles (eds.), Objects of High Redshift, 145–152.

of the subject, one should refer to that paper. Different approaches
have been tried:
    i)   Multi-Binning Analysis;
    ii) Power Spectrum Analysis;
    iii) Source Counts.

The MBA was performed by dividing the survey into bins of areas
increasing from 0.45 to 62 square degrees. The comparison between the
observed source distribution and that expected from a random population
(Figures 4, 5 and 6 of paper I) does not show any significant difference, suggesting that no kind of clustering is revealed by this type of
analysis.

The PSA was applied to the B2 catalogue in the way developed by
Webster (1976). Without going into details, I only remind that the power
spectrum $Q(\lambda)$ gives a measure of the departure from random of a point
(in our case radiosources) distribution, in this sense: if sources are
completely at random, Q will be equal to $\underline{1}$ for all λ; if all the sources
are in clusters of population $\underline{q}$ and size $\overline{\lambda}_c$, then Q will be equal to $\underline{q}$
for all $\lambda \gg \lambda_c$ and equal to $\underline{1}$ for all $\lambda \ll \lambda_c$. Intermediate cases will
be in between. We make use of the ordinary Fourier Transform instead
of the spherical harmonics adopted by Peebles and co-workers in his
analyses of galaxies and galaxy clusters. This makes computations
easier and does not sensibly affect the results, provided one makes an
appropriate equal area projection of the celestial sphere. This projection might eventually distort the clustering shape, but this is not
important in the present analysis.

The power spectrum of the B2 catalogue, shown in Figure 1, does
not deviate significantly from that of a random distribution, in agreement with the MBA.

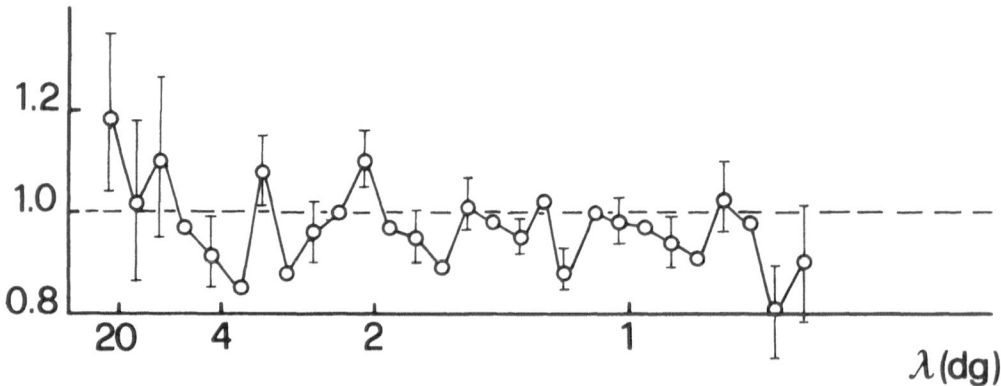

Figure 1. Power spectrum of the B2 radiosources

The results of Fanti et al. (1978) agree with the analyses performed by Webster (1976b, 1977) on catalogues of comparable depth at
different frequencies. As far as I know, the only significant
exceptions are:

a)   A small but significant excess found in the 4C catalogue by
     Seldner and Peebles (1978) which Massons (1978), on the other
     hand, explains as due to zones of incompleteness in the 4C
     catalogue and to the presence of SNRs which, being a low galac-
     tic latitude, are obviously not at random positions;

b)   A source surface density ~ 30% greater than that of B2 cata-
     logue in the MC2 and MC3 catalogues (Mills, 1973).   Recent
     observations of these regions with the Bologna Cross by Fanti
     et al. (1979) show ~ 15% difference in the flux scales.   This
     is enough to bring the discrepancy between the source counts
     within the statistical uncertainties.

Concluding, I think I can still make the conservative statement
that there is no convincing evidence so far that in fairly deep surveys
sources are clustered at any level.

On the other hand, most of the radiosources in a catalogue are
believed to be radiogalaxies, and since galaxies do cluster, there
ought to be clustering in the radiosources' positions at some level.

Actually, Seldner and Peebles (1978) detected a faint but significant cross-correlation between the source positions of the 4C catalogue
and the Lick galaxies counts, suggesting that at least the 4C sources
associated with galaxies do cluster.   This effect is much stronger when
they cross-correlate the 3C radiogalaxies with the Lick galaxies.

On these lines I have analyzed the spatial distribution of a few
complete samples of radiosources optically identified.   In particular,
I have applied the PSA to the five samples of QSOs and to that of radiogalaxies listed below.

		Freq	N (obj)	S_{lim} (Jy)	References
	3CR	178	33	9.0	M. Schmidt, 1968
	GV	408	122	0.9	G. Grueff & M. Vigotti, 1972, 1973, 1979
QSO	Olsen	178		2.5	M. Schmidt, 1974 & references therein
			127		
	4C	178		3.0	D. Wills & R. Lynds, 1978
	B2	408	58	0.25	R. Fanti et al. 1979
	PKS	2700	60	0.35	D. Wills & R. Lynds, 1978
GAL	GV	408	136	0.9	G. Grueff & M. Vigotti, 1972, 1973, 1979

In no case has any significant deviation from randomness been found
for QSOs (Figure 2a).   Again, I have to mention that in the identification programs by Hazard (1977) and Hunstead (1979) of the Molonglo radiosources, they find density fluctuations in different areas of the sky

much larger than expected. As the authors themselves state, further
investigations are needed before one can assume a real QSO anisotropy.
Nevertheless, this is a point to be kept in mind.

For radiogalaxies the PSA has been performed for several limiting
magnitudes (Figure 2b). While for the full sample there might be only
some suspicion that the distribution is not random, for the radiogalaxies brighter than m_r = 15 there is no doubt that their spatial distribution significantly deviates from randomness. For the faintest galaxies,
however, there is no indication at all.

This is roughly consistent with what is expected from the knowledge
of the power spectrum of a galaxy sample of the same magnitude limit
(for example, the Zwicky galaxies) and of the radioluminosity function
of elliptical galaxies. One should find about half a radiosource in
excess of random in the bright sample, and ~ 0.02 in the faint sample,
over angular scales $\gtrsim$ 20°.

On the assumption that the same hierarchical properties still hold
at large redshifts, one can try to extrapolate these considerations to
unidentified radiosources, which are usually thought to be radiogalaxies.

Even lower values of clustering are expected if we assume a density
evolution as strong as $(1 + z)^6$ or $(1 + z)^8$.

It is therefore not surprising that clustering of radiosources
associated with clusters or superclusters of galaxies similar to those
existing at the present epoch is not detected in surveys of radiosources.

At very large scales, from the analysis of the B2 it is possible to
put an upper limit of 2 to 4% on the r.m.s. fluctuations in the source
counts over linear scales up to ~ 40°. Assuming an average redshift of
2, this means a fluctuation around the mean density which is less than
5 to 10% over linear scales of 1 Gpc (H = 50 km s^{-1} Mpc^{-1}).

In Figure 3 these upper limits are compared to the actual values
of the covariance function for galaxies.

To improve these limits it would be necessary to analyze surveys
either more extended or deeper in order to increase the source number.
However, calibration uncertainties, catalogue completeness, etc., begin
to be very important at this stage. Moreover, in very deep surveys,
intrinsically faint sources begin to be very common. Also, one starts
to split sources into several components. These effects might mask any
cosmological clustering and it would not be straightforward to disentangle the various contributions.

Finally, one can look for anisotropies in the radiosource evolution
by computing the log N–log S relation in different regions of the sky.
This was already done on the 4C catalogue by Golden (1974) with completely negative results.

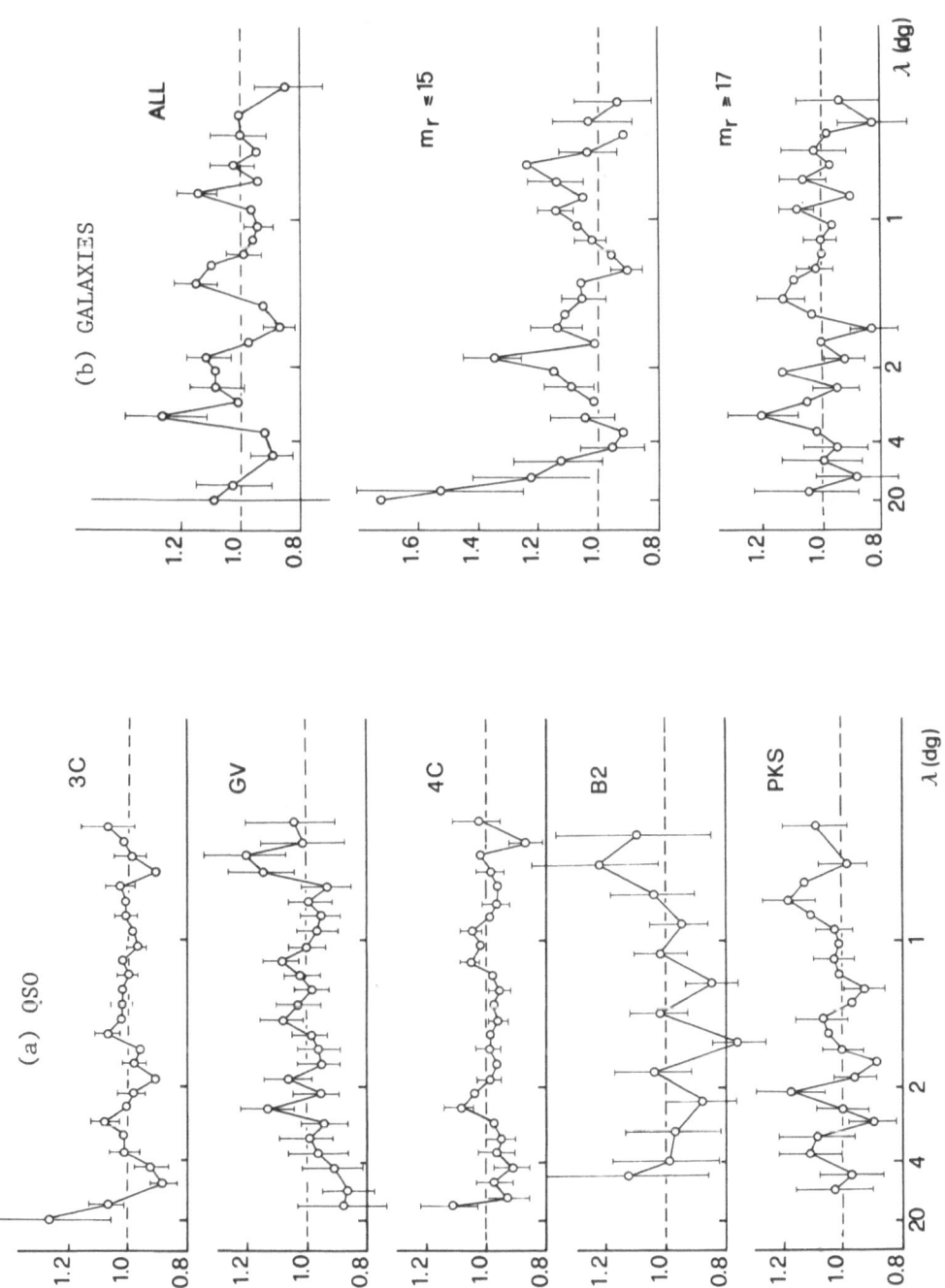

Fig. 2. (a) Power spectrum of QSO's and (b) radiogalaxies samples

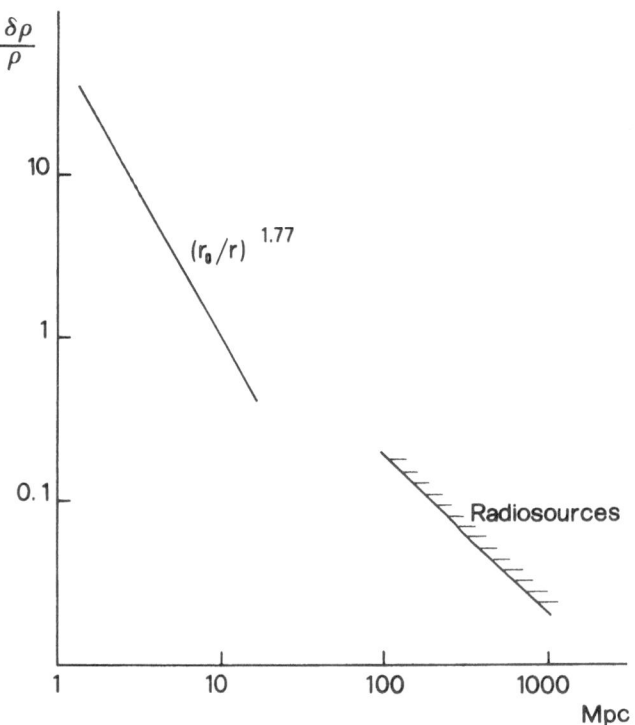

Figure 3.   Covariance function for galaxies compared to the limits
            obtained from the B2 catalogue

We repeated the test at a deeper flux level on the B2 survey in a
slightly different way. For each of the networks used in the MBA we
have considered patchwork regions built up with all the bins which have
the same occupation number in the intermediate flux interval 0.5 to 1.0
Jy.

In each of these regions we computed the log N-log S relation using
only the flux intervals adjacent to the 0.5 to 1.0 Jy interval.   In
Figure 7 of paper I is plotted the slope of the log N-log S against the
occupation number for each network. No kind of correlation is present.

One concludes, therefore, that not only is condensed matter very
uniformly distributed in the unverse, but that also the physical conditions which determine the phenomenon "radiosource" have to be spatially
very uniform in spite of the strong dependence with time shown by the
steepness of the source counts. The density and evolution of radiosources are similar in regions of the universe which were causally quite
unconnected at the epoch of the radiosources formation.

REFERENCES

Fanti, C., Lari, C., and Olori, M.C.: 1978, Astron. Astrophys., <u>67</u>,
     175.

Fanti, C., Ficarra, A., Gregorini, L., and Mantovani, F.: 1979 (in
    preparation).
Fanti, R., Feretti, L., Giovannini, G., and Padrielli, L.: 1979, Astron.
    Astrophys., 73, 40.
Golden, L.: 1974, Monthly Notices Roy. Astron. Soc., 166, 383.
Grueff, G., and Vigotti, M.: 1972, Astron. Astrophys. Suppl., 6, 1.
_____ 1973, Astron. Astrophys. Suppl., 11, 41.
_____ 1979, Astron. Astrophys. Suppl., 35, 371.
Hazard, G.: 1977, "Radioastronomy and Cosmology," IAU Symposium No. 74
    (ed. D.L. Jauncey), p. 157, D. Reidel and Co.
Jauncey, D.L. (ed.): 1977, "Radioastronomy and Cosmology," IAU Sympo-
    sium No. 74, D. Reidel and Co.
Massons, C.R.: 1978, Monthly Notices Roy. Astron. Soc., 185, 9p.
Mills, B.Y.: 1977, "Radioastronomy and Cosmology," IAU Symposium No.
    74, D. Reidel and Co.
Peebles, P.J.E.: "The Large Scale Structure of the Universe," IAU Sym-
    posium No. 79 (M.S. Lonair and J. Einastro, eds.), 217, D. Reidel
    and Co.
Schmidt, M.: 1968, Astrophys. J., 151, 393.
_____ 1974, Astrophys. J., 193, 505.
Seldner, M., and Peebles, P.J.E.: 1978, Astrophys. J., 225, 7.
Webster, A.: 1976a, Monthly Notices Roy. Astron. Soc., 175, 61.
_____ 1976b, Monthly Notices Roy. Astron. Soc., 175, 71.
_____ 1977, Monthly Notices Roy. Astron. Soc., 179, 511.
Wills, D., and Lynds, R.: 1978, Astrophys. J., 36, 317.

DISCUSSION

Murdoch: The Molonglo flux density scale has been carefully revised
        over the whole available sky. Whilst this reduces previously
claimed anisotropies in radiosource density to ~ 2σ, the revision is not
as great as suggested by the Bologna results.

        More interesting is the comparison of QSO surface density
        (number per square degree) between the north galactic hemi-
sphere region of MC2 and the Southern selected area of 0.66 sterad at
~ $-20°$ (mentioned earlier by Richard Hunstead). The surface density of
QSOs in the south is only 1/3 of that in the north galactic hemisphere
region of MC2, 3 at an optical limit of $19^m_.5$ and a radio limit of 0.95
Jy (on the revised scale for both areas). The root-mean-square difference is 3.2 σ.

Fanti: About the first point of the comment, I only said I have found
        a difference between the two flux scales which is enough to
remove the discrepancy in the source counts without pretending to attrib-
ute all this difference to the Molonglo data. The fact that the flux
scale revision is not as great as suggested by our analysis probably
means that Bologna and Molonglo are not yet on the same absolute flux
scale.

Seldner: Jim Peebles and I have re-examined the auto-correlation func-
            tion for the 4C radiosource catalogue in light of Colin
Masson's suggestion that the incompleteness in the regions around the
brightest sources has caused the observed effect. We have generated
several random catalogs, placing no sources in the holes, and including
anti-correlation at small angular separation to account for confusion.
Mean pair counts from these random catalogues are used to normalize the
data to obtain the correlation function. We also found no north-south/
east-west asymmetry in the pair distribution which might have been
caused by varying sensitivity across the declination strips.   The net
result is that there are 3 ± 1% excess pairs in the 4C catalogue at
angular separations less than 3°. Elimination of the region $|b| < 10°$,
where galactic sources could contaminate the sample, has almost no effect
on the result.

Fanti: I have not mentioned in my talk that above 1 Jy there is a
          very marginal ($< 2\sigma$) evidence of clustering on an angular
scale of a few degrees at a level of ~ 10%. Half of this value can
easily be explained by the clustering of the radiogalaxies brighter than
$m_r = 15$ diluted in the total sample. It is reasonable, therefore, to
think that also the excess you find in the 4C catalogue is due to nearby
radiogalaxies.

Masson: Although power spectrum analysis is convenient for analyzing
           the clustering of galaxies, I think that covariance function
analysis is more appropriate for radio sources.   This is because confusion, which affects all terms of the power spectrum, only alters the
covariance function on small angular scales.   Thus, it is easier to
separate the effects of confusion and clustering.

          I have recently carried out a covariance function analysis
          of the Cambridge 6C catalogue. The weaker sources (s < 200
mJy) are uniformly distributed, within statistical errors, but there is
an indication of clustering among the stronger sources on angular scales
of a few arcminutes.

Fanti: For the B2 catalogue, confusion is not a major problem since,
          as described in detail in Fanti et al. (1978), any source
areas in which confusion by side lobes may occur have been excluded in
a homogeneous and predictable way, and we can take count of the effect
of these "excluded" areas in our analysis. Also, the confusion effect
which results by considering a close couple of sources as a single one
is negligible because we stopped the analysis well above the confusion
limit. However, I agree that, in general, covariance function may be
more appropriate for radio catalogues.

          About the second point, it would be interesting to know if
          you are detecting real clustering on small angular scale or
if, perhaps, you are not counting complex sources.

21-CM PROBES OF REDSHIFTED CLOUDS

A. M. Wolfe
University of California, San Diego and
University of Pittsburgh

21 cm studies, carried out with the 305 m telescope at Arecibo, of absorption-line clouds in QSOs have resulted in the following discoveries: (a) A search for 21 cm absorption in redshift systems selected for exhibiting Mg II/Fe II absorption in radio-bright QSOs resulted in the detection of a $z \approx 0.4$ 21 cm line in PKS 1229-02; (b) the detection of 21 cm absorption at $z \approx 1.8$ in MC3 1331+170; and (c) the discovery of 21 cm emission from regions ~ 100 kpc distant from the ScI galaxy NGC 628, that occurred during a search for outlying H I in a complete sample of spiral galaxies.

A. INTRODUCTION

I shall discuss results of recent experiments carried out during the past year whose objective was to understand the nature of absorption-line clouds in QSOs. I shall also discuss some theoretical work that has grown out of these studies. Because a detailed version of an identical paper will be published elsewhere (Wolfe 1980), I shall summarize the results very briefly.

B. DETECTION OF 21 CM ABSORPTION IN 1331+170

The very strong L_α absorption feature at $z \approx 1.8$ in 1331+170 (Carswell et al. 1975) makes this redshift system a prime candidate for 21 cm absorption. A large H I column density, $N(H\ I) \sim 10^{21}\ cm^{-2}$, is indicated because the rest-frame equivalent width of L_α, $W_\lambda \sim 20\ \text{Å}$, is much larger than equivalent widths of the heavy-element lines identified with the two redshift systems comprising this absorption complex; and this is the case if L_α is on the damping portion of the curve-of-growth. Recent spectral scans of the L_α absorption trough (Hegge,

G.O. Abell and P. J. E. Peebles (eds.), Objects of High Redshift, 153–159.

Liebert, and Strittmatter 1979) show that most of the absorption arises in system A (z_A = 1.7763) rather than in system B (z_B = 1.7851).

M. M. Davis and I recently detected a 21 cm absorption line in system A (Wolfe and Davis 1979). This weak absorption feature has an optical depth τ (21) = 0.02 and the line profile can be fit by a Gaussian function with velocity dispersion σ = 8.5 km s^{-1}. The detection of 21 cm absorption and Lα absorption in the same material allows a determination of the hydrogen spin temperature T_s in a high-z object for the first time (Wolfe 1979). We find that T_s < 980 K: the upper limit accounts for the possibility that the absorbing cloud doesn't cover the entire radio source. When combined with the upper limits on collisional de-excitation set by the absence of C II λ 1335.7 absorption from the J = 7/2 fine-structure level, this limit on T_s requires that system A be further than d $\approx$ few kpc from 1331 + 170; otherwise 21 cm continuum excitations would raise T_s above 980 K. This limit would rule out radiative ejection at the implied velocity of u $\approx$ 0.1 c from this z = 2.08 QSO.

System A could be closer to 1331 + 170 provided that another source of hyperfine excitation dominates the 21 cm continuum. Lα pumping of the ground hyperfine levels via 2p excitations is an interesting possibility since atomic recoil causes $T_s \rightarrow T_k$, where T_k is the local kinetic temperature (Field 1959): T_k is expected to be less than 980 K in the H I region that gives rise to 21 cm absorption. Furthermore, recombination Lα produced through photoionization by the QSO Lyman continuum results in a pumping rate that is many orders of magnitude larger than the excitation rate due to 21 cm radiation emitted by 1331 + 170. The problem is that Lα is produced in the H II region that faces the QSO, while 21 cm absorption arises in the outward facing H I region which has an optical depth in Lα of τ (Lα) ~ 10^8. Thus it is not clear whether Lα remains strong enough to cause pumping after propagating into this extremely opaque medium.

J. J. Urbaniak and I (Urbaniak and Wolfe 1979) have studied the transfer of Lα that arises in the H II region and propagates across an ionization front into the H I region. The important parameters in this problem are v, the relative velocity between H II and H I regions, and σ and N(H I) of the H I region. For all plausible values of v, that is 0 < $|v|$ < 250 km s^{-1}, we find that Lα pumping dominates the 21 cm continuum. Solutions with larger values of $|v|$ cause larger pumping rates since larger amounts of radiation penetrate and emerge from the H I region. In all cases the emergent radiation displays a double-hump spectrum that is symmetric about the line center, with peak-to-peak separation $\Delta\lambda$ ~ 8 Å. Such a feature would be detectable if

$|v| > 100$ km s^{-1}, and if the fraction of the sky subtended by system A at the QSO is given by $\Omega/4\pi > 2 \times 10^{-4}$. Recently Hegge, Liebert, and Strittmatter (1979) detected a narrow ($\Delta v < 350$ km s^{-1}) feature centered at $\lambda = (1 + z_A)(1215.7$ Å). Its central location indicates that this feature does not arise according to the above scenario which predicts a negligible fraction of emergent radiation at the line center. The redshift, velocity width, and luminosity ($\sim 10^{43}$ ergs s^{-1}) are more easily understood if this feature originates in a galactic nucleus with redshift z_A. In these respects it resembles the star-like emission object associated with the 21 cm absorber in AO 0235+164 (Smith, Burbidge, and Junkkarinen 1977). Further observations are needed in order to confirm this very interesting discovery.

C. A SEARCH FOR 21 CM ABSORPTION IN Mg II/Fe II ABSORPTION SYSTEMS

At redshifts $z < 1.8$ where L_α is inaccessible the leading candidates for 21 cm absorption are redshift systems with Mg II ($\lambda\lambda$ 2796, 2803) and Fe II ($\lambda\lambda$ 2344 $\rightarrow$ 2600) absorption. These clouds are promising because Mg II and Fe II lines are present in previously detected 21 cm systems (cf. Wolfe 1979), and because these ionic stages dominate in galactic H I regions. However, the presence of strong Mg II absorption need not indicate 21 cm absorption. The reason is that low-resolution optical scanning devices preferentially detect saturated lines produced by highly turbulent material, whereas the high resolution of radio spectrometers selects material of low velocity dispersion and high column density. Therefore we need to find redshift systems in which highly opaque clouds of low velocity dispersion lurk in the midst of rapidly moving clouds of considerably lower optical depth.

F. H. Briggs and I have searched for 21 cm absorption in 16 redshift systems that are in front of 14 QSOs. The results of our observations are mainly negative: 15 out of the 16 redshift systems in the survey have no radio absorption lines. Our subsequent data analysis was motivated by two questions: (1) are Mg II systems with detectable 21 cm absorption optically distinct from the others?, and (2) why are most Mg II systems transparent to 21 cm radiation?

To answer the first question Briggs and I compared $\tau(21)$ with W_λ (Mg II λ 2796) for redshift systems in which Mg II equivalent widths are available. A plot, which includes data from previously detected 21 cm systems, shows no obvious correlation (see Wolfe 1979). Moreover, $\tau(21)$ is not correlated with DR, the Mg II doublet ratio. In fact the largest DR (= 1.54) is found in the $z = 0.395$ absorption system in PKS 1229-02 for which we have a 21 cm detection. This would be

surprising if the same gas produced both Mg II and 21 cm absorption
since the indicated Mg II optical depth, τ (Mg II) $\sim$ 1, corresponds to
τ (21) $\ll$ 1 unless the metal abundances are unacceptably low. But it
is clear that the two types of lines form in different gas. A curve-of-
growth analysis of the Mg II lines indicates $\sigma \approx 100$ km s^{-1}, while the
21 cm absorber has $\sigma = 4.8$ km s^{-1}. This supports the turbulent/
opaque model for Mg II/21 cm systems, as suggested above.

The absence of 21 cm absorption in most Mg II systems can be explained in a number of ways. First, let us assume that Mg$^+$ is associated with regions in which H is mostly neutral. If Mg$^+$/H$^\circ$ = X(Mg)$_\odot$,
our null detections require that $T_s > 30$ K (recall that $\tau(21) \propto N(H\ I)/$
$(\Delta v\ T_s)$). This is not a prohibitive restriction for clouds ejected from
QSOs, nor for clouds in intervening galaxies. However, we would not
expect gas in galactic disks nor in galactic halos to have solar abundances of Mg, so these limits on T_s probably do not pertain to intervening galaxies. Indeed, if Mg$^+$/H$^\circ$ assumes the "Copernicus" abundance,
then $T_s > 500$ K in many cases. Since spin temperatures this large are
rarely observed in the galactic plane, one would consign the Mg II
regions to galactic halos, a site previously suggested by Bahcall and
Spitzer (1969). Perhaps the simplest explanation for the absence of
21 cm absorption is that the Mg II-producing region is associated with
H that is mostly ionized. But the similar kinematics of the 21 cm and
optical absorption profiles in the 21 cm absorber in AO 0235 + 164
(Wolfe et al. 1978) suggests that the "turbulent" Mg II clouds contain
H that is mostly neutral.

From a statistical point of view, the incidence of 21 cm absorption
in Mg II redshift systems is compatible with the intervening galaxy hypothesis. In a recent study of a complete sample of QSOs Weymann
et al. (1979) find that the incidence of Mg II absorption is about 13 times
that expected if Mg II lines form in galactic disks that extend to one
Holmberg radius. Therefore galactic Mg II-filled halos with radius
$R \sim 3.5\ R_{H_0}$ (≈ 70 kpc) are required by the intervening galaxy model.
Since 21 cm absorption occurs in $\sim$ 10% of the Mg II systems (including
the 21 cm absorber in AO 0235 + 164), galactic disks with $R \sim 1\ R_{H_0}$ are
adequate to explain the 21 cm data.

D. A SEARCH FOR OUTLYING H I IN A COMPLETE SAMPLE OF GALAXIES

If the Mg II lines are produced by 70 kpc halos, then galaxies
should be surrounded by large halos of hydrogen. Since there is a good
chance that Mg$^+$ is associated with hydrogen that is mostly neutral, one
would expect 21 cm emission from the outskirts of a significant fraction

of spiral galaxies. To test this hypothesis F. H. Briggs, N. Krumm, E. E. Salpeter and I began a systematic search for 21 cm emission from the outlying regions of a complete sample of spiral galaxies. The idea is to search for 21 cm emission from selected areas around each galaxy. By placing the narrow (4' diameter) beam of the Arecibo 21 cm antenna at two opposite points located at $R \sim 3\,R_{H_0}$ along the major and minor axes, one effectively probes the gaseous cross-section required to produce the Mg II lines. We integrated for an hour at each point so that our 3-σ upper limits on N(H I) would place meaningful upper limits on W_λ (Mg II).

Our sample consists of 27 galaxies that (a) are outside the local group, (b) are later than SO, (c) have optical major-axis diameters greater than 7', and (d) are in the Arecibo Dec. range. So far we have investigated 13 galaxies at beam locations comprising $\sim 1/3$ of the total sample. Our results, while obviously not complete, are still interesting enough to report on (see Briggs et al. 1979). Of the 33 beam locations already investigated 70% are null detections, 21% are marginally significant (i.e., signals that exceed the 3-σ noise limit but are confused by side lobe contributions from the main galaxy), and only 9% showed clear evidence for 21 cm emission. We note that isolated galaxies and galaxies with companions obey the same statistics.

The absence of H I halos at a sensitivity level of N(H I) $< (2 \rightarrow 3) \times 10^{18}$ cm^{-2} for bandwidths $\Delta v \simeq 16$ km s^{-1} is a new result with interesting implications. By adopting a velocity dispersion σ for gas at a given beam location and a ratio Mg^+/H^0 we can calculate W_λ (λ 2803) from upper limits on N(H I) that are appropriately smoothed. We may then compare the resultant W_λ vs σ curves with the range of these parameters inferred for QSO absorption-line clouds: the statistics of Weymann et al. (1979) are based on systems with $W_\lambda \gtrsim 0.5$ Å, and $\sigma = 100$ km s^{-1} is a conservative upper limit for Mg II systems (Briggs and Wolfe, 1979). We find that gas in beam locations with null or marginal detections cannot produce observable Mg II lines unless $Mg^1/H^0 > 0.3\,X(Mg)_\odot$. Thus halo gas with population II abundances, $X(Mg) \approx 0.04\,X(Mg)_\odot$ (Searle and Zinn 1978), or with interstellar abundances, $X(Mg) \approx 0.03\,X(Mg)_\odot$ (Spitzer 1978), cannot produce observable Mg II lines, if the Mg^+ is associated with H I regions. We emphasize that this conclusion is valid even if the Mg^+ is associated with ionized material. Ionization equilibrium arguments (Briggs and Wolfe 1979) show that in this case Mg^+/H^0 cannot exceed a conservative upper limit of $\approx 4\,X(Mg)$, so that $N(Mg^+)$ is less than 1/2 of the required value for population II abundances. Consequently, intervening galactic halos like those in our sample will not give rise to observable Mg II absorption lines unless gaseous material with population I metal abundances is present.

The discrepancy between the required Mg II cross-section of inter-vening galaxies and the average cross-section, $\langle A \rangle$, implied by our survey can be stated in the following manner: The frequency of detect-ing 21 cm emission at our beam locations indicates that $1.8 < \langle A \rangle / A_{H_0}$ < 5.0, where A_{H_0} $(= \pi R_{H_0}^2 / 2)$ is the average cross-section of a Holm-berg disk. We are not certain whether a null detection at a radius R means that H I is confined to R_{H_0}, or whether it extends to just within 1 beamwidth of R, and this accounts for the range in $\langle A \rangle$. In any case the value of $\langle A \rangle$ required by Weymann et al. (1979) is $\langle A \rangle = 13\ A_{H_0}$, so there is a factor of ~ 4 discrepancy.

The only "isolated" galaxy to exhibit 21 cm emission at $R > R_{H_0}$ is NGC 628. The emission coincides with four beam locations surrounding the S-W extension of the outer major axis (p.a. = 80°), which are ~ 100 kpc from the nucleus. Briggs (1980) has recently mapped the entire galaxy and finds H I in every quadrant at $R \sim 3.5\ R_{H_0}$, with a pro-nounced asymmetry toward the S-W. Because the galaxies nearest to NGC 628 are two dwarfs displaced by ~ 350 kpc on the sky, it is improb-able that tidal interactions are the cause of the outlying H I. Rather it appears to be primordial in origin. On the other hand it is difficult to see how a primordial asymmetry could survive the effects of differential rotation which would smear it out in 1 or 2 rotation periods of $\sim 10^9$ y.

REFERENCES

Bahcall, J. N. and Spitzer, L.: 1969, Ap. J. Lett. 156, L63.

Briggs, F. H.: 1980, in preparation.

Briggs, F. H. and Wolfe, A. M.: 1980, in preparation.

Briggs, F. H., Wolfe, A. M., Krumm, N., and Salpeter, E. E.: 1980, submitted to the Ap. J.

Carswell, R. F., Hilliard, R. L., Strittmatter, P. A., Taylor, D. J., and Weymann, R. J.: 1975, Ap. J. 196, 351.

Field, G. B.: 1959, Ap. J. 129, 551.

Hegge, E. K., Liebert, J. L., and Strittmatter, P. A.: 1979, private comm.

Searle, L. and Zinn, R.: 1978, Ap. J. 225, 357.

Smith, H. E., Burbidge, E. M., and Junkkarinen, V. T.: 1977, Ap. J. 218, 611.

Spitzer, L.: 1978, "Physical Processes in the Interstellar Medium" (New York: Interscience).

Weymann, R. J., Williams, R. E., Peterson, B. M., and Turnshek, D. A.: 1979, preprint.

Wolfe, A. M.: 1979, "Active Galactic Nuclei" (eds. C. Hazard and S. Mitton), p. 159 (Cambridge University: Cambridge).

Wolfe, A. M.: 1980, to appear in Physica Scripta.

Wolfe, A. M., Broderick, J. J., Condon, J. J., and Johnston, K. J.: 1978, Ap. J. 222, 752.

Wolfe, A. M. and Davis, M. M.: 1979, A. J. 84, 699.

DISCUSSION

G. Burbidge: Would you agree that there is really very little evidence for extended halos containing metals in galaxies?

Wolfe: I would agree that we haven't found any evidence. But the point I was trying to make was that the upper limits we have set to 21-cm emission from the outskirts of galaxies do not rule out MgII absorption at the required level if the Mg/H abundance is population I; this is true if the gas is neutral, and virtually true if it is ionized. Whether population I gas exists in halo regions is a separate question. We know that the stellar content of the halo has low population II abundances. One possibility is a galactic wind. But whether cool regions can persist in a necessarily hot (T ~ 10^6 K) wind, out to R ~ 100 kpc is problematic.

Spinrad: Your results imply the absence of MgII-producing regions in galactic halos. Yet Boksenberg and Sargent find CaII absorption in the QSO/galaxy pair 3C 232/HGC 3067 where the QSO lies outside the optical image of the galaxy. How can these results be reconciled?

Wolfe: Yes, there is a paradox if the radial distance, R, of the absorbing gas from the nucleus of NGC 3067 is greater than 3 x R_{HO} (R_{HO} = Holmberg radius). But all we know is the impact parameter to 3C 232 which equals 1 R_{HO}. But the 90 km s^{-1} width of the CII lines rules this out. So it is possible that the reason why the absorbing gas in this galaxy is above our upper limit on column density is that it is at a radius smaller than that of our survey points.

Wehinger: Have you looked for 21-cm absorption in Seyfert and/or N galaxies?

Wolfe: No.

D. Roberts: I'd like to report another negative result in the search for 21-cm absorption associated with MgII optical absorption. Bennett, Lawrence, and Burke, at MIT, have used the NRAO 300-foot telescope to observe the double quasar 0957!561 at the optical redshift of 1.3914. A total of 2 hours integration yields a limit of 0.02 K, where the total source temperature is 2 K. However, most of the flux at 594 MHz comes from the extended ratio source associated with the north component, and as a result the optical depth limit is only $\tau_{21 cm}$ $\lesssim$ 0.2 (assuming that the absorbing cloud completely covers both of the compact sources).

RADIO STRUCTURE OF SEYFERT GALAXIES

A. S. Wilson
Astronomy Program, University of Maryland

A. G. Willis
Radiosterrenwacht Westerbork

We have recently mapped about a dozen Seyfert galaxies with the Very Large Array at 5 GHz and obtained less complete structural information on about a dozen more. Most sources are heavily resolved at resolutions near or below 1 arc sec, with linear scales in the range several hundred parsecs to a few kiloparsecs. For 3 galaxies (Mark 3, NGC 1068 and NGC 5548) the structure is double with two components more or less symmetrically placed on opposite sides of the optical nucleus. A third component, when present, coincides with the optical nucleus. This result provides strong evidence that "double radio source machines" also reside in the nuclei of active spiral galaxies as well as ellipticals. Other sources show a more diffuse morphology, but usually also possess a compact radio source associated with the optical continuum nucleus. A close relation between the extended radio emission and the thermal gas in the forbidden line region is indicated since (a) they have similar extents, (b) the radio and forbidden line powers are correlated, (c) the relativistic plus magnetic ($B^2_{eq}/4\pi$) and thermal pressures ($n_e k T_e$) are similar and (d) the kinetic energy of the thermal gas and the minimum energy for synchrotron radiation are comparable. For the double sources, the radio emitting plasma is probably ejected from the compact nucleus and slowed by the large quantities of thermal gas in the forbidden line region. Alternatively, radio sources with more diffuse morphology may derive their luminosity from cosmic rays accelerated "in situ" by shock waves associated with the high velocity thermal gas and a magnetic field from a compressed (accreted?) interstellar medium.

G.O. Abell and P. J. E. Peebles (eds.), Objects of High Redshift, 161–163.
Copyright © 1980 by the IAU.

DISCUSSION

Wolfe: Many radio galaxies studied by Osterbrock have extended
                  forbidden-line regions and broad permitted lines resembl-
ing Seyfert nuclei. Yet the radio structures are compact, rather than
being extended doubles. Since the emission-line characteristics of
these ellipticals are so similar to those of the Seyferts, one would
expect, on the basis of your model, that the radio structure would be
extended. Could you explain this apparent contradiction?

A.S. Wilson: Yes, a small fraction of galaxies with Seyfert-type spec-
                  tra have radio structures dominated by a very powerful
compact core. In some cases (e.g., NGC 1275), extended radio emission
is also seen. Since the cores are so much more powerful and compact
than for typical Seyferts, one feels something different is going on.
Alternatively, the radio emission from the core may be so strong in
these cases because we view the galaxy almost along the direction of a
relativistic beam (see Readhead's paper, to follow). This would also
account for their lack of double radio sources. Note also, however,
that even the radio-weak Seyferts I have described have, in general,
weak, unresolved radio core coincident with their optical nuclei as well
as the extended emission. Probably an isotropic radio component from
the core is also present.

J. Roberts: Is it correct that all of your Class 1 sources are ellip-
                  tical galaxies? What is the percentage of spirals in
your Classes 2 and 3?

A.S. Wilson: Although many of the galaxies in Class 1 are classified
                  merely as N, the answer to the first question is almost
certainly yes, since almost all powerful double radio sources are
associated with ellipticals, and never with spirals. The number of
galaxies known in Class 2 is rather small and their classification is
sometimes controversial (e.g., 3C 120), so no firm statements on their
morphological types can be made. Since almost all optically-selected
Seyferts lie in Class 3, their morphological breakdown follows that for
Seyferts as a whole, for which the great majority are spirals (see
Adams, T.F.: Ap. J. Suppl. 33, 19, 1977).

Gaskell: An interesting result has been obtained by Bill Keel of
                  Lick Observatory (which has been submitted to the Publi-
cations, Astronomical Society of the Pacific). He measured the orien-
tations of all known nearby Seyfert I and Seyfert II galaxies on the
sky survey plates (by measuring the axial ratios of the discs). He
also measured a control sample of disc galaxies. He found that orientation of the Seyfert II galaxies was the same as that of the control
galaxies, but that the Seyfert I galaxies were preferentially face on!
This could be either because there is some beaming along the pole of
the galaxy, or because obscuration in the discs of Seyfert I galaxies
is hiding their nuclei from us when they are near edge on.

A.S. Wilson: This sounds interesting and I look forward to learning more. However, I find it hard to see how the line emission can be beamed! Also, Type 1 Seyferts can now be discovered via their X-radiation, which should not be affected by absorption above a few keV. Thus, there are severe constraints on such interpretations.

VLBI MAPPING OF THE NUCLEI OF RADIO GALAXIES AND QUASARS

A.C.S. Readhead
Owens Valley Radio Observatory
California Institute of Technology

It is now possible, by means of VLBI hybrid mapping, to make maps of radio sources with a resolution of ∿ 1 milliarcsecond. This enables us for the first time to compare the morphologies of the small- and large-scale structures of extragalactic radio sources, and they are strikingly different.

1. INTRODUCTION

There are two major classes of extragalactic radio sources namely the extended (> 1 arc sec) and compact (< 1 arc second) objects. These are fairly well delineated by their radio spectra. In Figure 1 is

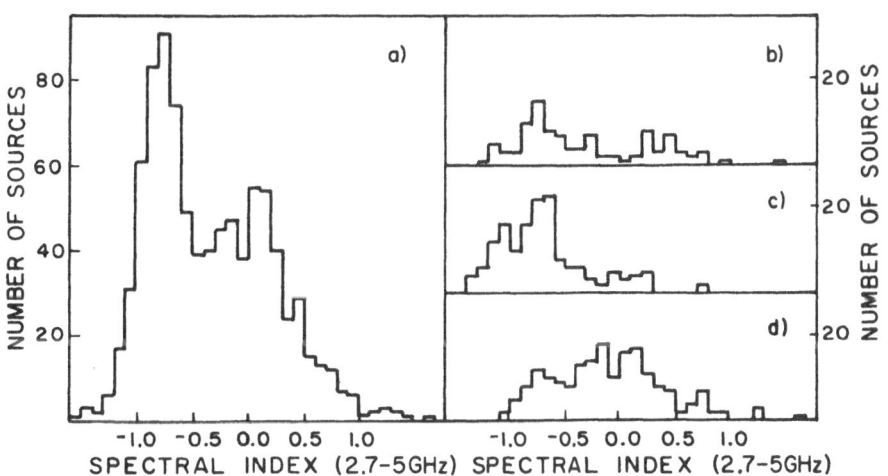

Figure 1. The spectral index distribution of sources in the SIV survey: a) complete sample, b) empty fields, c) radio galaxies, d) quasars (from Pauliny-Toth et al. 1978). Most objects with α > -0.5 are compact.

G.O. Abell and P. J. E. Peebles (eds.), Objects of High Redshift, 165–176.
Copyright © 1980 by the IAU.

shown the distribution of radio spectral indices from the SIV survey
(Pauliny-Toth et al. 1978), and we see immediately that over 60% of
these objects have flat spectra (i.e. they are compact), and only about
40% have steep spectra (i.e. they are probably extended). Many hundreds
of these extended objects have been mapped by aperture synthesis at
Cambridge, Westerbork, Greenbank and recently the VLA. Only $\sim$ a dozen
compact objects have been mapped thus far. Nevertheless the results of
mapping this small sample have been illuminating, and it is these objects
which I will discuss here.

The spectral index classification is not a completely reliable
guide to angular size, since there are some steep spectrum objects which
are compact (e.g. 3C147, 3C380) and which look quite different to the
"classical double" extended sources. A more useful classification is
given by the radio morphology at a given frequency and the disposition
of the radio structure relative to the optical objects. It is useful
to define "symmetric" objects as those in which the dominant radio
emission at 5 GHz comes from two regions on opposite sides of, and
roughly equi-distant from the optical object; and to define "core"
objects as those in which the dominant emission at 5 GHz comes from a
component coincident with the optical object, and in which the extended
radio emission region, if any, is asymmetric.

2.   THE NEED FOR HIGH RESOLUTION (<< 1 ARC SECOND) MAPS

In order to understand the physical processes responsible for the
creation and evolution of extragalactic radio sources we need to map
these on the angular scales $10^{-4} \rightarrow 1$ arc second, i.e. on VLBI scales,
for the following reasons:
1) As we have just seen, the compact objects comprise more than half
the total population of radio sources at high frequencies and they are
too small (<< 1") to be mapped by conventional methods.
2) Many compact objects vary on time-scales of days $\rightarrow$ decades. Thus
the study of variation in structure adds another dimension to the conventional mapping of extragalactic objects, and it enables us to study
the physics of these objects in a completely different way.
3) It is now commonly believed that the energy of the extended objects
originates in the nucleus of the galaxy or quasar. VLBI observations
enable us to map these centres of activity with a resolution $\sim$ 1 pc
at the Hubble radius.
4) Searches for interstellar scintillation in extragalactic objects
(Condon & Backer (1975), Condon & Dennison (1978), Readhead et al.
(1980b))have shown that the angular sizes of the most compact regions
are >> 10^{-5} arc seconds. Thus it is possible with high frequency VLBI
to study the smallest detectable features in these objects.

3.   HYBRID MAPPING

The hybrid mapping technique which we have used has been described in detail by Readhead & Wilkinson (1978). It is superior to the old model fitting methods for three reasons:
1) The visibility phases are derived from the closure phase (Jennison 1958) and amplitude data. The inclusion of the phase data is essential for making reliable maps, and removes the 180° ambiguity inherent in amplitude only reconstructions. This is particularly important when comparing maps at different epochs of a strongly varying source. In addition, with limited data it is sometimes possible to have quite different models which fit the amplitude data equally well. In our experience the closure phase data is a powerful discriminant in such cases.
2) Four or more telescopes are used for making hybrid maps. The CIT/JPL processor can now correlate all pairs of a five station network simultaneously, thus making observations on large networks more tractable.
3) Continuous tracking of the source for 9 to 12 hours, depending on the declination, ensures that we get all the possible visibility information on a given network.

In Figure 2 are shown three maps of 3C 147 at different frequencies. In the cases of 609 and 1671 MHz maps it was possible to use the core as an internal phase reference, so that the closure phase relations do give the visibility phases directly. The agreement of the three maps is excellent and gives confidence in the hybrid mapping procedure, especially when you consider that the maps were made on different networks at the three frequencies. In addition the fits to the data are very good in each case. Figure 3 shows the fit at 1671 MHz. This is typical of the fits obtained by hybrid mapping

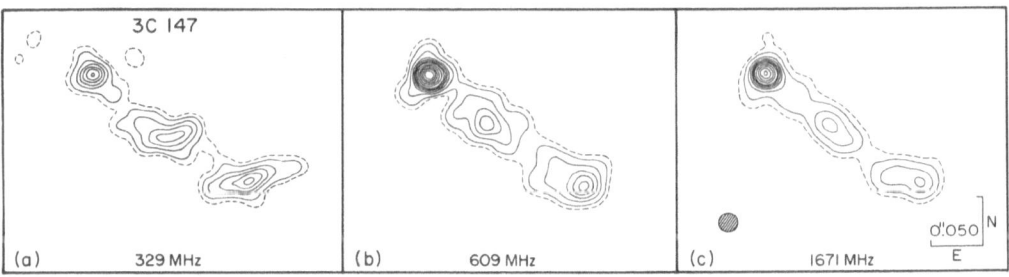

Figure 2. Hybrid maps of 3C 147:   a) Simon et al. (1980), b) Wilkinson et al. (1977), c) Readhead and Wilkinson (1980).

The morphology of 3C 147 is typical of core objects. The most striking characteristics are:

1) It is a one-sided jet,
2) There is a very compact, flat spectrum core at the end of an extended steep spectrum jet,

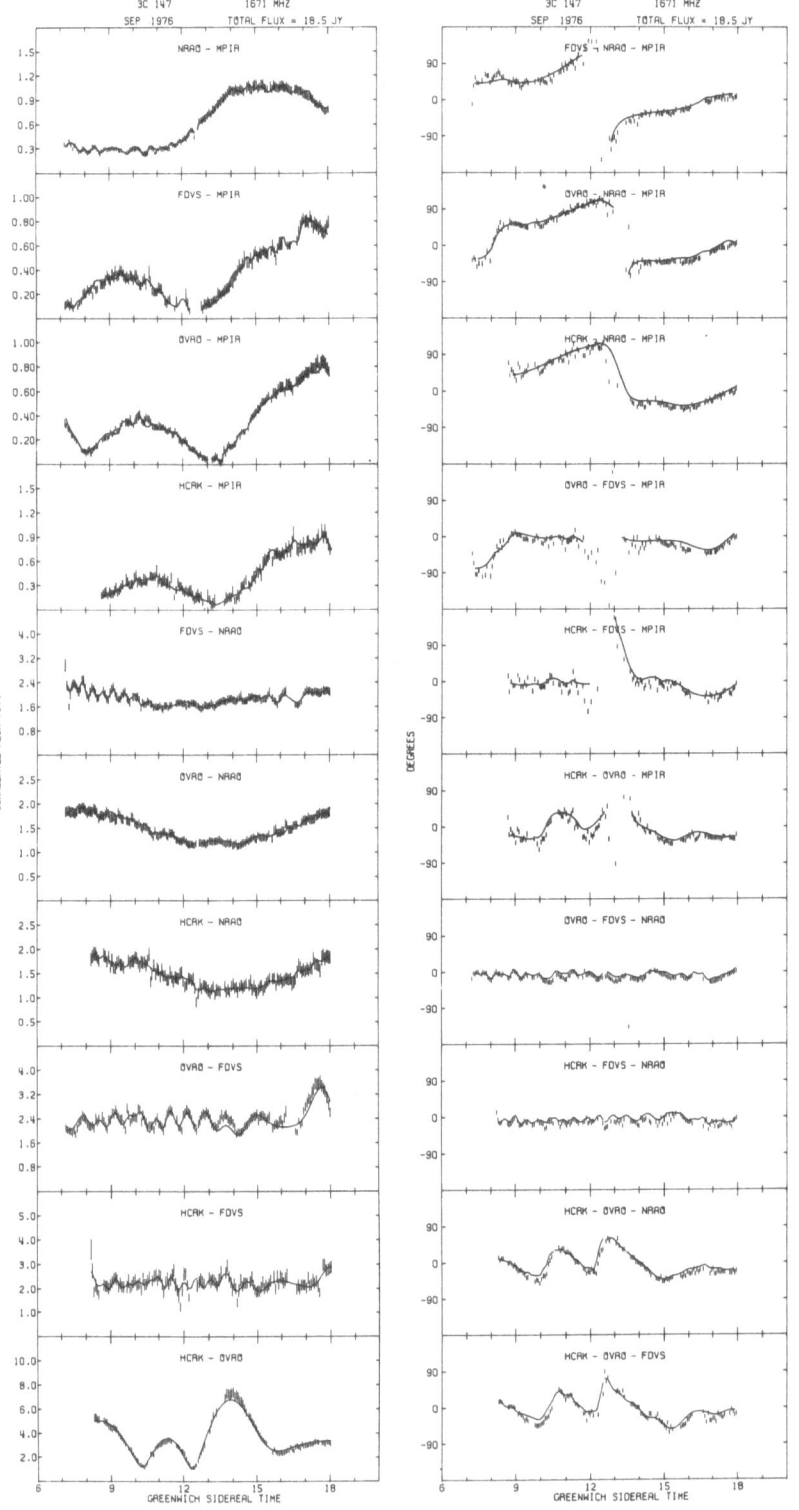

Figure 3. The fit to the amplitudes and closure phases of the delta functions which constitute the map shown in Figure 2c. (Readhead and Wilkinson, 1980). The fit is excellent, and gives confidence in the hybrid mapping procedure. At times of very deep minima in the correlated fluxes the closure phase flips through ~180°. In successive iterations the direction of the phase flip varied, on some baselines, and we have not tried to fit the closure phases at these times. This has a negligiable affect on the hybrid map.

3) The position angle of the jet is different to that of the larger
scale ($\sim$ 1") structure.

The objects which we have mapped thus far by this technique are
shown in Table I. Apart from 3C 84, nearly all of these objects are
one-sided jets with a flat spectrum core and a steep spectrum jet.
The only objects which are not clearly asymmetric are CTA 21 and 3C 119,
which have only been observed at one frequency. We have found that it
is often necessary to have maps at two frequencies before asymmetry is
clearly shown, because of the different spectral indices of the core
and jet.

TABLE I
Objects Which have been Mapped by VLBI

Object	Mapped at Frequencies (GHz)	Asymmetric	Flat Spectrum Core Steep Spectrum Jet	References
CTA 21	.609	?	?	43
3C 84	.609, 5.011, 10.650	Complex	Complex	19, 22, 28, 43
3C 111	10.650	√	?	15
3C 119	1.671	?	?	21
3C 120	5.011, 10.650	√	√	29
3C 147	.327, .609, 1.671, 4.885, 15.035	√	√	27, 32, 40, 42
3C 273	.609, 5.011, 10.650	√	√	24, 43
3C 286	.327, .609, 1.671	√	√	21, 40, 43
3C 345	.609. 5.011, 10.650	√	√	29, 43
3C 380	1.671, 5.011	√	√	22, 28, 32
3C 454.3	.609, 1.671	√	√	21, 43
CTA102	.609, 1.671	√	√	21, 43
NGC6251	10.650	√	√	5

4.   THE MORPHOLOGY OF THE NUCLEAR COMPONENTS

Among the objects mapped thus far there are symmetric objects,
steep spectrum core objects and flat spectrum core objects. Taken in
this order, the relative dominance of the central component is increasing, and I have chosen one example from each group to illustrate
their major characteristics

4.1 Symmetric Objects

A hybrid map, and a model of NGC 6251 are shown (Cohen & Readhead

1979) in Figure 4. It is a one-sided jet with a flat spectrum core and steep spectrum jet, and the jet is well aligned with the large scale (>> 1") structure (Readhead et. al. 1978a).

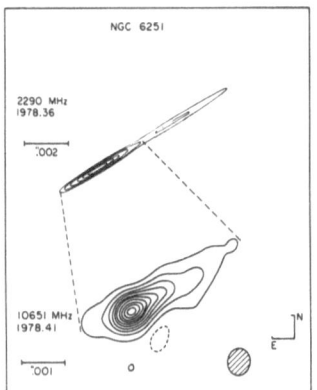

Figure 4. Hybrid map (10,651 MHz) and model (2290 MHz) of the nucleus of the symmetric radio galaxy NGC 6251 (Cohen and Readhead, 1979).

Thus far such good alignment has always been found in symmetric objects (Readhead et. al. 1978b). No significant structural variations have been seen, (i.e. over a period of 1 year).

4.2 Core Objects with Steep Spectra (at 5 GHz)

In Figure 5 is shown a hybrid map of 3C 380 (Readhead & Wilkinson 1980).

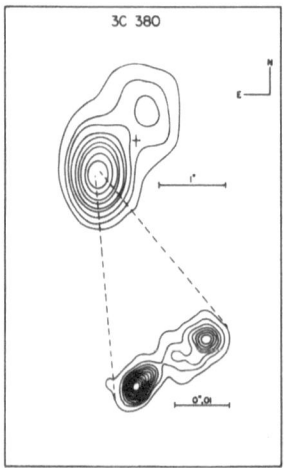

Figure 5. Top: Map of 3C 380 at 15 GHz made with Cambridge 5 km telescope (Scott, 1977). Bottom: Hybrid map at 1671 MHz (Readhead & Wilkinson 1980). The cross marks the position of a compact component The optical object coincides with the SE component of the 15 GHz map.

In many respects this source is very like 3C 147. The object is curved, i.e. the smallest and largest features are not aligned. This is typical of core objects (see §5). No significant structural variations have been seen in objects of this type.

4.3 Core Objects with Flat Spectra (at 5 GHz)

A hybrid map of 3C 273 (Readhead et al. (1979)) is shown in Figure 6. Objects in this class often show significant structural variations on time-scales ⪝ 1 year.

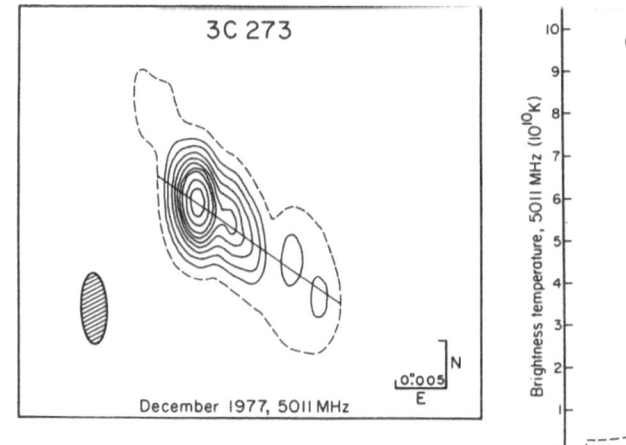
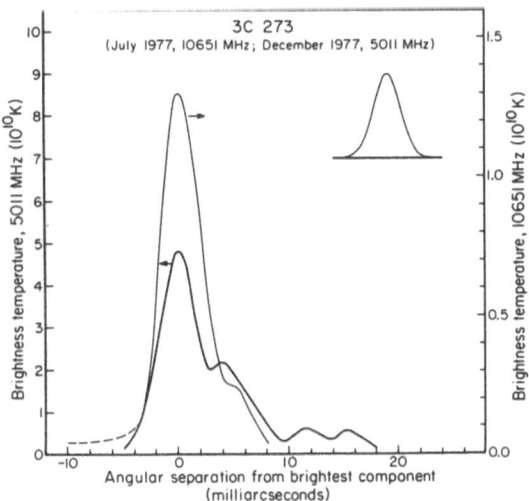

Figure 6. Left: Hybrid map of 3C 273 at 5011 MHz. Right: Sections through the 5011 MHz and 10,651 MHz maps of 3C 273, showing that the core has a flat spectrum and the jet a steep spectrum (Readhead et al. 1979).

An important point which emerges from this comparison is that the morphology of central components of symmetric objects is very similar to that of the core objects. R. Linfield (private communication) has recently made a hybrid map of the central component of 3C 111 and finds that, like NGC 6251, it is a one-sided jet.

5. SYMMETRIC AND/OR ASYMMETRIC JETS?

One of the most interesting questions in extragalactic radio astronomy is whether there are two mechanisms for producing jets - one symmetric and one asymmetric, or whether there is only one, symmetric mechanism.

Most extended (>1 arc sec) objects are symmetric, and until recently it appeared that asymmetric objects were rare, the best examples being 3C 273 and M87, and a few D2 quasars (Miley, 1971). It now appears that asymmetry might be as ubiquitous as symmetry in extragalactic objects. In particular, even in the symmetric objects 3C 111 and NGC 6251 the radio nuclei are asymmetric. These are the only nuclei of symmetric objects mapped thus far, but it is unlikely that the first two objects mapped will be entirely atypical. These jets could be instantaneously one-sided, but alternating between one side and the other. In that case, if the typical switching time were longer than $\sim 10^4$ years, the morphology of the objects in any complete sample of steep spectrum radio sources would change from asymmetric to symmetric at an overall size of a few tens of kiloparsecs - which is not observed. However NGC 6251 has an asymmetric jet 200 kpc long (Waggett et. al. 1978), implying a switching time $>10^5$ years. Thus either both jets are on quasi-continuously, and we just do not see the counter-jet, or NGC 6251 is atypical.

The occurence of asymmetric nuclei in symmetric objects strongly suggests that objects such as 3C 273 and M87 may not be as different from typical symmetric objects as they appear, and that the fundamental mechanism is symmetric but it only appears asymmetric to us. There is other evidence that this may be the case: In Table II are summarized some of the major differences between core and symmetric objects. There are three striking correlations between morphological class and other important physical characteristics:
(i) Alignment between small scale and extended structure (ΔPA) - symmetric objects are well aligned whereas core objects are often curved.
(ii) Overall Size - the core objects are consistently smaller than symmetric objects.
(iii) Distance Class - core objects are on average more distant than symmetric objects.

TABLE II
Objects in which both very compact ($\ll 1''$) and extended ($>1''$) Structure have been mapped. (H_0 = 50 km/s/Mpc, q_0 = 0.5)

Name	z	Size (kpc)	Type	Change in p.a. (deg)	References
NGC315	0.0167	1500	Symm	< 4	2, 15
NGC6251	0.023	2900	Symm	4.5	5, 25, 41
3C111	0.0485	270	Symm	< 4	14, 15
3C371	0.050	4	Core	30	22, 23, 28
3C390.3	0.0561	320	Symm	< 5	14, 15, 24
3C405	0.0565	190	Symm	< 4	11, 13
3C236	0.0989	5700	Symm	< 3.5	36, 44
3C273	0.158	70	Core	40	18, 26, 29, 35, 39
3C279	0.54	87	Core	~60	8, 16, 18
3C147	0.545	5	Core	25	10, 27, 32, 40, 62
3C345	0.594	25	Core	45	26, 29, 37, 38
3C380	0.691	40	Core	20	22, 28, 32, 34
4C39.25	0.698	25	Core	~40	22, 28, 37, 38, 39
3C454.3	0.860	50	Core	34	9, 21, 37, 39, 43

These effects can all be explained in terms of the simple model shown in Figure 7 (Readhead et. al. (1978b)). On this model both core and symmetric objects are produced by twin beams along which the particles travel with relativistic bulk velocity v. In core objects the angle, θ, between the line of sight and the beam is small ($\theta < \frac{1}{\gamma}$, where $\gamma = (1 - (v/c)^2)^{-1/2}$). In symmetric objects this angle is large. The observed asymmetry in the nuclei of both core and symmetric objects is due to relativistic beaming of the radiation from the jets. If the intrinsic bending of the beams is small ($\lesssim 5°$) this would account for the good alignment observed in symmetric objects, while projection and beaming effects in core objects would account for both the large apparent curvature and the small overall size. In addition, one has to observe a large number of objects, and hence a large volume of space in order to find a significant number of objects pointing almost along the line of sight. This helps to explain the larger average distance of the core objects.

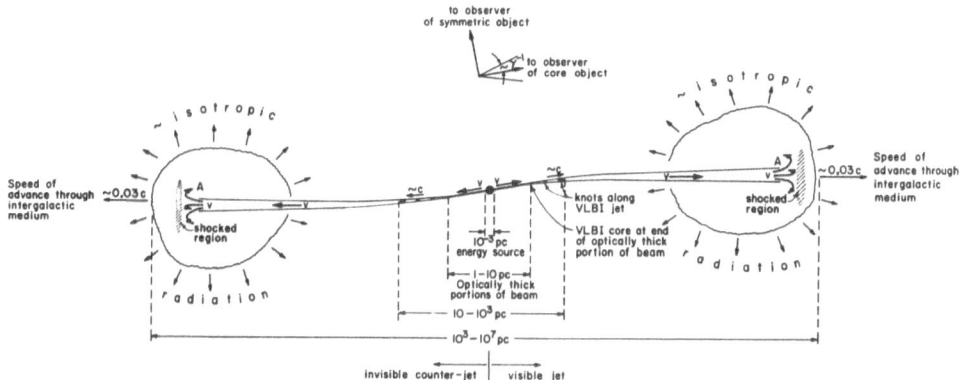

Particle bulk velocity along jets = v(~c). At positions "A", at heads of jets, there may be some beaming of radiation in directions almost parallel to the beam, but most of the radiation from the head regions is ~ isotropic.

Figure 7. Simple model which accounts for the most obvious differences between core and symmetric objects (Readhead et. al. (1978b), Scheuer & Readhead (1979), and Blandford & Konigl (1979)).

This model also accounts naturally for
iv) the superluminal velocities observed in some core objects (e.g. Cohen et. al. 1977, Cohen et. al. 1979), and
v) the ratio of radio-quiet to radio quasars (Scheuer & Readhead, 1979).

A detailed physical discussion of this model has been given by Blandford & Konigl (1979). I must emphasize that this is not the only plausible hypothesis which can account for some of these effects (Readhead et. al. 1978), but it is the only one which gives a natural explanation for _all_ of them.

Regardless of whether or not this simple model is either wholly or partially correct, the fact that the first VLBI maps are nearly all asymmetric is of great significance and seems destined to play as crucial a role in our understanding of these objects as the symmetry of the "classical double" sources.

REFERENCES

1. Blandford, R.D. and Konigl, A.: 1979, Astrophys. J., (in press).
2. Bridle, A.H., Davis, M.M., Fomalont, E.B., Willis, A.G. and Strom, R.G.: 1980, Astrophys. J., (submitted).
3. Cohen, M.H., Kellermann, K.I., Shaffer, D.B., Linfield, R.P., Moffet, A.T., Romney, J.D., Seielstad, G.A., Pauliny-Toth, I.I.K., Preuss, E., Witzel, A., Schilizzi, R.T. and Geldzahler, B.J.: 1977, Nature, 268,p405.
4. Cohen, M.H., Pearson, T.J., Readhead, A.C.S., Seielstad, G.A., Simon, R.S. and Walker, R.C.: 1979, Astrophys. J., 231,p293.
5. Cohen, M.H. and Readhead, A.C.S.: 1979, Astrophys. J. Letters, (in press).
6. Condon, J. and Backer, D.C.: 1975, Astrophys. J.,197,p31.
7. Condon, J.J. and Dennison, B.: 1978, Astrophys. J., 224,p835.
8. Cotton, W.D., Counselman, C.C., Geller, R.B., Shapiro, I.I., Wittels, J.J., Hinteregger, H.F., Knight, C.A., Rogers, A.E.E., Whitney, A.R. and Clark, T.A.: 1979, Astrophys. J. Letters, 229, L115-L117.
9. Davis, R.D., Stannard, D. and Conway, R.G.: 1977, Nature, 267,p596.
10. Donaldson, W. and Smith, H.: 1971, Monthly Notices Roy. Astron. Soc., 151,p253.
11. Hargrave, P. and Ryle, M.: 1974, Monthly Notices Roy. Astron. Soc., 166,p305.
12. Jennison, R.C.: 1958, Monthly Notices Roy. Astron. Soc., 118,p276.
13. Kellermann, K.I., Clark, B.G., Niell, A.E. and Shaffer, D.B.: 1975, Astrophys. J. Letters, 197, L113-L116.
14. Laing, R.: 1980, Monthly Notices Roy. Astron. Soc., (in press).
15. Linfield, R.P.: 1979, (private communication).
16. Lyne, A.: 1972, Monthly Notices Roy. Astron. Soc., 158,p431.
17. Miley, G.K.: 1971, Monthly Notices Roy. Astron. Soc., 152,p477.
18. Niell, A.E., Kellermann, K.I., Clark, B.G. and Shaffer, D.B.: 1975, Astrophys. J. Letters, 197, L103-L112.
19. Pauliny-Toth, I.I.K., Preuss, E., Witzel, A., Kellermann, K.I., Shaffer, D.B., Purcell, G.H., Grave, G.W., Jones, D.L., Cohen, M.H., Moffet, A.T., Romney, J., Schilizzi, R.T. and Rinehart, R.: 1976, Nature, 259,p17.
20. Pauliny-Toth, I.I.K., Witzel, A., Preuss, E., Kuhr, H., Kellermann, K.I., Fomalont, E.G. and Davis, M.M.: 1978, Astron, J., 83,p451.
21. Pearson, T.J., Readhead, A.C.S. and Wilkinson, P.N.: 1980, Astrophys. J., (submitted).
22. Pearson, T.J. and Readhead, A.C.S.: 1980, Astrophys. J., (submitted).
23. Perley, R.A. and Johnstone, K.J.: 1979, Astron. J., (in press).
24. Preuss, E., Kellermann, K.I., Pauliny-Toth, I.I.K. and Shaffer, D.B.: (in preparation).
25. Readhead, A.C.S., Cohen, M.H. and Blandford, R.D.: 1978a, Nature, 272,p131.

26.Readhead, A.C.S., Cohen, M.H., Pearson, T.J. and Wilkinson, P.N.:
     1978b, Nature, 276,p768.
27.Readhead, A.C.S., Napier, P.J. and Bignell, R.C.: 1980a, Astrophys.
     J., (submitted).
28.Readhead, A.C.S. and Pearson, T.J.: 1980, Astrophys. J., (submitted).
29.Readhead, A.C.S., Pearson, T.J., Cohen, M.H., Ewing, M.S. and Moffet,
     A.T.: 1979, Astrophys. J., 231,p299.
30.Readhead, A.C.S., Sargent, W.L.W. and Cohen, M.H.: 1980b, Astrophys.
     J., (submitted).
31.Readhead, A.C.S. and Wilkinson, P.N.: 1978, Astrophys. J., 223,p25.
32.Readhead, A.C.S. and Wilkinson, P.N.: 1980, Astrophys. J., (in press).
33.Scheuer, P.A.G. and Readhead, A.C.S.: 1979, Nature, 277,p182.
34.Scott, M.A.: 1977, Monthly Notices Roy. Astron. Soc.,197,p377.
35.Schilizzi, R.T., Cohen, M.H., Romney, J.D., Shaffer, D.B., Kellermann
     K.I., Swenson, G.W., Yen, J.L. and Rinehart, R.: 1975, Astrophys. J.,
     201,p263.
36.Schilizzi, R.T., Miley, G.K., van Ardenne, A., Baud, B., Beath, L.,
     Ronnang, B.O. and Pauliny-Toth, I.I.K.: 1979, Astron. Astrophys.,
     (in press).
37.Shaffer, D.B., Cohen, M.H., Romney, J.D., Schilizzi, R.T., Kellermann,
     K.I., Swenson, G.W., Yen, J.L. and Rinehart, R.: 1975, Astrophys. J.,
     201,p256.
38.Shaffer, D.B., Kellermann, K.I., Purcell, G.H., Pauliny-Toth, I.I.K.,
     Preuss, E., Witzel, A., Graham, D., Schilizzi, R.T., Cohen, M.H.,
     Moffet, A.T., Romney, J.D. and Niell, A.E.: 1977 Astrophys. J., 218,
     p353.
39.Shaffer, D.B. and Schilizzi, R.T.: 1975, Astron. J., 80,p753.
40.Simon, R.S., Readhead, A.C.S., Wilkinson, P.N., Moffet, A.T. and
     Anderson, B.: 1980, Astrophys. J., (submitted).
41.Waggett, P.C., Warner, P.J. and Baldwin, J.E.: 1977, Monthly Notices
     Roy. Astron. Soc., 181,p465.
42.Wilkinson, P.N., Readhead, A.C.S., Purcell, G.H. and Anderson, B.:
     1977, Nature, 269,p764.
43.Wilkinson, P.N., Readhead, A.C.S., Anderson, B. and Purcell, G.H.:
     1979, Astrophys. J., (in press).
44.Willis, A.G., Strom, R.G. and Wilson, A.S.: 1974, Nature, 250,p625

DISCUSSION

A.S. Wilson: Preliminary results from new 610 MHz observations at
               Westerbrook by R.G. Strom, A.G. Willis and me show the
probable existence of a very faint counterjet (i.e., to the SE) in
NGC 6251. This result supports the idea that both beams do exist
simultaneously.

Readhead: I am delighted to hear it. The ratio of the surface
               brightnesses in the two jets will place useful limits on
γ in our model.

Penston: I thought I noticed that the optically violently variable
                   objects fell among the compact sources. Could your model
accommodate such a difference in the optical properties?

Readhead: Yes, given that the intrinsic liminosities of core and
                   symmetric sources are probably very different, I would not
be surprised if other physical properties, such as optical and X-ray
flux, showed some differences between the two classes.

Schmidt: Is your hypothesis concerning the ratio of radio-quiet to
                   radio-noisy quasars compatible with a different evolution
of the two types of quasar?

Readhead: Taken in its simplest form, with a single value of γ, the
                   hypothesis is probably not compatible with a different
evolution of the two types of quasar. However, as Peter Scheuer and I
have pointed out (Scheuer and Readhead, 1979), it is clear that a range
of values of γ is needed, and it may be that the symmetric objects have
predominantly low values of γ, while the core objects have predominantly
high values of γ. Moreover, if relativistic beaming is important in
core objects, the intrinsic luminosities of these objects are much lower
than they appear. Thus, the power of the core objects is probably much
less than the symmetric objects, and it may well be that there is a different cosmological evolution for weak and powerful quasars.

                   As Martin Rees has pointed out, even without the model it
                   might be difficult to reconcile the different evolution
of the two quasar types with the strong evolution of the parent population of optically selected quasars. However, I would agree that the
model seems at first sight to aggravate the problem.

                   We need to know much more about the radio emission from
                   complete samples of optically selected quasars, and hence
limits on the values of γ for the two quasar types, before we can decide
if there is a serious confict.

NEW OBSERVATIONS WITH THE VLA

D. S. Heeschen
National Radio Astronomy Observatory

(Abstract)

The VLA, now under construction in New Mexico, is an aperture synthesis array of twenty-seven 25-meter diameter antennas, with over-all dimension of about 35 km. At 6-cm wavelength it will have resolution of 0.6 arcseconds and sensitivity of 0.1 mJy, and should be a superb instrument for radio studies of objects of large redshift. A more detailed description of the VLA has been given by Heeschen (1975).

At this meeting (August, 1979) the partially completed array is in use about one-third time for scientific observations, with up to 17 antennas. In this paper recent observations by a number of investigators were presented. The various programs discussed, and the investigators involved with them, are listed here:

The Structure of 3C 449: R. A. Perley, A. G. Willis and J. S. Scott
The Structure of 3C 31 and NGC 315: E. B. Fomalont
Structure and Polarization of 2349+32: J.F.C. Wardle and R. I. Potash
Structure in Some High-z Radio Sources: P. P. Kronberg
Asymmetric Double Sources: R. A. Perley, K. S. Johnston, and E. B. Fomalont
Radio Quiet QSO's: R. A. Sramek and D. W. Weedman
The Double QSO 1038+528: F. N. Owen, B. J. Wills, and D. Wills
The Double QSO 0957+561: D. H. Roberts, B. F. Burke, and P. E. Greenfield
Structure of the QSO 1442+101: M. J. Reid and M. S. Roberts

Results of these investigations will be published elsewhere by the authors listed.

Reference

Heeschen, D. S. 1975, Sky & Telescope, 49, 344.

G.O. Abell and P. J. E. Peebles (eds.), Objects of High Redshift, 177–178.
Copyright © 1980 by the IAU.

DISCUSSION

Longair: What fraction of the observing time at the VLA is unusable
              because of tropospheric fluctuations at 21 cm, 6 cm, 2 cm and
1 cm?

Heeschen: We don't yet have enough observing experience to give a
              quantitative answer. My present guess is that, averaged over
a year, tropospheric fluctuations make observations uncertain or impossible up to perhaps 30% of the time at 1.3- and 2-cm wavelengths, 10% at
6 cm, and essentially none at 21 cm. Much of this "unusable" time
occurs during July and August, when 1.3- and 2-cm observations are
generally not scheduled. The fact that the VLA can be rapidly switched
from one observing wavelength to another very greatly reduces the actual
time lost due to weather.

VLA OBSERVATIONS OF THE DOUBLE QUASAR 0957+561:
GRAVITATIONAL DOUBLE IMAGE OR BINARY QUASAR?

D. H. Roberts, P. E. Greenfield, and B. F. Burke
Department of Physics
Research Laboratory of Electronics
Massachusetts Institute of Technology
Cambridge, Massachusetts 02139 U.S.A.

The radio source 0957+561 was identified by Walsh et al. (1979, Nature 279, pp.381-384) with a pair of quasars, 6".1 apart on the sky, whose optical emission and absorption spectra are nearly identical. Walsh et al. suggested a gravitational lens interpretation in which a single object is split into two images by an intervening massive object. Using the Very Large Array of the NRAO we have made a 6-cm wavelength radio map of 0957+561. The map shows unresolved sources of 36 and 30 mJy coincident with the optical N and S quasars, and a complex extended source of $\sim$130 mJy (Roberts, Greenfield, and Burke: 1979, Science August 31). The extended emission lies on an arc containing the N quasar, and consists of two resolved sources containing 75 and 28 mJy, located 5".8 and 3".6 NE of the N quasar, a weak source of $\sim$10 mJy about 5".5 SW of the N quasar, and a suggestion of a bridge connecting the NE and SW sources and the N quasar. There is no evidence of radio emission from a massive object between the two quasars. Although the existence of the extended source does not rule out a gravitational lens model for 0957+561, the underlying source required would have an unusual morphology. In addition, if the refracting object is at the redshift of the absorption seen in both quasars, its mass would have to be at least 2×10^{14} solar masses. Further observations at the VLA could rule out the gravitational lens model for 0957+561 if a second image corresponding to the NE extended component is not found.

A more natural interpretation of 0957+561 is that there are two separate but physically related quasars, one of which is undergoing an active phase, ejecting relativistic plasma which gives rise to the extended emission. The linear size and radio luminosity of the extended component are typical of those seen in quasar-jet sources such as 3C273. In this model the most intriguing features of the source are the near identity of the optical absorption spectra of the quasars and their proximity in space. The National Radio Astronomy Observatory is operated by Associated Universities, Inc., under contract with the United States National Science Foundation.

G.O. Abell and P. J. E. Peebles (eds.), Objects of High Redshift, 179–180.
Copyright © 1980 by the IAU.

DISCUSSION

Wolfe: One crucial test of whether the northern extended radio
                emission is associated with the double QSO is to make
detailed VLB maps of this source.   If these VLB maps reveal a jet pointing from the compact component in the direction of the extended component, then the evidence for association would be strong.   Have VLB maps
been made, and if so, what do they reveal?

D. Roberts: I agree that the detection of such a jet in the north
                component would be good evidence for association of the
extended emission with the north quasar.   Conversely, if VLB maps of
the two components showed them both to be elongated in a direction
perpendicular to their separation, it would strongly support the gravitational double image interpretation.   The 6-cm flux ratio of 1.20 which
we measure predicts that the images would be elongated by a factor of
approximately 15 to 1.

A SUMMARY OF THE OBSERVATIONS OF THE TWIN QSOs, 0957+561 A,B

Frederic H. Chaffee, Jr.
Smithsonian Astrophysical Observatory
Mt. Hopkins Observatory

Ray J. Weymann
Steward Observatory, University of Arizona

Marc Davis and Nathaniel P. Carleton
Harvard-Smithsonian Center for Astrophysics

D. Walsh
University of Manchester
Nuffield Radio Astronomy Laboratories
Jodrell Bank

R.F. Carswell
Institute of Astronomy, Cambridge

ABSTRACT

An analysis of all observations of the "twin" QSOs, 0957+561 A,B, to date does not yet allow us to distinguish between their being two nearly identical QSOs or a single QSO split into two images by an intervening gravitational lens. The more identical the two objects are found to be, the more difficult any explanation which postulates the existence of two distinct QSOs becomes. Jodrell Bank and VLA observations reveal additional radio structure to the northeast of the northern QSO image which, if physically associated with a single QSO doubly imaged by a gravitational lens, would itself be imaged weakly to the southwest. More detailed radio mapping should be able to test the existence of such an image.

The VLBI map of Porcas and his collaborators reveals that the radio images corresponding to the optical ones are point sources separated by 6.175 arcsec having an angular extent to less than 20 milliarcseconds, whereas all further radio structure is resolved out.

Optical spectroscopy of the twins reveals two nearly identical sources with indistinguishable emission line redshifts and with absorption line redshifts identical to within 15 km/sec. It is the identity of these optical characteristics which makes all non-gravitational lens hypotheses most difficult.

The most compelling test of the lens hypothesis is the measurement of time variations of the two images at as many wavelengths as

G.O. Abell and P. J. E. Peebles (eds.), Objects of High Redshift, 181—184.
Copyright © 1980 by the IAU.

possible. If brightness variations of one image are repeated by the other after a time interval determined by the details of the observer-lens-QSO geometry (such an interval could be of the order of many months or years) the lens hypothesis would be confirmed. Several observations indicate prior variations of the images, and programs to monitor their relative brightness in the future will be of great importance.

DISCUSSION

D. Roberts: If the extended radio emission in the map I presented is either foreground or background, then one might expect it to have a compact component. Do you know if Walsh et al. have seen such a source in their VLBI? Also, is it not true that the object which seems to have undergone the greater reddening shows the shallower absorption lines, contrary to what one might have expected?

Chaffee: The VLBI map of Porcas et al. shows only the two QSOs as compact sources. No compact component is associated with the structure seen to the northeast.

It is true that the redder object seems to have shallower absorption lines.

H.E. Smith: What is the shortest time lag between the objects' variability that you would consider reasonable? (I.e., if William Liller were to see only one outburst, with a fairly short duration, could that eliminate the lens hypothesis?)

Chaffee: The time lag depends sensitively on the distance to the deflecting mass and to its alignment. If we place the mass at z = 0.7 (half the redshift of the QSO), the time lag is a few months. The farther away from us the mass is, the longer the time lag. The brightening of one image without the subsequent brightening of the other does not rule out the lens hypothesis because if the mass is close to the QSO that lag can become very long. The most convincing proof of the lens hypothesis would be for one image to vary in intensity in some characteristic way (a flickering of some kind, say) and some months later have the other image mimic that signature. It remains to be seen if nature will cooperate.

Epstein: Comment regarding the presence of tertiary optical images: At Montreal there was shown a radio map indicating the presence of a tertiary image. Dr. Barnothy pointed out that tertiary images are not unexpected under the gravitational lens hypothesis.

How did you arrive at the estimate of the time scale of variability to be expected if there is a gravitational lens present?

Chaffee: The time scale of variations depends on a large number of
                parameters -- most importantly, the distance to the
deflecting mass, its mass distribution and its angular displacement
from the line of sight to the QSO. Our estimate is based on arbitrarily
placing the mass at half the redshift of the QSOs, making reasonable
assumptions concerning the mass distribution (i.e., the velocity dispersion) in the deflector and using the estimate that the deflector is
0.4 arc sec off the line of sight, which follows directly from the
relative brightness of the two images.

Marscher: I'd like to comment that the detection of a single (rather
                than a double) X-ray source would not convincingly rule
out a gravitational lens. If the X-ray flux is highly variable -- sometimes detectable, sometimes not -- then the differential light travel
time could cause each source to "blink" out of phase with the other.

B. Wills: Concerning the interpretation of the division of the two
                spectral scans of 0957+561 A and B: the fact that the
emission lines divide out completely means that, to quite a high degree
of accuracy, the <u>equivalent</u> <u>widths</u>, not the intensities, are the same
in both components. This means that if the objects are due to a gravitational lens, the differences in the continua must be due to some kind
of extinction -- equally affecting the emission lines.

                Do the deep plates show any other features of interest in
                the region of the double QSO (e.g., at the positions of
other features in the radio maps or which could be attributed to galaxies)?

Chaffee: The video camera observations with the Kitt Peak 4-meter
                telescope by Adams and Boroson, which will be published
in <u>Nature</u> later this fall, show eight non-stellar objects in a 75 x 75
arc sec field centered on the twins. None of their positions corresponds to the excess radio emission, presented by Roberts in the preced-
ing paper, to the northeast of the twins.

M. Burbidge: Dr. Wills, how much differential extinction would that
                correspond to?

B. Wills: The differential extinction amounts to $A_V = 0\overset{m}{.}5$, and is
                almost independent of the distance of the absorber
$(1 \leq z \leq 1.4)$.

G. Burbidge: Is there any bright galaxy nearby? The difficulties that
                you raised with the idea that it is two separate objects
at a cosmological distance, namely the large energy and the similarity
of the absorption over ~ 70 kpc, are reduced by a large factor if the
objects are at a distance about 50 times less than the cosmological
distance.

Chaffee: There is a bright irregular galaxy 10 arc min to the
               southeast of the twins.

               Naturally, many of the problems I have discussed in
               explaining the existence of the identical QSOs would be
eased if they were closer than cosmological distances from us.

Koo: In light of the provocative presentation by Dr. G.
               Burbidge, has a search been made for a nearby galaxy and
additional QSOs?

Chaffee: A bright galaxy does exist 10 arc min to the southeast,
               and a number of faint ones have been detected by Adams
and Boroson nearer the QSOs. No search has yet been made for other
QSOs in the field nearby.

               No optical polarization measurements have yet been made.

THE RADIO PROPERTIES OF BL LACERTAE OBJECTS

Kurt W. Weiler
Max-Planck-Institut für Radioastronomie
Auf dem Hügel 69
5300 Bonn 1
West Germany

A recent study of the properties of BL Lacertae Objects at $\lambda\lambda 6$ and 21 cm by K.J. Johnston and myself has shown:

I. SPECTRA OF COMPACT COMPONENTS ($\lambda 6$ cm -- $\lambda 2$ cm)
    A) 11 of 34 (32%) have "normal" negative slope $\alpha < -0.05$.
    B) 10 of 34 (29%) have no slope $-0.05 < \alpha < +0.05$.
    C) 13 of 34 (38%) have "inverted" positive slope $\alpha > +0.05$.
    D) Half of the sources have $-0.15 < \alpha < +0.15$.
    E) Spectral indices are much flatter than a general sample of radio sources or quasars.

II. STRUCTURE ($\lambda 6$ cm)
    A) 23 of 42 (55%) have extended structure ($\theta > 1$ arcsec)
        -- usually containing < 50% of the total flux.
    B) 27 of 31 (87%) have very compact structure ($\theta <$ 2E-3 arcsec)
        -- usually containing > 50% of the total flux.

III. PHYSICAL PROPERTIES
    A) Extended structures
        1) Sizes $3 \lesssim D$ (kpc) $\lesssim 200$.
        2) Radio luminosities $\sim$ 1E41 -- $\sim$ 1E45 erg/sec.
        3) Particle energies $\sim$ 1E57 -- $\sim$ 1E60 erg.
        4) Magnetic fields $\sim$ 1E-5 -- $\sim$ 1E-6 Gauss.
    B) Compact structures
        1) Sizes $0.6 \lesssim D$ (pc) $\lesssim 14$.
        2) Radio luminosities $\sim$ 1E42 -- $\sim$ 1E46 erg/sec.
        3) Particle energies $\sim$ 1E52 -- $\sim$ 1E55 erg.
        4) Magnetic fields $\sim$ 1E-1 -- $\sim$ 1E-2 Gauss.

Briefly then, the BL Lacs closely resemble the other classes of strong sources in their radio properties.

A more detailed description of these results as well as a comparison of the optical properties of the BL Lacertae Objects with those of quasars and radio galaxies is in press in the Monthly Notices of the Royal Astronomical Society.

G.O. Abell and P. J. E. Peebles (eds.), Objects of High Redshift, 185–186.

DISCUSSION

A.S. Wilson: One hears the suggestion from time to time that BL Lac
                 objects are elliptical galaxies with double radio source
type relativistic beams directed almost exactly towards and away from
the observer. If that were the case, one might expect the large-scale
radio structure to be more or less spherically symmetric about the
core, rather than of double-radio-source type. What do your results
say about this?

Weiler: Our study was designed to provide the statistics of source
                 sizes rather than the structures of individual sources.
However, new results from the Jodrell interferometer reported at the
General Assembly by D. Stannard show a variety of structures, from classical doubles to core-halos. Thus, it seems that the radio structures,
as we also found for the radio sizes, energies, etc., are similar to
those seen in radio galaxies and quasars. This argues against a "pref-
erential alignment" explanation for BL Lacs.

X-RAY OBSERVATIONS OF OBJECTS AT COSMOLOGICAL DISTANCES FROM THE "EINSTEIN" OBSERVATORY.

Riccardo Giacconi
Harvard/Smithsonian Center for Astrophysics
Cambridge, Massachusetts 02138

INTRODUCTION

The launch of the Einstein X-ray Observatory on November 13, 1978 has brought about a qualitative change in the field of X-ray astronomy. A fairly detailed description of the Observatory has appeared in print (Giacconi et al, 1979a) and therefore will not be given here.

However, a summary of the relevant instrument parameters is useful. Figure 1 shows a schematic representation of the Observatory which was constructed under the management control of Marshall Space Flight Center of National Aeronautics and Space Administration. TRW, Inc. was responsible for the spacecraft and American Science & Engineering, Inc. for much of the instrumentation. The X-ray telescope, built by Perkin-Elmer, has a 0.6 meter aperture with a focal plane scale of 1 arc min/mm. It consists of 4 nested Wolter-type I paraboloid/hyperboloid grazing incidence mirrors. The telescope effective area is ~ 400 cm^2 at 0.25 keV and ~ 30 cm^2 at 4 keV.

The detectors are on a lazy-Susan arrangement which permits each instrument to be placed in the telescope focus. There are two types of imaging detectors: (a) an imaging proportional counter (IPC) with large field of view ($\sim 1^{\circ} \times 1^{\circ}$), moderate angular resolution (~ 2 arc min), and broadband spectral resolution from 0.25 to 4.0 keV, and (b) a high resolution imaging detector (HRI) which is limited by the telescope point response function to about 3 arc seconds resolution, with no spectral resolution. There are two types of spectrometers: (a) a non-dispersive Si(Li) crystal spectrometer, cooled to about 100°k, with high sensitivity and 150 eV resolution in the 0.8 to 3.5 keV range, and (b) a high resolution, but low sensitivity, Bragg Crystal Spectrometer. The imaging detectors were developed by the Center for Astrophysics, while the spectrometers were developed by Goddard Space Flight Center and Massachusetts Institute of Technology respectively.

G.O. Abell and P. J. E. Peebles (eds.), Objects of High Redshift, 187–205.
Copyright © 1980 by the IAU.

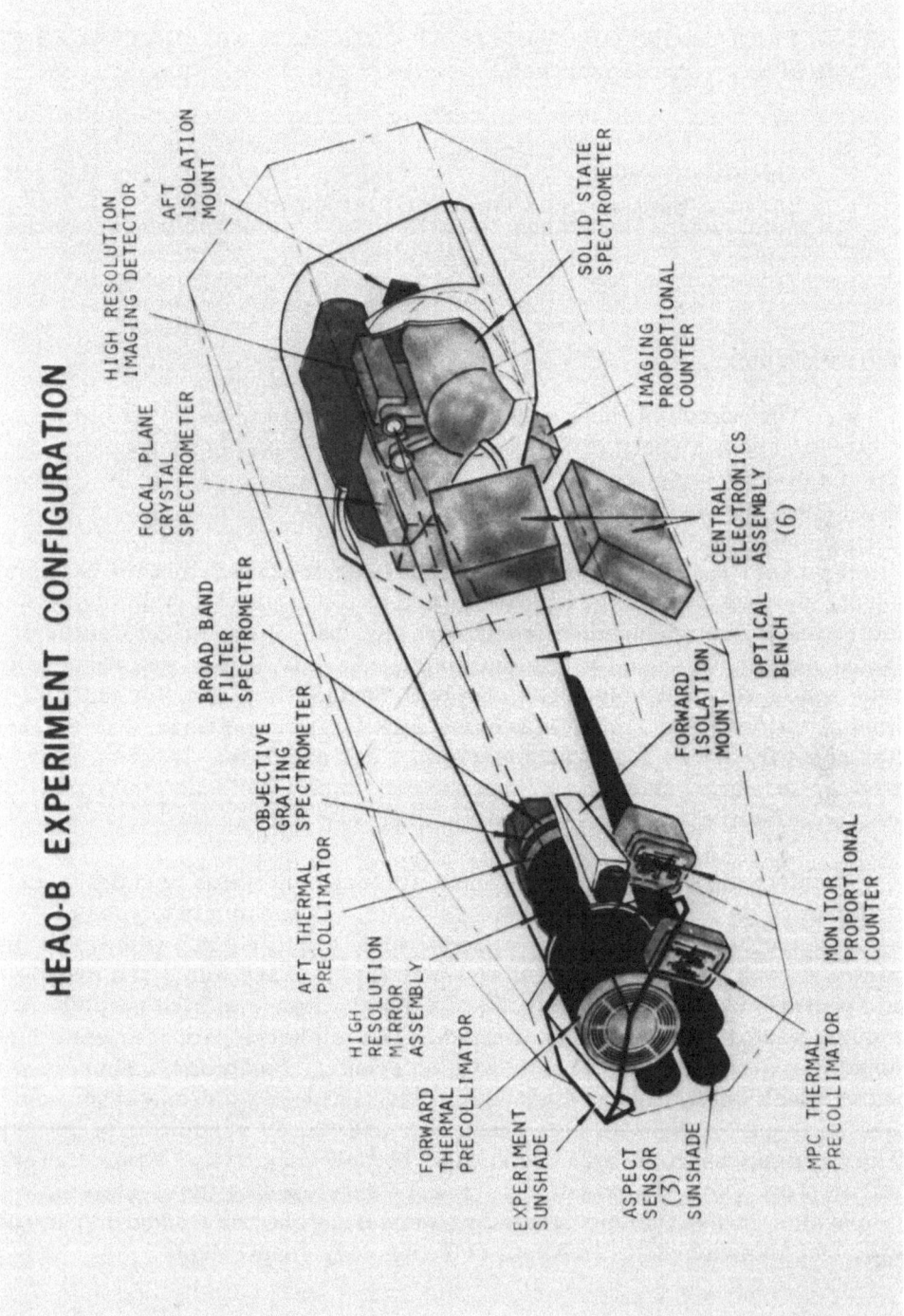

Figure 1

The use of focusing optics results in an increase of sensitivity of a factor of 1000 over any previous surveys and provides an angular resolution of a few arc seconds comparable to that usually achieved in visible light. These new observational capabilities have profoundly changed the scope of X-ray astronomy. X-ray emission can now be detected from intrinsically weak sources, such as the coronas of stars (10^{26} - 10^{33} erg sec^{-1}) or from the most distant known members of intrinsically bright classes of objects such as clusters of galaxies and QSO's (10^{43} - 10^{48} erg sec^{-1}). Every known astronomical object (save, perhaps, the planets) can now be studied in X-ray radiation.

The manner in which X-ray observations are carried out has also changed substantially. Although the project was conceived and initiated as a Principal Investigator experiment, and carried out under the scientific guidance of a consortium of institutions (Harvard/Smithsonian Center for Astrophysics, Columbia University, Goddard Space Flight Center and Massachusetts Institute of Technology), the capabilities of the Einstein Observatory have been made available to an ever increasing number of guest observers. For the first six months of the mission, 170 proposals for observations by guest investigators have been approved by a NASA Review Committee and are being carried out. This level of participation by astronomers not directly involved with the hardware construction is comparable to that of any major optical or radio astronomy observatory.

These qualitative changes particularly affect the ability of X-ray astronomy to contribute to the study of objects at cosmological distances. Although data reduction of the more than 10^8 bits of information that we receive daily is occurring in essentially real time, data analysis of the result is still in a preliminary phase. Therefore, I will be able today to present only tentative conclusions, more from the point of view of acquainting you with the nature and import of the observations than in the expectation of completely resolving any of the very difficult and long-standing questions that have been discussed at this meeting. I will limit my remarks to three subjects:

  X-ray studies of clusters of galaxies
  Detection and study of distant QSO's
  Preliminary findings of the deep X-ray surveys

CLUSTERS OF GALAXIES

Detection of X-ray emission from clusters of galaxies was first accomplished from UHURU in the early 1970's. Subsequent satellite investigations from OSO-8, Ariel V, ANS and HEAO-1 discovered a large number of cluster sources ($\gtrsim$ 50) and gave us a qualitative understanding of the main

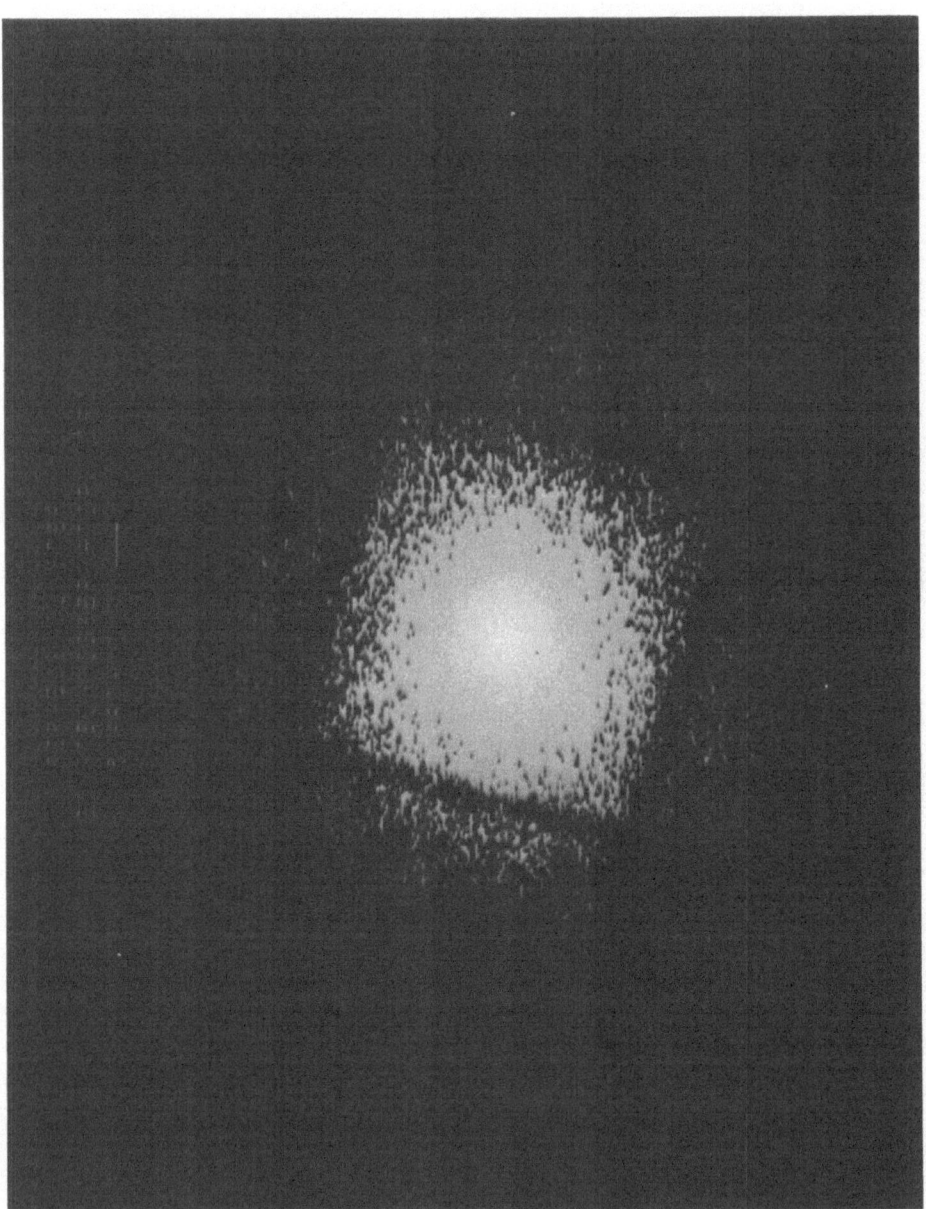

Figure 2

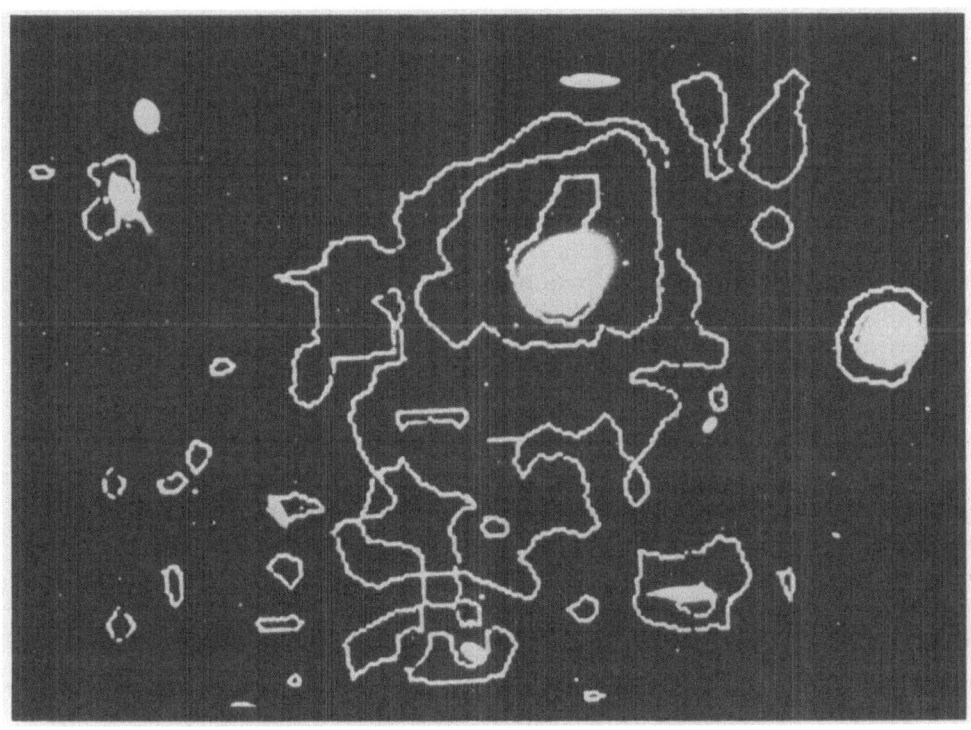

Figure 3

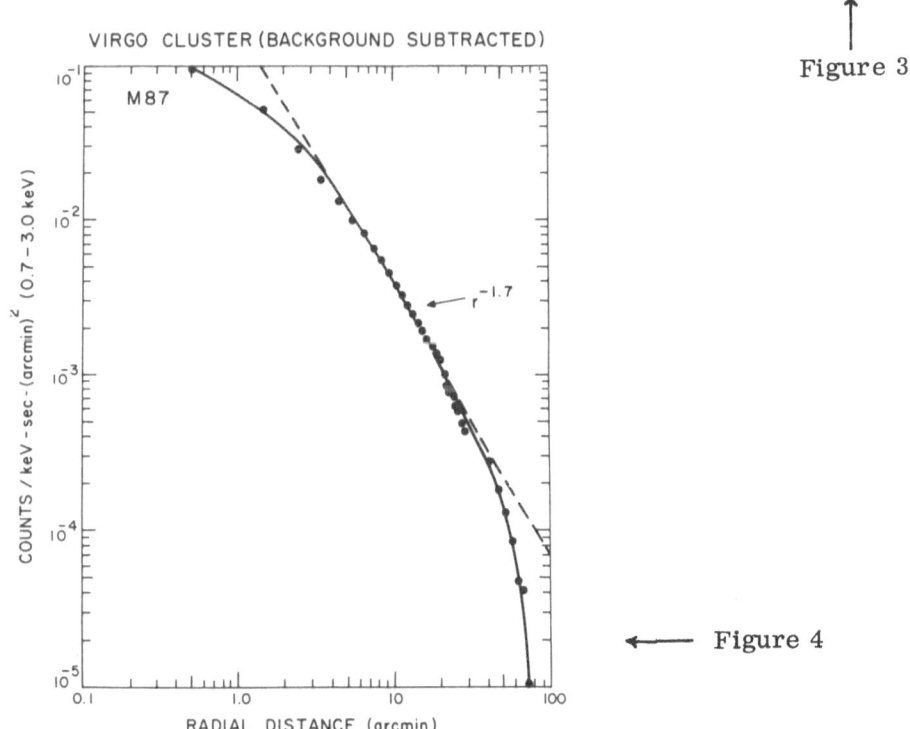

Figure 4

features of the physical processes involved. The bulk of the emission originates from a hot intergalactic medium pervading the cluster and containing a
total mass comparable to that contained in the visible galaxies. The emission
is due to thermal bremsstrahlung and K_α emission lines of iron at 6.7 keV
have been detected from several clusters. In some clusters evidence is found
for gas with at least two different temperatures. Given the presence of high Z
elements, the gas is believed to be material from the galaxies contained in the
cluster rather than primeval gas left over by inefficiencies in the process of
formation. Einstein observations contribute to the study of clusters in several
distinct areas.

The study of nearby clusters, such as Virgo, Perseus and Coma, can
be carried out in some detail. Images of the Virgo field obtained with the
IPC, for example, reveal a complex morphology which is being studied at
CFA (Forman et al, 1979). The X-ray emission is resolved in several components: emission from M87 itself which is being further studied at high
resolution to resolve the nucleus and jet contributions; emission from cool
gas (~ 2 keV) surrounding M87 itself, M85 and possibly other galaxies in
the cluster; finally, a higher temperature (~ 10 keV) gas component pervading the entire cluster and presumably following the cluster potential (Figures 2 and 3).

Gorenstein and Fabricant, of CFA, have carried out detailed studies on
the temperature and source brightness distribution of the gas surrounding M87.
In the central region and out to at least 25 arc minutes the emission from the
cooler component is predominant and we find essentially a constant temperature
of $2.5 \times 10^7 {}^0k$. Figure 4 shows the fit to the X-ray data with projection
effects not yet removed. Removing projection effects we find that the gas distribution in the region between a few and 25 arc minutes can be well described
by a power law $r^{-1.3}$. If we assume that the gas is in hydrostatic equilibrium,
we can derive the underlying mass required to explain the observed distribution. This mass would mainly reside not in the gas ($10^{12} M_\odot$), but in a dark
halo component such as the one suggested by Bahcall and Sarazin, and by
Matthews ($10^{13} M_\odot$). We find for this mass a distribution of r^{-2} and a total
mass (within 100 kp) of $10^{13} M_\odot$. We will be able to apply similar analysis
techniques to the study of M86, although at the moment we do not yet have
sufficient statistics to go into as much detail. The cool gas component surrounding M86 appears to have a temperature of ~ 1 keV and central gas density of
4×10^{-3} particles/cc. The total mass of the gas within 60 kpc is found to be
of order $6 \times 10^9 M_\odot$.

Since we know from previous experiments, as well as the spectrometer
data on Einstein, of the existence of an additional hot component of the gas,
we can study the conditions under which M86 will lose its gas halo when crossing the center of the Virgo cluster due to ram pressure stripping. Our data,

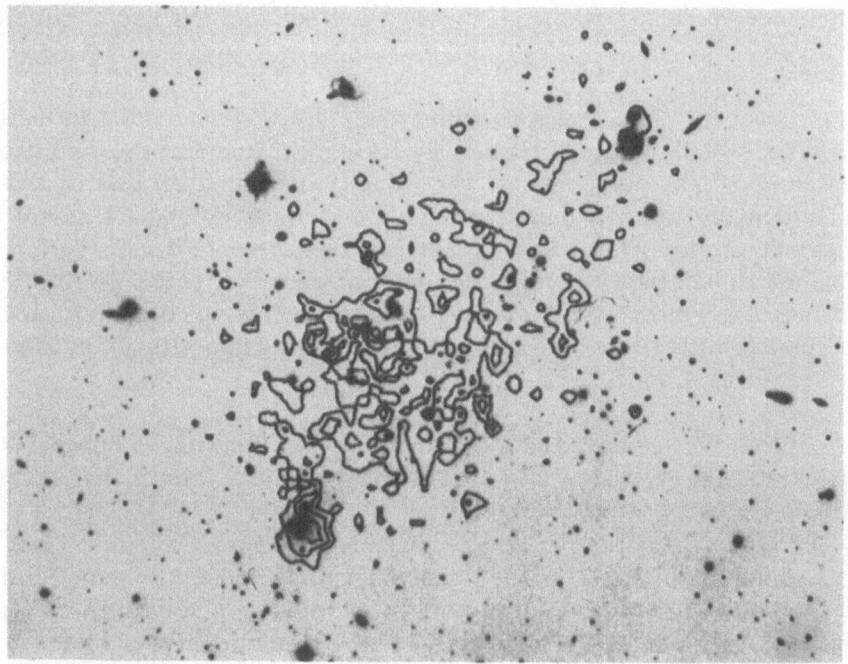

Figure 5

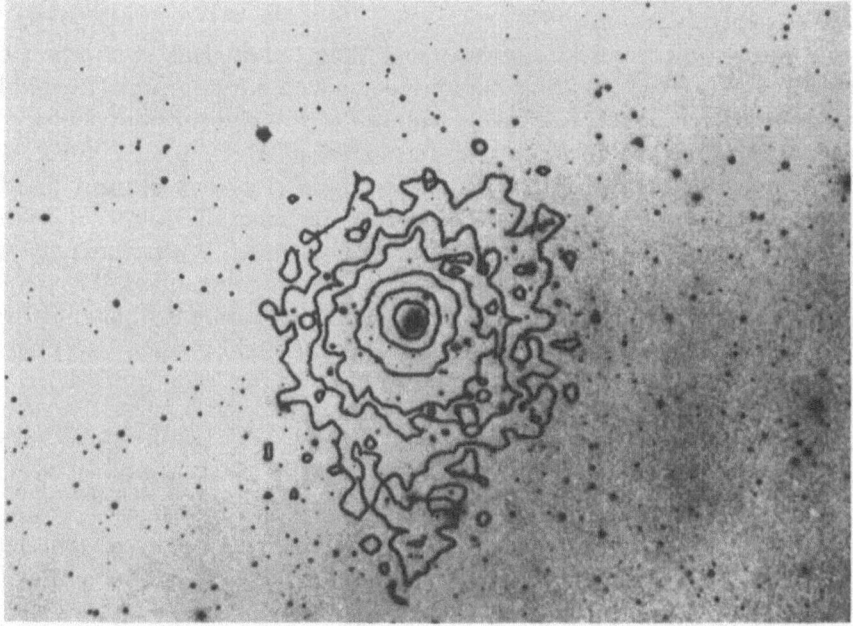

Figure 6

therefore, strongly suggests that at least some of the gas originated in the galaxies.

Probably M86 is approaching the cluster center now, as indicated by its velocity component towards us; taking into account the densities of the hot and cold gas components, the velocity of the galaxy and its mass one can show that the gas will quite probably be stripped during its next crossing (Forman et al, 1979). Many other sources are detected in Virgo; while some turn out to be simply foreground or background objects, we find that we detect all bright galaxies with $M_V \lesssim 12.5$. These galaxies could either be intrinsic emitters or be surrounded by glowing gas either lost from the galaxy or accreting onto it.

This type of detailed study of cluster properties is being carried out for Coma and Perseus, as well, although the analysis of these clusters is still in its early stage.

In addition, preliminary analysis of spectroscopic data obtained both with the non-dispersive solid state spectrometer and with the Bragg Crystal spectrometer confirms the existence of a two-temperature gas in general agreement with the qualitative description given above. The SSS spectrometer, for instance, resolves lines from both hydrogen and helium like atoms of silicon and sulphur and from their relative strength the temperature of the two components can be measured.

For more distant clusters, such detailed studies cannot yet be carried out. However, the imaging experiment reveals a complex morphology of the gas distribution which is closely related to the cluster morphology and probably to their evolutionary state (Jones et al, 1979). Figures 5 and 6 of A1367 and A85 exemplify two extreme cases of a very clumpy and broad , and a very centrally condensed and smoothly distributed cluster medium. Figure 7 illustrates our findings for a number of clusters. We tentatively identify at least four different types. Two extreme cases mentioned above are an inter-mediate case, exemplified by A2256, which is centrally enhanced but less peaked than A85, and finally the type exemplified by A2666, a poor group con-taining a cD galaxy.

Correlation between X-ray properties and morphological type and evolutionary state has been discussed by Jones et al and by N. Bahcall.

In rough outline it is clear that the observed X-ray properties fit well in the context of an evolutionary scheme such as that discussed by Peebles or Perrenod.

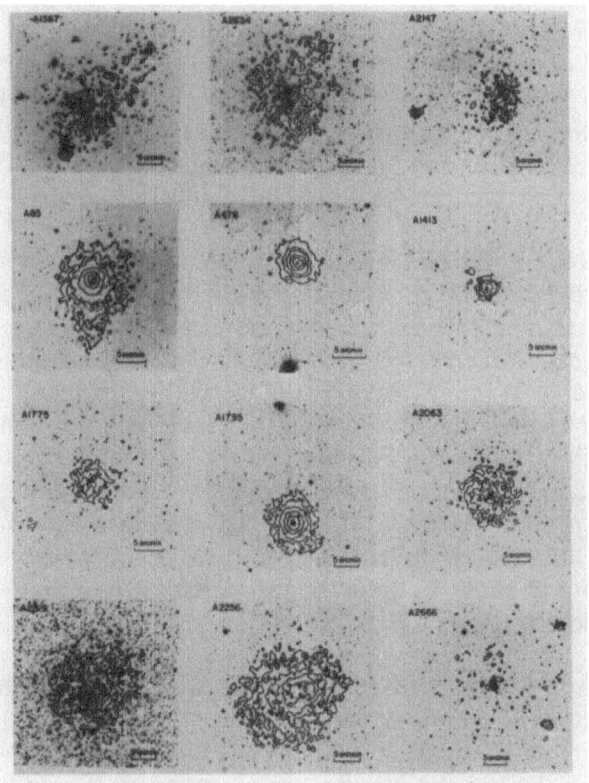

Figure 7

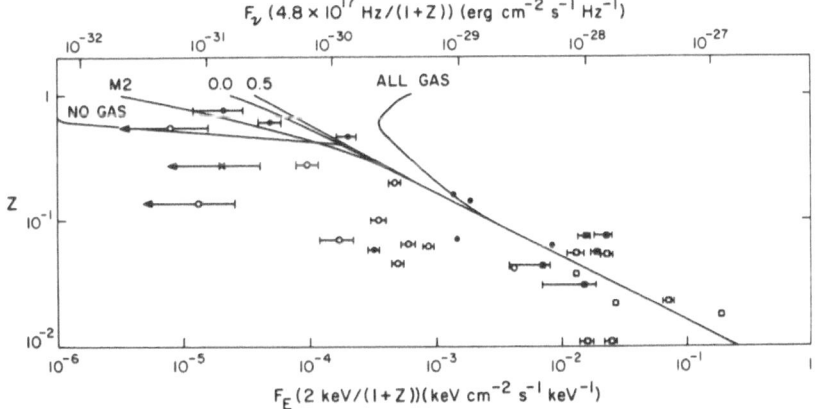

Figure 8

Time does not permit me today to dwell on this point; I would only like to remark that once the extensive survey of clusters which is planned in the Einstein mission is completed, the X-ray measurements will give us a solid quantitative base to discuss integral properties of clusters. It is clear that the gas distribution can be measured more readily than the galaxy distribution and the temperature of the gas is an easier quantity to measure than the velocity dispersion particularly for distant clusters.

It may at some point become more convenient to carry out searches and classification of clusters on the basis of their X-ray, rather than optical, properties. In this connection, we have extended the detection of clusters to the furthest known objects of this kind with a view toward studying possible evolutionary effects. Figure 8 summarizes our preliminary findings. It is clear that we can readily detect any cluster discovered in visible light observation. At the moment the large intrinsic luminosity dispersion and the presence of what are understood as different evolutionary stages at any given epoch makes the interpretation of the results extremely complex. Notwithstanding these problems which hopefully will be resolved by additional data, we seem to be able at least to discriminate against the most extreme hypothesis of cluster evolution.

I would like to make clear that although we have described the data within the simple evolutionary scheme, for instance of Perrenod, the data are yet too incomplete to rule out competing models. In fact, there already are indications that not all may be well in the simple evolutionary picture which I have used as a framework to describe the data. The X-ray detection of 3C295, indicating that there is sufficient intergalactic gas to cause stripping of the cluster spirals, although the colors of these galaxies imply that they have not been stripped, is, for instance, such an indication (**Dressler, 1978**).

In fact, we are just beginning the X-ray study of clusters in some detail and it may well be that our views of cluster formation and evolution will have to be considerably modified, as a result of these observations.

For the future, X-ray research promises to be a very powerful tool for cosmology. As an example, it is sufficient to mention the recent proposal for an absolute measurement of H_0 and q_0 through measurement of temperature and of X-ray surface brightness distribution in clusters and cooling of the microwave radiation (Zeldovich-Sunyaev effect) made independently by several authors. This research is currently in progress, but the main difficulty is the scarcity of reliable data in the microwave domain.

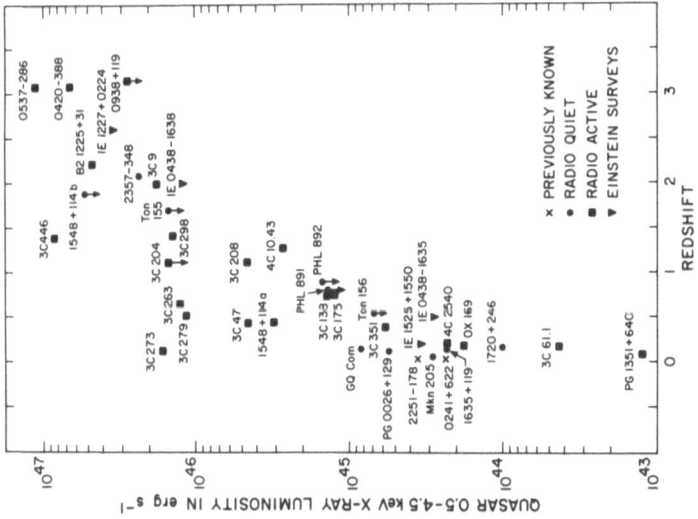

Figure 10

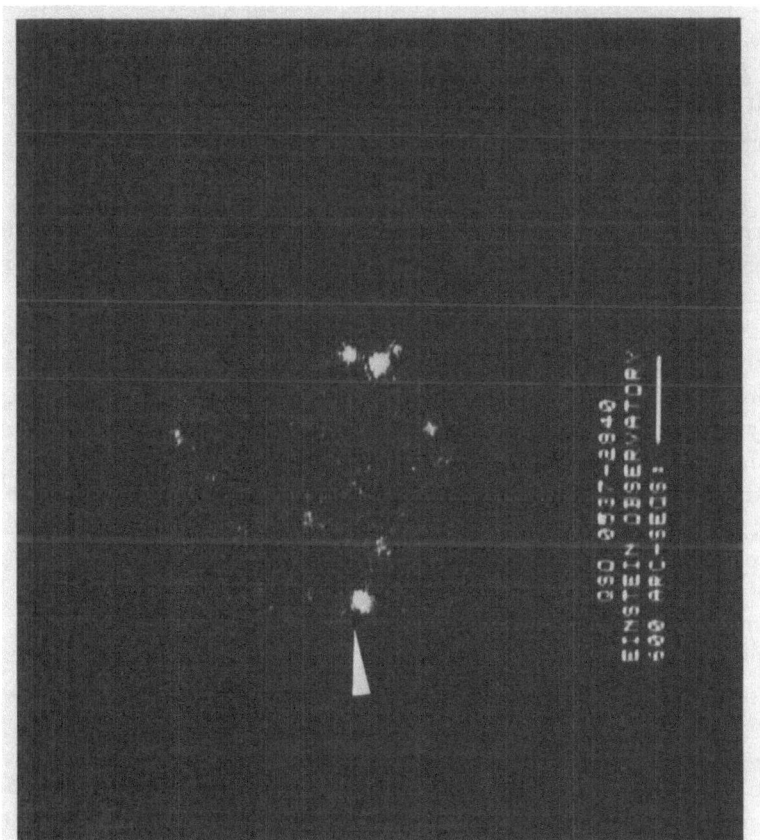

Figure 9

QUASARS

We are carrying out a program of observations to detect X-ray emission from known QSO's which have been selected to span a wide variety of optical and radio properties and also a significant range of redshifts (Tananbaum et al, 1979). Prior to Einstein, only a few quasars had been detected as X-ray sources, the furthest at redshift $z < 0.2$. We have now observed over 40 QSO's with the furthest at redshift $z = 3.1$. In addition, more than 20 new quasars have been discovered with redshifts ranging from 0.5 to 2.1 as a result of deep surveys and optical follow-ups.

Figure 9 shows the results of an observation of the QSO 0537-286 which was obtained with the IPC after a 10,000 second exposure. The quasar is indicated by the cursor and has a luminosity of 10^{47} erg s^{-1} in the 0.5 to 4.5 keV band (at the source and assuming $q_0 \approx 0$). As can be seen, there are several sources detected in this field ($\sim 1^0$ x 1^0). These include two main sequence F stars and several as yet unidentified sources which are fainter than 15th magnitude. The ease with which we have observed this quasar, and a second quasar (0420-388) also at $z = 3.1$ indicates that with the Einstein Observatory we should be able to detect still more distant QSO's. Thus, the cosmologically intriguing question of the formation epoch for QSO's may be within our ability to answer. Of course this depends upon assumptions regarding evolution of the quasars (both in density and luminosity) which may be present.

A summary of the X-ray properties of QSO's is shown in Figure 10. This is a plot of the 0.5 to 4.5 keV luminosity (at the source) versus redshift. The luminosity range is from 10^{44} erg cm^{-2} s^{-1} to 10^{47} erg cm^{-2} s^{-1}. We observe that there is not obvious redshift-luminosity correlation. The absence of low luminosity points at high redshifts is due to selection effects since the sensitivity of the observations is limited. Radio quiet and radio bright quasars are indicated by symbols. Thus far we have observed primarily radio bright QSO's. This may be a source of bias in any preliminary conclusions which may be drawn from the data. However, those radio quiet quasars we do observe, do not appear to be significantly different in their X-ray properties , suggesting little, if any, correlation of X-ray and radio properties.

Tananbaum et al (1979) have found a tentative relation between the optical and X-ray fluxes of quasars such that the slope of a power law joining the X-ray and optical flux densities is nearly constant with average slope $\alpha_{ox} \approx 1.30$. This is illustrated in Figure 11 where α_{ox} is plotted for each of the observed QSO's. Most significant is that α_{ox} shows no apparent dependence on redshift and again there is no evidence for a bias with regard to the quasar radio properties. Using the optical luminosity function for QSO's from Braccesi et al (1979) and the value of $\alpha_{ox} \simeq 1.30$, the contribution of QSO's to the X-ray

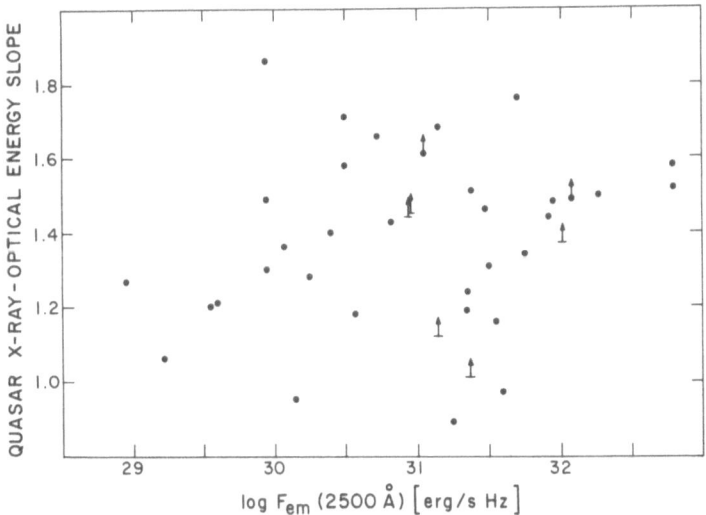

Figure 11

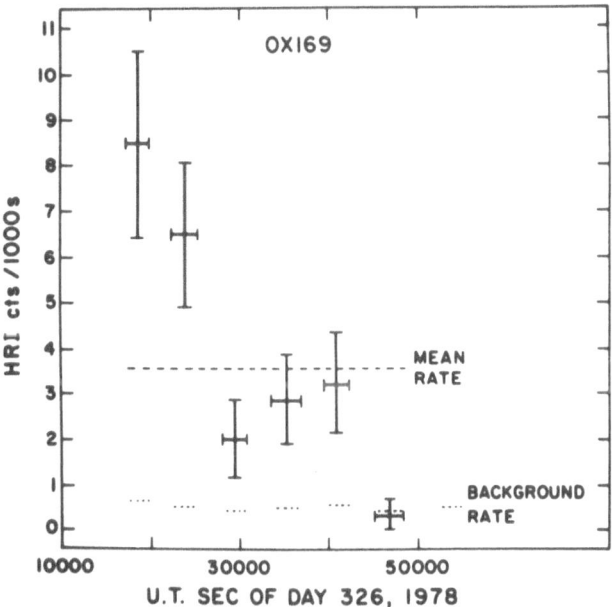

Figure 12

background can be calculated. The results of such a calculation depend strongly on the value of α_{ox} and the range of QSO luminosity which Tananbaum et al have taken as 10^{44} to 10^{46} erg s^{-1}. They find that from 10% to all of the X-ray background can be accounted for from quasars. Moreover, for some choices of the luminosity function and the evolution of quasars, that is, the value of n and z_{max} in the density evolution function (z) α (1+ z)^n and z $\leq$ z$_{max}$, the X-ray background is exceeded. Thus the X-ray data can be used to place constraints on quasar evolution. For example, current observations of quasars are not consistent with a local origin hypothesis unless they are at least 500 Mpc away, which is then essentially equivalent to a cosmological interpretation of the redshifts. This conclusion is based on the arguments originally made by Setti and Wolter (1973).

The emission mechanism for QSO's is a major astrophysical problem since we are concerned with processes capable of providing luminosities of up to 10^{47} erg s^{-1}. X-ray studies of quasars may provide unique insights into this question by studying time variability of QSO emission. Figure 12 shows the X-ray observations of QSO OX169 which is at redshift z = 0.2. The source is variable with a decrease in luminosity from 2.0 to 0.6 x 10^{44} erg s^{-1} within 100 minutes. This behavior can be interpreted as a signature of the underlying energy source of the quasar. For a process involving conversion of mass to energy (which seems likely on the basis of the large amounts of energy involved), rapid variability requires high efficiency. This supports models involving release of gravitational energy through accretion onto a compact object. If such a model for OX169 is assumed, then a black hole with mass between 10^6 to 10^8 M$_{\odot}$ is required so that neither the Eddington luminosity limit, nor the size scale from variability, is exceeded. We also note that the short time scale requires the X-rays to be produced near the central source, which implies that X-ray observations may ultimately provide the means for understanding mechanisms for powering quasars and active galaxies in general.

DEEP X-RAY SURVEY

The increased sensitivity of the Einstein Observatory has made possible for the first time a truly deep survey in limited areas of the sky. Of primary concern in conducting these surveys has been the long-standing question of the origins of the X-ray background. Both diffuse and discrete models have been proposed and are discussed extensively in the literature (cf. a recent review by Schwartz, 1978). We can approach the problem directly with Einstein observations through deep surveys by attempting to image the background at the limit of sensitivity. The expected source density, if the background is discrete, is high enough so that the 2 arc min resolution of the IPC is near the source confusion limit. The HRI, of course, is virtually free from this problem.

Without going into details of our observing program, which is described in the deep survey paper by Giacconi et al (1979b), I will present the results which are already of substantial significance. Figure 13 shows an example of the observing pattern for a deep survey. This is for the region in the constellation Eridanus. We have conducted a similar survey in the constellation Draco. Observing times vary from 10,000 to 50,000 seconds. The flux limit of the survey is 1.3×10^{-14} erg cm^{-2} s^{-1} or about 10^{-7} Sco X-1. In the two surveys thus far completed, we detect 43 X-ray sources in about 1/2 square degree of sky. Our positional accuracy in locating sources is dependent on the detector used and the source intensity. In the case of the IPC we have adopted a positional uncertainty of 60 arc sec radius mainly due to systematic uncertainties in converting detector coordinates to sky positions. For the HRI sources the location accuracy varies from ~ 5 to ~ 20 arc sec radius, which depends mainly on the statistical uncertainty from the small number of source counts.

Since we are concerned with the contribution of discrete sources to the extragalactic X-ray background, we first limit our band to the 1 to 3 keV range (for conversions of count rate to flux we assume that the sources have the spectrum of the background) and then we separate the contribution of stars from the objects detected. This is done by direct measurement of the redshifts of candidate objects where possible, or by the optical and radio morphology of potential counterparts. To accomplish this we have been assisted by various optical and radio astronomers whose contributions to our study are greatly appreciated. We have also measured the B magnitude and B-V colors for optical candidates as well as their proper motions over a 25-year baseline. In making these observations we have identified new quasars, compact galaxies, and possible BL Lac objects. In some instances we have found X-ray error boxes which appear completely empty at the plate limit of available optical material. In one such case additional observations using CCD cameras at Mt. Hopkins (Gursky and Schild, 1979) and Mt. Palomar (Kristian, Westphal and Young, 1979) have revealed the presence of objects fainter than $M_R \sim 21.5$. We conclude that about 1/3 of the X-ray sources are associated with stars and we eliminate these from further consideration.

Of the remaining source we select, for the purposes of estimating the background contribution, only those within the central 32 x 32 arc min region of the survey field and which are observed at the 5 sigma level of significance or higher. These constraints reduce possible errors due to statistical uncertainty in source intensity and geometrical corrections to the flux. The results are shown in Figure 14 which is a plot of the number-flux observations for X-ray sources. Included in this figure are the previous distributions from the UHURU survey (Murray, 1977), and the HEAO-1 flux limit, our new result and the Einstein flux limit for an exposure of 5×10^5 seconds. We also indicate the background limit for a number-flux relation which has the form $N(> S) S^{-1.5}$.

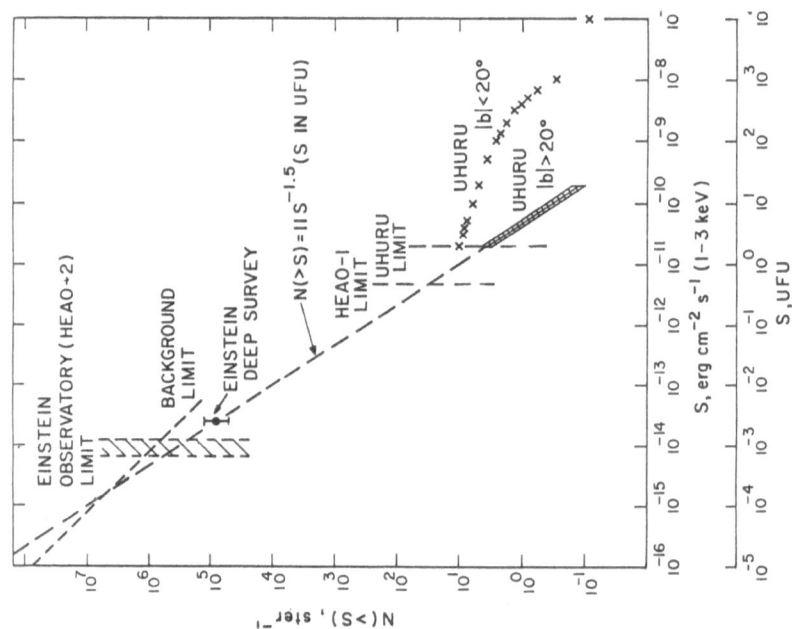

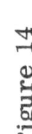

Figure 14

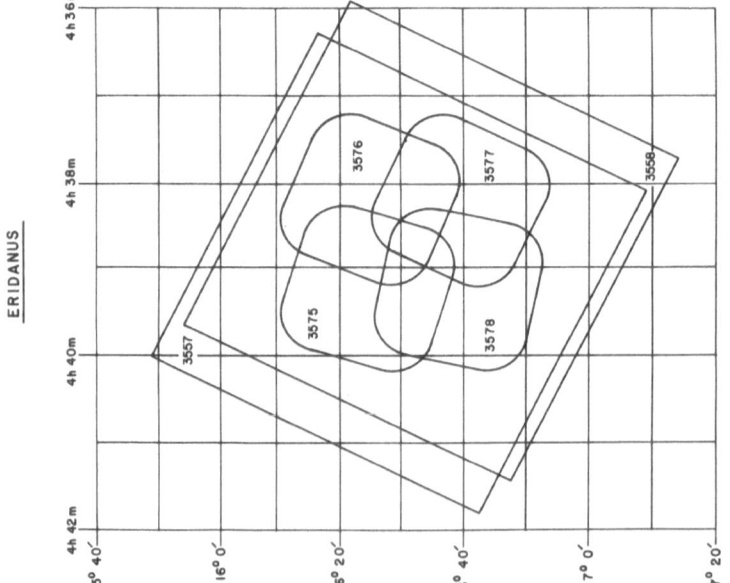

Figure 13

The Einstein deep survey point is at a flux of 2.6×10^{-14} erg cm^{-2} s^{-1} and corresponds to 6.3×10^4 sources ster^{-1}. This lies just below an extrapolation of the UHURU results over almost 3 orders of magnitude in flux. This agreement indicates that this simple form for N($>$S) is valid over a large dynamic range in flux. This can be explained either in terms of a uniform distribution of objects in space which would have to be nearby, or in terms of cosmological density evolution compensating for cosmological distance-volume effects. This first case would occur if we are detecting only objects within a small range of redshifts and a large range of intrinsic luminosities. This is unlikely in view of the optical data and the general considerations of Schwartz (1978). The second occurs if we are observing sources at large redshift ($z \gtrsim 1$) as indicated by the optical data. For example, the number-flux relationship for active galaxies, QSO's (assuming a strong density evolution) and nearby classes of X-ray sources, such as rich clusters, can approximate the observed power law slope of 1.5 over our flux interval. In this case we would expect the slope of the number-flux relationship to begin steepening at or beyond the present flux limit as quasars dominate the sources.

Taking the number-flux data point from the deep survey, and assuming a power law index of 1.5, we compute the contribution of discrete sources to the total extragalactic background to be $(25 \pm 11)\%$ at the flux limit of 2.6×10^{-14} erg cm^{-2} s^{-1}. An extrapolation of a factor of two in flux is made based on our observation of sources to this faint level. This will give a contribution of 37% of the background. Even this value is conservative since in selecting sources to include in the calculation of the number-flux relation we have not included sources with observed flux greater than 2.6×10^{-14} erg cm^{-2} s^{-1} if they were below the 5 sigma significance level. Thus, real sources have been left out of the computation. This may be as much as a 30% effect, which is an underestimation of the source counts. Thus, we feel quite confident in concluding that a substantial fraction, if not all, of the extragalactic X-ray background is due to discrete sources. We note that while previous theoretical discussions of the contribution of discrete sources to the background have yielded estimates comparable to our result, they were based on an extrapolation of at least 3 orders of magnitude in flux. The Einstein deep surveys have given us the first direct observation of these sources at fluxes low enough to yield a substantial contribution.

CONCLUSION

The Einstein X-ray Observatory has given X-ray astronomy an observational capacity not too different from that available in the optical and radio wavelengths, especially for the study of objects at cosmological distances. X-ray observations have unique capabilities in the study of high energy processes and for probing the most distant reaches of the Universe. Observations with

Einstein have already shed new light on the distribution of quasars and their contribution to the X-ray background. Studies of variability should prove to be a powerful tool in understanding the underlying energy source for these objects. The X-ray study of clusters promises to be extremely interesting, even on the basis of our preliminary results. The deep survey has extended our knowledge of X-ray sources to the point where much of the X-ray background is resolved. We have reached a sensitivity which allows us to begin to observe cosmological and evolutionary effects.

REFERENCES

Bahcall, J. and Sarazin, C.L., 1978 Ap. J. 219, 781.

Bahcall, N. 1979 Joint Discussion on Extragalactic High Energy Astrophysics IAU Montreal - to be published.

Braccesi, A., Zitelli, V., Bonoli, F., and Formiggini, L. 1979 submitted to Astronomy and Astrophysics.

Dressler, A., 1978 Ap. J. 226, 55, 113-E6.

Fabricant, D., Lecar, M., and Gorenstein, P., 1979 Joint Discussion on Extragalactic High Energy Astrophysics, IAU Montreal - to be published.

Forman, W., Schwarz, J., Jones, C., et al, 1979 Ap. J.(Letters) in press

Giacconi, R., Branduardi, G., Briel, U. et al, 1979a Ap. J. 230: 540-550.

Giacconi, R., Bechtold, J., Branduardi, G. et al, 1979b Ap. J.(Letters) in press

Gursky, H. and Schild, R., 1979. Private communication

Jones, C., Mandel, E., Schwarz, J., et al 1979 Ap. J.(Letters) in press

Kristian, J., Westphal, J., and Young, P. 1979. Private communication

Mathews, W.G., 1978 Ap. J. 216, 413.

Murray, S.S. 1977, CFA Preprint #680.

Peebles, P.J.E. 1970 A.J., 75, 13.

Perrenod, S. 1978 Ap. J. 226, 566.

Schwartz, D.A. 1979 COSPAR X-RAY ASTRONOMY, edited by W. A. Baity and L.E. Peterson; Pergamon Press Oxford and New York 1979.

Setti, G. and Woltjer, L., 1973,"X-and Gamma-Ray Astronomy", IAU Symposium No. 55 (eds. H. Bradt and R. Giacconi); D. Reidel Publishing Co.

Tananbaum, H., Avni, Y., Branduardi, G., et al. 1979 Ap. J. (Letters) in press.

DISCUSSION

Perrenod: How many square degrees have your surveys covered? I take
it that you've seen no clusters.

Giacconi: We have not yet detected clusters in the deep surveys. They
only covered less than 1° of the sky.

Silk: The clusters that exhibit the Butcher-Oemler effect appear
to show a radial gradient in the color distribution of the
galaxies, with the bluer galaxies being preferentially concentrated in
the center regions. Have you made a detailed comparison of the extent
of the X-ray emitting region with the blue galaxy distribution? This
is evidently necessary before one can decide on the relevance of ram
pressure stripping models for these clusters.

Giacconi: We have not yet analyzed the results from that point of view.

Stockton: You showed the X-ray variations in OX-169 and mentioned that
3C 273 also showed variations at about 10%. These two QSO's
are the only two I know of that show bright optical jets. Have you
looked for variations in other QSO's?

Giacconi: Yes, but we haven't found any.

Green: Has any correlation been found between X-ray (not X-ray
optical) spectral index and redshift in quasars?

Giacconi: Not yet; we are studying the question.

Peebles: Can you use the X-ray luminosities of the main sequence M
stars to place limits on the hypothesis that galaxies have
massive halos of low mass main sequence stars?

Giacconi: A galaxy containing 10^{11} to 10^{12} M-type stars will produce
10^{38} to 10^{40} ergs per sec in X-rays. Massive halos from
nearby galaxies could therefore be detected. Conversely, lack of
detection may be useful in setting upper limits on the number density
of M-stars.

Abell: Following up on Jim Peebles' question, how many M dwarfs
have now been observed for X-ray emission, and what fraction
of them show the surprising ratio of L_x/L_v of 10^{-1}?

Giacconi: About a dozen M dwarfs have been observed; most of them show
this ratio.

THE X-RAY BACKGROUND: ORIGIN AND IMPLICATIONS

Martin J. Rees
Institute of Astronomy,
Cambridge, England.

This paper will be concerned with three topics relevant to the
X-ray background: (i) X-ray emission mechanisms in quasars; (ii) the
contributions to the X-ray background from quasars, clusters of galaxies,
intercluster gas, young galaxies, etc; and (iii) the use of X-ray background observations as a probe for large-scale density irregularities
in the Universe.

1. X-RAY EMISSION FROM QUASARS

In the new era of X-ray astronomy initiated by HEAO 2 (the "Einstein
Observatory"), remote quasars and active galactic nuclei can be studied
in detail in the X-ray band as well as at optical (and sometimes radio)
wavelengths. These exciting developments are reviewed by Giacconi elsewhere in this volume. My main topic is the X-ray background – its
implications as a probe for the evolution and large-scale structure of
the Universe. Quasars and related objects apparently make a major contribution to this background; it may therefore be appropriate briefly
to consider the mechanisms whereby X-rays can emanate from such objects.
At the moment, such inferences as we can draw concerning physical conditions in the emitting volumes suggest that a wide variety of processes
could generate X-rays.

Optical data on the broad emission lines imply that these arise
from clouds moving at speeds of several thousand km s^{-1}. Shocks
involving this kind of velocity automatically yield gas at temperatures
of $\gtrsim 10$ keV: indeed, the clouds giving the optical emission lines
($T_e \simeq 10^4$ $^\circ$K) may be in pressure balance with a hot and more rarified
medium capable of emitting thermal X-rays via bremsstrahlung, or
comptonisation of soft photons. We also observe directly that quasars
emit non-thermal continuum radiation, implying that relativistic electrons
are present. (These particles could have been accelerated via shocks, or
by some electromagnetic process near a massive central object.) These
can give X-rays via synchrotron or compton emission. The various possible
processes can be summarised in a flow diagram reproduced in Figure 1.

G.O. Abell and P. J. E. Peebles (eds.), Objects of High Redshifts, 207–225.
Copyright © 1980 by the IAU.

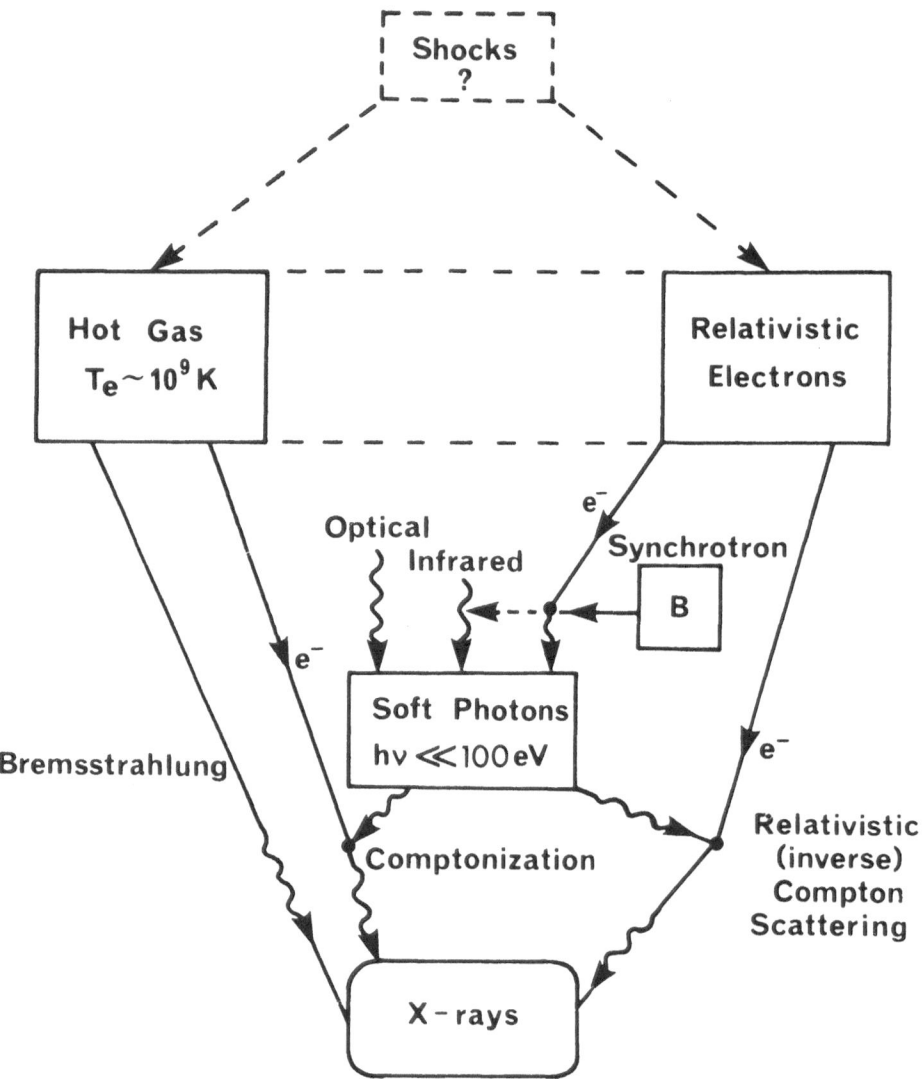

Figure 1.   Possible pathways for X-ray emission in galactic nuclei
(from Fabian and Rees (1978a), in which a fuller discussion of these
various mechanisms can be found).

        The gas responsible for the broad emission lines – even for
Lyman α – has densities $n_e \lesssim 10^{11}$ cm^{-3}; photoionization models then
suggest that this gas lies $\gtrsim 10^{18}$ cm from the central continuum, and
that it consists of enormous numbers of fast-moving clouds of individual

dimensions $\leq 10^{15}$ cm (Blumenthal and Mathews 1979; Carswell and Ferland 1979). There is no physical reason why, still closer to the central non-thermal continuum, there should not be even denser ($n_e >> 10^{11}$ cm^{-3}) and more opaque clouds. Indeed, in a wide class of accretion models (Bergeron 1979; Maraschi et al. 1979), infalling clouds could exist, with ($n_e T_e$) $\propto r^{-5/2}$, in pressure balance with a "hot-phase" medium at the virial temperature. At a distance r from the centre, the equivalent black body temperature is

$$T_{bb} \simeq 5 \times 10^4 L_{47}^{\frac{1}{4}} (r/10^{15} \text{ cm})^{-\frac{1}{2}} \text{ K}$$

(the non-thermal continuum luminosity, $10^{47}L_{47}$ erg s^{-1}, being assumed to come from a central region of size $r_{cont} < r$). In the high density limit, clouds located where $T_{bb} \gtrsim 10^4$ K would have $T_e \simeq T_{bb}$ and would reprocess the incident continuum into approximately black body radiation. The form of the reprocessed continuum depends on the configuration and dynamics of the clouds – specifically, on how the covering factor depends on radius for $r_{cont} \leq r < r_2$, where r_2 is the radius such that $T_{bb}(r_2) \simeq 10^4$ K. Recent IUE data on 3C 273 (Ulrich et al. 1979) reveal evidence for such a "thermal" continuum feature. The dense clouds at $r << 10^{18}$ cm may be embedded in a hot (almost relativistic plasma) which radiates non-thermally or by comptonisation of the soft photons generated by the cool clouds (cf. Liang 1979).

To discriminate among the various mechanisms illustrated in Fig. 1, further extended spectral observations covering the hard X- and γ-ray bands are needed. These, combined with variability timescales, may eliminate some emission mechanisms. Highly variable sources seem most likely to emit via non-relativistic Compton scattering, the timescales for this process being shorter than for bremsstrahlung from the same electrons. Relativistic Compton scattering seems unable to produce short timescales without multiple scattering being important, thereby creating an energy problem and predicting dominant γ-rays. The bulk of the material in the emitting region may be mildly relativistic ($T_e = 10^9 - 10^{10}$ °K). Observations of X-rays from galactic nuclei should stimulate much needed study of processes such as: radiation processes in 'transrelativistic' plasmas where electron-electron bremsstrahlung and relativistic corrections are important; production of $e^+ - e^-$ pairs, and their effects on cooling and opacity; effects of major differences between electron and ion temperatures; comptonisation of spectra in inhomogeneous gas clouds; and acceleration of ultrarelativistic particles by mildly relativistic shocks.

X-ray observations – along with studies of the non-thermal optical continuum – offer the most direct evidence on physical conditions close to the central "power house" (dimensions $\leq 10^{15}$ cm). Relative to the continuum source, even the broad-line region is diffuse "fuzz" out on the periphery, $10^2 - 10^3$ times further removed from the centre.

2. ORIGIN OF X-RAY BACKGROUND

2.1 The Contribution from Quasars

Preliminary data from the first deep surveys carried out with the Einstein Observatory, covering only of order one square degree, and reported by Giacconi elsewhere in these proceedings, show that most quasars down to $\sim$ 19th magnitude are detectable as X-ray sources. The limiting sensitivity is $\sim 10^{-14}$ erg cm^{-2} s^{-1} in the 1 - 3 keV band, $\sim 10^3$ lower than earlier surveys. This is not a surprising result - it was predictable from the scanty previous data if the ratio of X-ray and optical luminosities for the typical quasars was similar to that of 3C273 and Seyfert nuclei. Setti and Woltjer (1973, 1979) and others have pointed out that only a modest extrapolation from the Einstein Observatory limiting sensitivity would account for all the X-ray background (at $\leq$ 5 keV): indeed, the argument can now be inverted, and the X-ray background used to set constraints on how far, and how steeply, the quasar counts can be extrapolated to fainter magnitudes (Fabian and Rees 1978b, Setti and Woltjer 1979). See Figure 2 and caption for further explanation. This sets constraints on the evolution with z of the quasar population, and on extrapolations of the luminosity function towards fainter objects (Seyfert galaxies, etc.).

The parameter α_{ox} - denoting the spectral index that is obtained by interpolating a power-law spectrum between the optical (2500 Å) and X-ray (2 keV) flux densities - shows no obvious correlation with luminosity or redshift (Tananbaum et al. 1979; Avni et al. 1979). This gives us some confidence in assuming that the X-ray counts do indeed have the same slope as the optical counts. However it will be important, both for cosmological applications and for our physical understanding of quasars and their evolution, to test this by studying the counts and redshift distribution for an X-ray selected sample. (There should soon be a large enough sample to permit an analysis similar to Schmidt's (1970) well-known comparison of an optically-selected and a radio-selected sample.)

The Einstein Observatory data refers to relatively <u>low</u> energy X-rays (below $\sim$ 4 keV). One needs some spectral information before one can estimate how much quasars contribute to the $\geq$ 10 keV X-ray background.

At the time of writing, the best evidence on the X-ray spectra of active galactic nuclei comes from the work of Mushotsky et al. (1979) using the HEAO-A2 experiment, which covers the energy range 2 - 50 keV (Holt 1979). Of the seven Seyfert nuclei in the sample, all are roughly fitted, <u>above 5 keV</u> by power laws with (energy) spectral index $\alpha \simeq 0.6$. NGC4151 has a flatter spectrum, which may continue all the way up to γ-ray energies, but there is otherwise no correlation between α and L_x, even though the sample contains a range of $\sim$ 100 in L_x. <u>Below 5 keV</u> there are two other features: a (possibly variable) cut-off due to absorption; and a low energy "excess", with $\alpha \simeq 2$, which may also be variable. The 5 "BL Lacs" in the sample show two components: one "hard", with $\alpha \simeq 0.2$; the other "soft", with $\alpha \simeq 2$.

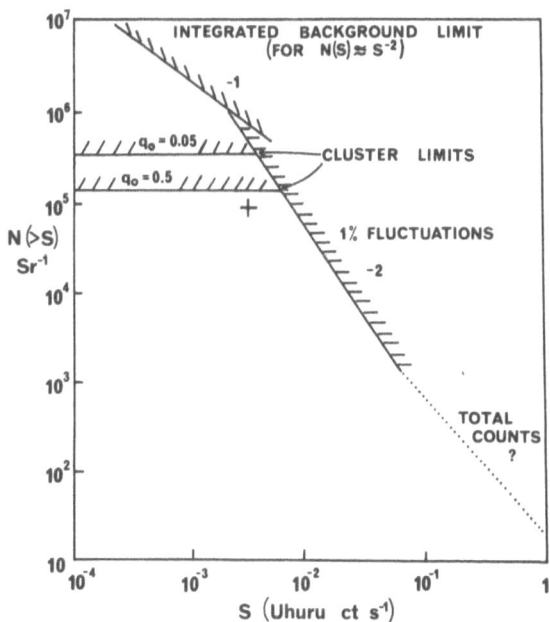

Figure 2.   Integrated background and fluctuation constraints plotted in
Uhuru flux units ($\simeq 1.1 \times 10^{-11}$ erg cm^{-2}s^{-1} at 1 - 3 keV). These
constraints depend in detail on the actual source counts, and thus are
only approximate; the lines drawn (logarithmic slopes -1 and -2
respectively) do however indicate the limits set by the integrated
background (assuming $N(> S) \propto S^{-2}$), and the absence of detected fluctuations in the background, respectively (Fabian and Rees 1978b). The
total surface density for rich clusters out to $z \simeq 3$, assuming a number
density 4×10^5 $(c/H_o)^{-3}$, is indicated for $q_o = 0.5$ and $q_o = 0.05$: it
is clear that rich clusters, however they may evolve, cannot reach the
integrated background line without contradicting the fluctuation constraints. The cross denotes a point on the source counts estimated by
Giacconi et al. (1979) from the Einstein Observatory's first deep
surveys. Most of these sources may be quasars. This point lies on an
extrapolation, with logarithmic slope -1.5, of the counts at high flux
densities. This is, however, fortuitous: most of the Uhuru-level
extragalactic sources are clusters, whose counts probably have a slope
flatter than -1.5; the quasar counts may be as steep as -2, in which
case only a modest further extrapolation to fainter fluxes suffices to
account for the entire X-ray background.

Evidence on the hard X-ray spectra of bona fide quasars is even more sparse, being restricted essentially to 3C 273 (Worrall et al. 1979). This object shows $\sim$ 40% variations on timescales of 6 months, with α in the range 0.4 - 0.7 between 5 and 50 keV. The spectrum seems steeper both at lower and at higher energies (there is a measurement by CosB of $\sim$ 100 Mev γ-rays from 3C 273 (Swanenberg et al. 1978)).

According to Avni et al. (1979) α_{ox} varies between 0.94 and 1.86. The mean is 1.27 (though this refers to a sample biased in favour of radio-selected quasars). The fact that the X-ray spectral index is flatter than α_{ox} has the astrophysically-interesting consequence that the luminosity of quasars is concentrated in the optical/UV band or in the hard X-ray band. As far as the X-ray background is concerned, the X-ray spectral index of quasars seems flat enough to ensure that - given that they are a major contributor at $\leq$ 5 keV - they may also contribute most of the background at higher energies: indeed the fact that the γ-ray background is not stronger allows us already to exclude the possibility that most active nuclei have spectra like NGC 4151.

2.2 Clusters of Galaxies

The X-ray properties of clusters are reviewed by N. Bahcall else-where. In clusters such as Coma there now seems little doubt that most of the X-rays below $\sim$ 10 keV are thermal bremsstrahlung from metal-enriched gas in the core of the cluster. We know much less about the gas in the outlying parts of clusters because its emissivity ($\propto n_e^2$) is low - it may not be in equilibrium and its temperature and composition are uncertain. The properties of small groups and "unrelaxed" clusters such as Virgo are more complex. The X-ray luminosity of rich clusters seems positively correlated with virial velocity, cluster richness and central density; but negatively correlated with the fraction of spirals in the cluster. The latter is in accord with ideas that, in "dynamically old" clusters, the gas has been stripped from individual galaxies.

Some consideration of how clusters evolve is a prerequisite not only for understanding their present properties, but also for inter-preting the data on large-redshift clusters that HEAO-B is now providing, and for pinning down the cluster contribution to the X-ray background.

Two contrasting hypotheses have been explored:

A. Galaxies (and dark matter) may have condensed from primordial gas before clusters assemble. Galaxies then gradually aggregate into clusters in a manner that can be simulated by gravitational N-body computations. The present gas in clusters then results from either: infall of the small fraction of material ($\sim$ 10%) that escaped incorpora-tion into the first generation of gravitationally-bound systems; or ejection from galaxies, via stellar winds and planetaries, supernovae, "sweeping" of interstellar matter by ram pressure of pre-existing cluster gas.

B. Protoclusters (i.e. gas clouds of $\sim 10^{14}$ M$_\odot$) may have been the first objects to have condensed out of the expanding universe; they then cool and fragment into galaxies. In this scheme, advocated particularly by the Moscow group (Doroshkevich et al. 1974), the present intracluster gas represents material that has not (yet) cooled and condensed.

Obviously a type-A hypothesis yields clusters whose X-ray luminosities increase with time; on the other hand, hypothesis B would predict that clusters were brightest where they had just virialised but were still predominantly gaseous, but that the X-ray luminosity would thereafter decline as the initial gas gradually cooled and condensed into galaxies.

Perrenod (1978) and others have investigated type-A models. The general findings are that L_x may increase by a factor up to ~ 10 between the epoch corresponding to $z \simeq 1$ and the present. This is a combined consequence of the cumulative build-up of gas, and the deepening of the cluster potential well as virialisation proceeds.

The X-ray emission from clusters may turn on more suddenly than in Perrenod's schemes: for instance, gas could perhaps be retained in galactic halos (at temperatures $kT \lesssim 1$ keV) until the galaxies fall together into a cluster and the halo material is stripped away and shock-heated to the cluster virial temperature (Norman and Silk 1979). One difficulty stems from the fact that the hypothesised halo gas has a short cooling time; on the other hand, the Butcher-Demler (1978) evidence for a sharp evolutionary change in galaxy colours (indicating a sudden quenching of star formation) supports this general idea, at least qualitatively.

Observations of distant clusters with the Einstein Observatory offer our best hope of discriminating between these various schemes. In practice, a broad range between the extremes of A and B seem possible within the limits of current observation. Preliminary data on a small sample of clusters with $z \simeq 0.5$ can merely rule out extreme evolution of either sign (Henry et al. 1979).

Rich clusters probably contribute 5 - 10% of the $\lesssim 5$ keV X-ray background - one can rule out the possibility that they contribute all the background - in whatever fashion they may evolve - because the small angular scale fluctuation limits imply that the total number of contributing sources must exceed the number of rich clusters out to $z \simeq 3$ (Fabian and Rees 1978b).

2.3 Ultra-hot Intercluster Gas?

The gas temperatures in clusters are ~ 5 keV, implying an exponential fall-off in their contribution to the background at higher energies. One possible contributor to a genuinely diffuse hard X-ray background might be thermal bremsstrahlung from a very hot gas between the clusters. As has recently been emphasised by the Goddard group (Marshall et al. 1979) the background spectrum is, empirically, very closely fitted by a 35-40 keV bremsstrahlung spectrum. The excellence of this fit may well be fortuitous: indeed the fit could be destroyed by merely subtracting off the contributions from other sources of background (e.g. quasars and clusters), which are known to be substantial below 10 keV and do not have this spectrum.

Field and Perrenod (1977) discussed the possibility that intergalactic gas is indeed responsible for the background. If such gas is reheated at redshift z_{heat} to a temperature such that $kT = 35(1+z_{heat})$keV, then it can emit the entire X-ray background if its density corresponds to

$$\Omega_{gas} \simeq \frac{1}{3} \left((1+z_{heat})^{1.6} - 1 \right)^{-\frac{1}{2}} h_{100}^{-3/2} \qquad (1)$$

Of course, less gas would be needed if it were clumped; but at such high temperatures it would be uninfluenced by the gravitational field fluctuations arising from clusters of galaxies.

The main difficulty with this hypothesis, as Field and Perrenod realised, is the energy requirements for heating such a large mass of gas. Subsidiary problems relate to the Compton distortions of the microwave background spectrum that this ultra-hot gas would cause (Wright 1979); also, conductivity and pressure balance considerations might prejudice the existence and survival of low-density HI clouds around galaxies.

2.3   Other Possible Classes of Extragalactic X-ray Sources

Compton scattering in remote radio sources. Inverse Compton scattering of microwave background photons by relativistic electrons was long ago suggested as a contributor to the X-ray background (Felten and Morrison 1966). This process would tend to be more efficient at large redshifts, for two reasons: the energy density of the microwave background varies as $(1 + z)^4$; and the many suppliers of relativistic electrons - radio source, quasars, etc. - were more prolific at early epochs. As a corollary of this effect, Compton scattering may "snuff out" extended sources at large redshifts (Rees and Setti 1968).

X-rays observed in the range 1 - 10 keV are generated predominantly by electrons with Lorentz factors in the range $(1 - 3) \times 10^3$. The radio sources with the highest inverse Compton X-ray luminosities are therefore those with the largest energy content in the form of such electrons. These sources will not necessarily be those with the highest radio luminosities: the maximum stored energy is inferred to exist in very extended "giant" sources and in the low surface-brightness "bridges" joining the components of some strong double sources. On the assumption of equipartition, the energy stored in $\gamma \simeq 10^3$ relativistic electrons in NGC 6251 and 3C 236 is $\geq 10^{60}$ erg (Waggett et al. 1977; Willis et al.1974).

If the magnetic field is weaker than its equipartition strength, the inferred energy stored in relativistic electrons is even larger. Some of the X-rays seen from Centaurus A are probably inverse Compton emission from the inner radio lobes (Schreier et al. 1979).

A radio source with a typical spectra index ~ 0.7 containing 10^{60} ε_{60} erg of relativistic electrons with $\gamma \gtrsim 10^3$, and at redshift z, will emit an inverse Compton X-ray power of $\sim 3 \times 10^{43}$ $(1+z)^4$ ε_{60} erg/s. We do not know the appropriate values of ε_{60}, but it would seem quite probable that there may be diffuse objects at $z \simeq 2$, a few hundred kpc in extent, emitting $\sim 10^{46}$ erg of 1 - 10 keV X-rays. These sources would generally not be particularly powerful radio sources (objects such as Cygnus A or 3C 9 are powerful radio sources because they have a strong magnetic field, rather than because they have an exceptional energy content), but they may feature in deep radio surveys. Several Westerbork radio sources have already been identified in Einstein Observatory deep surveys, though it is not yet clear whether the X-rays come from the radio lobes or from the active nucleus (Giacconi et al. 1979).

Young galaxies. According to some theories, galaxies pass through a bright early phase when the rate of star formation, supernova outbursts, etc. is ~ 100 times higher than in a present-day galaxy (e.g. Meier 1976, Ostriker and Thuan 1975). The properties of such systems, and their potential detectability, depend on many uncertainties; but we can readily see that a young galaxy where supernova outbursts were frequent could give rise to thermal or non-thermal X-rays with a much higher power than a present-day galaxy.

Winds from young galaxies. A possible mechanism that could generate a 40 keV bremsstrahlung-type spectrum without demanding such a high mass of gas might be supernova-driven winds in young galaxies. Mathews and Baker (1971), Mathews and Bregman (1978) and others have discussed galactic winds. The mass is supplied by supernova ejecta, with characteristic velocities of $\sim 10^4$ km/sec (~ 100 keV per particle), and by more gentle processes such as stellar winds and planetary nebulae. Supernovae give the main energy input, though not necessarily the main mass supply. Two simple requirements for a wind are

$$\dot{M}_{SN}/(\dot{M}_{SN} + \dot{M}_{other}) \gtrsim \frac{\text{(escape energy, per proton, from galaxy)}}{100 \text{ keV}} \quad (2)$$

and

$$t_{cool} \gtrsim t_{outflow} \quad (3)$$

(Note that these constraints could perhaps be eased in a more complex - though maybe more realistic - case when the gas has a multiphase structure.)

For a given value of the ratio (2), condition (3) is more easily
satisfied now than in the past. However, galaxies might have experienced
an initial bright phase when supernovae provided the main mass loss.
There would then have been a very fast ($\sim 10^4$ km/sec) hot (~ 100 keV)
wind emanating from young galaxies. For instance the supernova rate
within the core region of a young galaxy may be high enough to sustain,
for $\sim 10^7$ yrs, a wind with kinetic energy output $\sim 3 \times 10^{46}$ erg/sec.
Even though (3) is fulfilled by an unduly wide margin, such a wind
would yield $\sim 5 \times 10^{44}$ erg/sec, per galaxy, in hard X-rays (Bookbinder
et al. 1979). If their upper-main-sequence stars have a Population I
composition, young galaxies may be stronger X-ray sources than present-
day galaxies because metal-poor stars spend longer in the part of the
H-R diagram that permits strong radiation-driven winds, so the lifetime
of a typical X-ray binary is correspondingly prolonged.

2.4 The Background Spectrum

The well-known problem of accounting for any sharp break or
feature in the background spectrum will be reassessed by di Zotti in
his contribution. If various categories of source contributing to the
background have different spectra, there is of course no reason why the
degree of isotropy should be similar at different X-ray energies (cf.
Rees 1973).

3. THE BACKGROUND AS A PROBE OF LARGE-SCALE INHOMOGENEITY

The X-ray background obviously holds important clues to the
evolutionary history and spatial distribution of its sources. I shall
mention here just one aspect of this subject: the sensitivity of the
X-ray background as a probe for density irregularities on scales larger
than superclusters. Whatever the precise origin of the background may
be, it is plausible to suppose that any large scale inhomogeneities in
the overall distribution of gravitating matter will give rise to
associated inhomogeneities in the spatial density of X-ray sources.
(This relation may not be a strict proportionality - the X-ray source
density or emissivity may depend on a higher power of the overall
density - but this would do no more than change a numerical coefficient
of order unity in the following expressions, where we consider only
small fractional perturbations.)

Although the covariance function data become imprecise on scales
exceeding 20 Mpc, the universe definitely appears more homogeneous on
scales $\gtrsim 100$ Mpc than on any scale $\lesssim 10$ Mpc. There is good radio, and
optical evidence that the distribution on the sky of all kinds of
luminous objects becomes smoother as we look deeper.

Fluctuations on larger scales ($\lambda \gtrsim 100$ Mpc, say) are thus only
of small amplitude (certainly $\delta\rho/\rho \lesssim 1$). Their influence may nonetheless
be significant. This is because the velocity perturbations that they

induce are of order

$$V_{pec} \simeq c\Omega \left(\delta\rho/\rho\right) \left(\lambda/\lambda_H\right) \tag{4}$$

and the gravitational field perturbations are

$$\Delta\phi \simeq c^2\Omega \left(\delta\rho/\rho\right) \left(\lambda/\lambda_H\right)^2 \tag{5}$$

This means that peculiar velocities are dominated by the largest scales unless $\delta\rho/\rho$ falls off more steeply than λ^{-1} ($\propto M^{-1/3}$). Anisotropies in the microwave background on small angular scales are due to gravitational and doppler effects on the 'cosmic photosphere' (the gravitational effect dominating for scales exceeding the horizon size at that epoch (Sunyaev and Zeldovich 1970)).

Constraints are placed on large-scale inhomogeneities by the iso-tropy of any class of extragalactic discrete source; but these are of limited value. Inferences drawn from optical counts of galaxies are bedevilled by the possible effects of patchy Galactic obscuration; for radio sources, absorption is negligible but the problem here is the broad luminosity distribution (such that intrinsically faint nearby sources and powerful remote ones appear in surveys in comparable numbers at the same flux density). As discussed by Fanti in these proceedings, the best limits amount to $\leq$ 5 % on scales of $\sim$ 1/3 the Hubble radius.

Given that the X-ray background is predominantly from cosmological distances ($z \gtrsim 1$), and that the Galactic contribution, away from the plane, is only a small contamination, X-ray data can provide useful evidence on the distribution of matter, if we assume that the X-ray emission per unit volume (at a given epoch) scales with the overall matter density. The present isotropy limits have a precision of about 1 per unit on large angular scales. Warwick et al. (1979), in an analysi of Ariel V data, (2 – 18 keV) obtained an upper limit of better than 1 per cent to any 24h effect, after subtracting off a component depending on galactic latitude. The limits on angular scales of $\leq$ 20° are no better than a few percent, but there are prospects of significant improve ments from larger-area detectors, and from measurements at higher energie where the Galactic disc contributes relatively less.

Suppose that the input into the X-ray background at a redshift z amounts to a power per unit comoving volume of $\mathcal{E}(z)$. The background intensity then involves an integral of the form

$$\mathcal{I} = \frac{c}{4\pi} \int \mathcal{E}(z) \left(1 + z\right)^{-\alpha} \frac{dt}{dz} dz \tag{6}$$

where α is the effective spectral index, and dt/dz depends on the cos-mological model. It is convenient to define a number

$$x = \frac{c}{4\pi} \, \mathcal{E}(0) \, H_o^{-1} / \mathcal{J} \tag{7}$$

The significance of this number is that a static non-evolving Euclidean universe of radius cH_o^{-1} would emit x times the actual X-ray background. The value of x is of order unity: the $(1+z)^{-\alpha}$ term in (6) tends to reduce it below unity; on the other hand, any evolutionary effect which enhanced the contribution from large z would tend to raise it.

Clumping of the sources, on any scale, would cause a corresponding anisotropy or 'graininess' in the observed X-ray background. Suppose, for illustration, that the clumping has a characteristic (comoving) scale λ (<< the present Hubble radius) and that the amplitude of the variations is $\delta\rho/\rho$. In general $(\delta\rho/\rho)$ will be a function of z. The inhomogeneities nearest to us will cause anisotropies in the X-ray background. The precise amplitude depends on the configuration of the irregularities around us, but the characteristic expected amplitude is obviously

$$\left(\frac{\Delta I}{I}\right)_x \simeq x\left(\frac{\delta\rho}{\rho}\right)\left(\frac{\lambda}{\lambda_H}\right) \tag{8}$$

Additionally, there will be fluctuations on small angular scales $\sim (\lambda/\lambda_H)$ due to similar irregularities at redshifts $z \gtrsim 1$. (Turner and Geller (1979) have already been able to show that the X-ray background is uncorrelated with observed bright galaxies, to a precision which implies sufficient source evolution to make x twice as large as its non-evolutionary value.)

If the irregularities in the X-ray emissivity are related to inhomogeneities in the distribution of matter, then the gravitational effects should induce peculiar velocities, which themselves show up as '24h effects' in the X-ray and microwave background. This relationship is complicated in the general case where the perturbations have become non-linear. Things are simpler, however, if we restrict attention to perturbations whose gravitational influence makes merely a small fractional change in the Hubble flow, as seems to be the case for all scales $\gtrsim 20$ Mpc (the scale of the Local Supercluster). These linear perturbations are nevertheless interesting because the dominant contribution to our peculiar motion could arise from perturbations with $\lambda \gg 20$ Mpc and $(\delta\rho/\rho) \ll 1$ (cf. (4)).

The peculiar velocity induced at the boundary of a lump of scale λ and amplitude $\delta\rho/\rho$ actually depends on the value of Ω. The problem is discussed by Silk (1974), who presents graphs for V_{pec} more accurate than the crude relation (4), as a function of $\delta\rho/\rho$ for different choices of Ω. When $\lambda \ll \lambda_H$, our peculiar velocity will give a 24 hour X-ray anisotropy of amplitude

$$\left(\frac{\Delta I}{I}\right)_x = (3 + \alpha) \; \frac{V_{pec}}{c} \tag{9}$$

(when $\lambda \simeq \lambda_H$ the situation is more complicated because the sources of a substantial fraction of the X-ray background are themselves given a peculiar velocity: see below). The microwave background, if observations are made on the Rayleigh-Jeans portion of the spectrum ($\alpha = -2$), would yield

$$\left(\frac{\Delta I}{I}\right)_{mic} = \frac{V_{pec}}{c} \tag{10}$$

Comparison of (8) - (10) shows that these measurements yield an estimate of Ω: if the microwave background anisotropy is induced by inhomogeneities on scale λ ($<< \lambda_H$), then the X-rays should show an effect $(3 + \alpha)$ times larger due to our motion (equation 9), and an effect due to the clumping of sources themselves which is larger by a factor $\sim x\Omega^{-1}$. In principle, therefore, comparison of X-ray and microwave data can yield constraints on Ω (Fabian and Warwick 1979): if Ω is very small the Galactic peculiar velocity of ~ 600 km s^{-1} reported by Smoot et al. (1977) and Corey and Wilkinson (1976), and discussed by Smoot in these proceedings, could not be induced by a large-scale inhomogeneity on any scale between ~ 100 Mpc and the Hubble distance, without the corresponding anisotropy in the X-ray source distribution exceeding what is observed.

Figure 3 shows the relative sensitivity of the microwave, X-ray and other limits in restricting the inhomogeneity of the universe on large scales. The amplitude of the inhomogeneities on a scale λ is expressed in terms of the present value of $(\delta\rho/\rho)$. The large scales ($\lambda > 100$ Mpc) would have grown since entering the horizon unimpeded by radiation pressure effects, so for them $(\delta\rho/\rho)$ is related to the curvature fluctuation by

$$\epsilon \simeq \left(\frac{\delta\rho}{\rho}\right) \left(\frac{\lambda}{\lambda_H}\right)^2 \tag{11}$$

(Note that Ω does not enter into this expression.) The upper limit on any 24 h component in the X-rays sets a limit shown as a line in Fig. 3. The peculiar velocity induced by the corresponding inhomogeneity shows up as an extra contribution depending on Ω.

We see from this diagram that X-ray isotropy observations with ~ 1 per cent precision provide our best limits on the spectrum of inhomogeneities on scales between ~ 100 Mpc and ~ 1000 Mpc; on scales exceeding 1000 Mpc the most sensitive limits come from the microwave background - more specifically, from limits to the gravitational potential fluctuations around the surface of last scattering. On scales $\gtrsim \lambda_H$ other gravitational effects must be allowed for (cf. Rees and Sciama 1968; Dyer 1976).

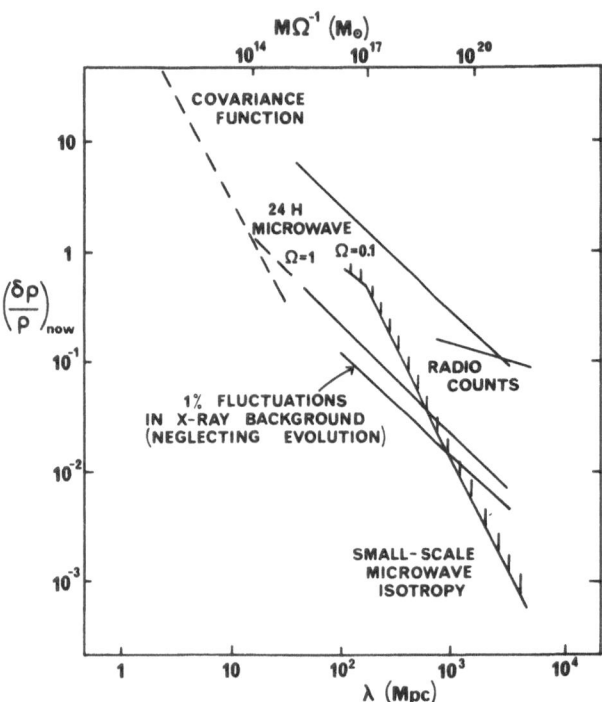

Figure 3. Constraints on density perturbations ($\delta\rho/\rho$) on various length scales λ exceeding ∿ 10 Mpc (adapted from Fabian <u>et al</u>. 1979). The lines labelled 'covariance function' and 'radio counts' assume that the galaxy and radio source distributions mimic the underlying mass distribution. The 'microwave velocity' lines give δv_G = 600 km s^{-1} and assume that we lie at the edge of such a perturbation. The radio count limits can perhaps be improved to 3 percent on scales of 1 Gpc, but depend upon the radio luminosity function etc. The microwave background observations imply upper limits to the Doppler and gravitational perturbations at the epoch of last scattering, and these yield approximately the limits indicated. Limits of ∿ 1 % on the isotropy of the X-ray background would yield, apart from an evolutionary correction (cf. equation 8), the line shown: note that this is potentially the most sensitive constraint on scales 100 – 1000 Mpc; on larger scales the microwave limits are likely to remain the best.

(The interpretation of the experimental limits to $\Delta T/T$ on __smaller__ angular scales depends on the sharpness of the last scattering surface or 'cosmic photosphere', the influence of early reheating etc. However, these uncertainties do not enter on large scales, and on the smaller scales the X-ray limits may be better in any case.)

In compiling Figure 3, inhomogeneities of characteristic amplitude $(\delta\rho/\rho)$ were assumed to pervade the whole universe — $(\delta\rho/\rho)$ measures the amplitude of the Fourier components over a particular range of length scales. If, on the other hand, the universe contains isolated 'lumps' on large scales, embedded in a much smoother general background, then, as Fabian and Warwick (1979) have pointed out, there may be detectable consequences for the X-ray background, for observations with presently attainable sensitivity, even though the influence of the 'lump' may be undetectable as far as the microwave background is concerned. Of particular interest is the possibility that there may be isolated inhomogeneities on scales fully comparable with the observed part of the universe.

The naive discussion leading to (8) and (9) needs some modification for inhomogeneities with scales comparable with the Hubble radius λ_H ($\lambda \gtrsim 1000$ Mpc, say). A general treatment of cosmological models containing inhomogeneities on scales $\gtrsim \lambda_H$ would be prohibitively complicated. However, drastic simplifications ensue if we restrict attention to very small amplitudes — this is a justifiable restriction for observational cosmology because we know, from the microwave background, that the Universe is not far from being accurately "Robertson-Walker".

Provided that the curvature fluctuations are indeed small, we can consistently take a Friedmann model, with a certain overall curvature (or, equivalently, a well-defined mean density parameter Ω_0 at the present epoch t_0). The nature of the fluctuations can then be specified by defining the density at each point (comoving coordinate r) as $\bar{\rho}(1 + \delta(\underline{r}, t))$; at each location $\underline{r}$, δ increases with t according to the standard expression for the growth of linear density perturbations in a Friedmann model of density parameter Ω_0. Under the restriction that $\delta \ll 1$, one can extend relation (4) for the peculiar velocity to the case $\lambda \simeq \lambda_H$. For an observer at the edge of a "Swiss cheese" perturbation one obtains

$$V_{pec} = \frac{\delta}{2} c \left(\frac{\lambda}{\lambda_H}\right) \Omega \left(\frac{2(1-\Omega)^{\frac{1}{2}} - 1}{2(1-\Omega)^{\frac{1}{2}} - \Omega}\right) \tag{12}$$

The effect on the X-ray background can be analysed into three components;

(i) A 24h effect due to the enhanced density of sources in the 'lump'. (This effect would not exist if the lump lay beyond the redshift at which the X-ray background originates.)

(ii) A 24h effect due to our motion towards the 'lump'. (Reduced below (12) because the sources are themselves "falling" as well.)

(iii) A 12h quadrupole effect, with a minimum along the axis of
symmetry towards the 'lump', due to the shear induced within the
volume whence the bulk of the X-ray background originates (see Fig. 4
and caption).

Finally, we note that the X-ray background can set constraints on
cosmological anisotropies on scales $\gg \lambda_H$ (Wolfe 1970).

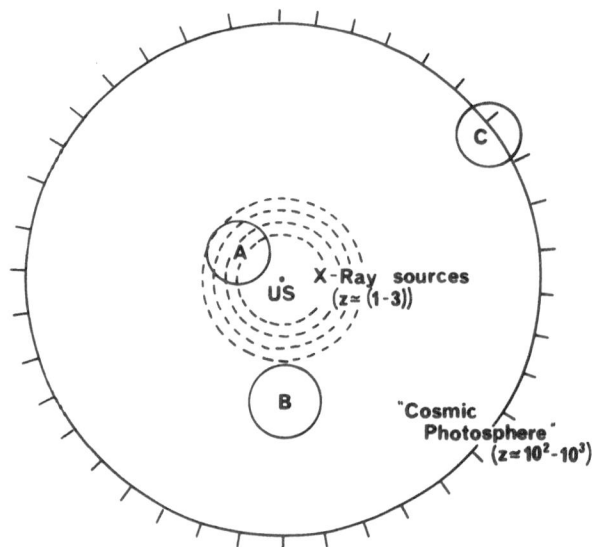

Figure 4. This diagram is intended to illustrate schematically the
effect of a single isolated 'lump' of dimensions comparable with the
Hubble radius, placed in various positions around us. The radial coordinate is the 'comoving r' of the Robertson-Walker metric; the 'cosmic
photosphere' is assumed to be at $z \simeq 1000$, and the sources of the X-ray
background at $z \lesssim 3$. (Note that, in terms of the coordinate r, $z \simeq 3$ is
about half-way to $z \simeq 1000$ if $\Omega = 1$. If $\Omega < 1$ it is less than half way;
moreover, volumes then increase faster than $\propto r^3$.) The effects on the
X-ray and microwave background are as follows (cf. Fabian and Warwick 1979)
    A. (i) 24h effects in microwave and X-ray backgrounds due to our
peculiar motion. (ii) 12h effect of similar magnitude in X-ray background
owing to shear induced by lump. (iii) 24h effect, Ω^{-1} times larger than
others, due to excess sources in region of lump.
    B. As compared with case A, effect (ii) is more important relative
to (i); but (iii) is now absent because the lump lies beyond the X-ray
sources.
    C. There will in this case be a small 12h effect in the X-rays; our
peculiar velocity shows up in the microwave background but not significantly in X-rays, since all the "sources" are falling at almost the same rate
as us. The dominant effect, however, would be in the microwave background,
due to the gravitational perturbation on the cosmic photosphere.

ACKNOWLEDGEMENTS

I am grateful to many colleagues, particularly A.C. Fabian, for helpful and relevant discussion.

REFERENCES

Avni, Y. et al.:1979, Astrophys.J.(Lett.) in press.
Blumenthal, G. and Mathews, W.G.: 1979, Astrophys.J., 233, p.479.
Bookbinder, J., Cowie, L.L., Krolik, J., Ostriker, J.P. and Rees, M.J.:
     1979, Astrophys.J., in press.
Butcher, H. and Oemler, A. 1978, Astrophys.J., 219, p.18.
Carswell, R.F. and Ferland, G.: 1979, Mon.Not.R.astr.Soc., in press.
Corey, B.E. and Wilkinson, D.T.: 1976, Bull.A.A.S. 8, p. 351.
Doroshkevich, A.G., Sunyaev, R.A. and Zeldovich, Y.B.: 1974 in "Con-
     frontation of Cosmological Theory and Observational Data" ed. M.S.
     Longair (Reidel, Dordrecht).
Dyer, C.C.: 1976, Mon.Not.R.astr.Soc. 175, p. 429.
Fabian, A.C. and Rees, M.J.: 1978a, Proc.IAU/COSPAR Symposium on X-ray
     Astronomy, ed. W.A. Baity and L.E. Peterson p. 381. (Pergamon,Oxford)
Fabian, A.C. and Rees, M.J.: 1978b, Mon.Not.R.astr.Soc. 185, p.109.
Fabian, A.C. and Warwick, R.S.: 1979, Nature 280, p.39.
Fabian, A.C., Warwick, R.S. and Pye, J.P.: 1979, Physica Scripta, in press
Felten, J.E. and Morrison, P.: 1966, Astrophys.J. 146, p.686.
Field, G.B. and Perrenod, S.C.: 1977, Astrophys.J. 215, p.717.
Giacconi, R. et al. 1979, Astrophys.J.(Lett.) in press.
Henry, J.P. et al.: 1979, Astrophys.J.(Lett.) in press.
Holt, S.S.: 1980, in "X-Ray Astronomy" ed. G. Setti, (Reidel, Holland)
     in press.
Liang, E.: 1979, Astrophys.J.(Lett.) 231, L.111.
Maraschi, L. et al.: 1979, Astron.Astrophys. in press.
Marshall, F.E., Boldt, E.A., Holt, S.S., Miller, R., Mushotsky, R.F.,
     Rose, L.A., Rothschild, R.E. and Serlemitsos, P.J.: 1979,
     Astrophys.J., in press.
Mathews, W.G. and Baker, J.C.: 1971, Astrophys.J., 170,p.241.
Mathews, W.G. and Bregman, J.N.: 1978, Astrophys.J. 244, p.308.
Meier, D.: 1976, Astrophys.J., 207,p.343.
Mushotsky, R.F., Boldt, E.A., Holt, S.S., Marshall, F.E., Pravdo, S.H.,
     Serlemitsos, P.J. and Swank, J.H.: 1979, Astrophys.J. in press.
Norman, C. and Silk, J.I.: 1979, Astrophys.J.(Lett.) 233, L1.
Ostriker, J.P. and Thuan, T.X.: 1975, Astrophys.J. 202, p.353.
Perrenod, S.C.: 1978, Astrophys.J. 226, p.566.
Rees, M.J.: 1973 in "X-ray and gamma-ray astronomy" ed. H.Bradt and
     R. Giacconi, p.250 (Reidel, Dordrecht).
Rees, M.J. and Sciama, D.W.: 1968, Nature 217, p.511.
Rees, M.J. and Setti, G.: 1968, Nature 217,p.326.
Schmidt, M.: 1970, Astrophys.J. 162, p.371.
Schreier, E.J., Feigelson, E., Delvaille,J., Giacconi, R., Grindlay, J.,
     Schwartz, D.A. and Fabian, A.C.: 1979, Astrophys.J.(Lett.) in press

Setti, G. and Woltjer, L.: 1973 in "X-ray and gamma-ray astronomy"
    ed. H. Bradt and R. Giacconi (Reidel, Holland).
Setti, G. and Woltjer, L.: 1979, Astr.Astrophys. 76, L1.
Silk, J.I.: 1974, Astrophys.J. 193, p.525.
Smooth, G.F., Gorenstein, M.V. and Muller, R.A.: 1977, Phys.Rev.Letters
    39, p.898.
Sunyaev, R.A. and Zeldovich, Y.B.: 1970, Astrophys.& Sp.Sci, 7, p.3.
Swanenberg, B.N. et al.: 1978, Nature 275, p.298.
Tananbaum, H. et al.: 1979, Astrophys.J.(Lett.) in press.
Turner, E. and Geller, M.: 1979, Astrophys.J. in press.
Ulrich, M.H. et al.: 1979, Mon.Not.R.astr.Soc. (in press)
Waggett, P.C., Warner, P.J. and Baldwin, J.E.: 1977, Mon.Not.R.astr.Soc.
    81, p.465.
Warwick, R.S., Pye, J.P. and Fabian, A.C.: 1979, Mon.Not.R.astr.Soc. in
    press.
Willis, A.G., Strom, R.G. and Wilson, A.S.: 1974, Nature 250,p.625.
Wolfe, A.M.: 1970, Astrophys.J. 159, p.61.
Worrall, D.M., Mushotsky, R.F., Boldt, E.A., Holt, S.S. and Serlemitsos,
    P.J.: 1979, Astrophys.J. 232, p.683.
Wright, E.L.: 1979, Astrophys.J. 232, p.348.

DISCUSSION

Abell: Another way to place a limit on the scale of inhomogeneities is
            from the counts of numbers of galaxies, n, brighter than visual
magnitude, m_V. The $n(m_V)$ curves found by Rainey (doctoral thesis, UCLA,
1977) in three widely separated directions in the sky are remarkably
isotropic for $m_V > 16.5$. Rainey showed that inhomogeneities along the
line of sight with density enhancements, $\Delta\rho/\rho \sim 0.5$ to 1, with a linear
extent of 300 Mpc or greater would easily show up in his data in the
magnitude range $16.5 \leq m_V \leq 19.5$, and can be ruled out. Incidentally,
Peebles' power-law relation for the covariance function extends to about
20 Mpc (rather than 10 Mpc) if $H_0 = 50$ km s^{-1} Mpc^{-1} (Peebles uses $H_0 =$
100 km s^{-1} Mpc^{-1} for ease in comparison between various data sets).

Rees: These results are very interesting, but I suspect that the
            X-ray background isotropy limits will soon be substantially
stronger (though these are admittedly not quite so unambiguous to interpret). As regards the covariance function, its value is unity for
~ 10 Mpc (taking $H_0 = 50$ km s^{-1} Mpc^{-1}); I agree with you that its power-
law form extends out to larger distances, as described in Groth's contribution.

Tyson: Faint number counts of galaxies suggest that there is a nearby
            supercluster of size 500 Mpc to 700 Mpc in the general direc-
tion of Virgo. More data from the south is needed to better define its
angular extent.

            In your schematic QSO model, perhaps we can learn something
            about the mechanisms at the bottom of your column by observing

polarization in the X-rays, which would be caused by synchrotron radiation -- assuming the magnetic field there is not hopelessly tangled.

Rees: Such a discovery would be extremely interesting in this context. If it represented an "overdensity" of, say, 50%, then it would induce a peculiar velocity much larger than observed unless Ω were well below 0.1 (or unless the bulk of the matter were in some unclustered relativistic form). Such an effort -- if real -- ought to show up in radio source counts and in the X-ray background.

The magnetic field in the continuum-emitting region may be as strong as 10^3 to 10^4 gauss. The most important and direct evidence on physical conditions in this region comes from the studies of polarization variations at optical wavelengths, particularly by Roger Angel and his collaborators. I suspect that it will be a long time before the same thing can be done by the X-ray astronomers.

Margon: What are the chances that the X-ray spectrum of the "typical" high redshift QSO will prove to be no more complex than something that mimics a 45 keV exponential, thus allowing a simple superposition to explain the observed spectrum of the diffuse background?

Rees: Even if quasars do have a standard spectrum, there would still be a problem in accounting for the sharp bend in the background spectrum at 20 to 40 keV, particularly if one allows for the spread in redshift of the quasar contributing to the flux. Dr. De Zotti will be addressing this problem more fully in his paper.

SOURCES OF THE X-RAY BACKGROUND: TEMPORAL STABILITY

Elihu Boldt
Laboratory for High Energy Astrophysics
NASA/Goddard Space Flight Center
Greenbelt, Maryland 20771 U.S.A.

The A2 experiment[1] on HEAO-1 was especially developed to make systematics-free measurements of the extragalactic X-ray background (Boldt et al. 1979) over the band (up to 60 keV) of maximum flux. The spectrum observed has a remarkably simple thermal form (Marshall et al. 1980) with a mean photon energy of about 40 keV, an order of magnitude above the high-energy limit of the Einstein Observatory (HEAO-2) telescope. If most of this hard X-ray flux is not diffuse, then the main sources of this background could be 1) unresolved objects of known classification (e.g. BL Lac type, quasars, active galaxies) at high redshift, 2) redshifted (z > 1) gamma-ray bursts and/or 3) a new class of X-ray objects peculiar to high redshifts. If we assume that the number of such sources that are highly variable is less than 10^6, then our first-cut analysis of the temporal stability measured for the X-ray background indicates that 1) their contribution is less than 15% if they are variable on scales less than 10^4 seconds, and 2) their contribution is less than 60% if they are variable on scales less than a half-year.

Boldt, E., Marshall, F., Mushotzky R., Holt, S., Rothschild, R., and Serlemitsos, P., in Proceedings of COSPAR/IAU Symposium on X-ray Astronomy (edited by W. Baity and L. Peterson) Pergamon Press: Oxford and New York, p. 443, 1979.

Marshall, F., Boldt, E., Holt, S., Miller, R., Mushotzky, R., Rose, L.A., Rothschild, R., and Serlemitsos, P., 1980, Ap. J., in press.

[1]The A2 experiment on HEAO-1 is a collaborative effort led by E. Boldt of GSFC and G. Garmire of CIT, with collaborators at GSFC, CIT, JPL and UCB.

DISCUSSION

Epstein: Why did you choose, in modelling the short-term variations, a time scale of less than 0.5 hours -- because of a priori information on the intrinsic nature of the sources or simply to have a time scale much less than your total observation time of 9 hours?

Boldt: Since the averages considered here are for bins of half-hour duration, we must deal with two regimes of temporal variability, viz., $\tau < 1/2$ hour and $\tau \gtrsim 1/2$ hour. The variance to be associated with the shorter regime is singled out since it represents the minimal effort to be expected among scales of variability less than the total 9-hour exposure.

Masson: In your models of the effect of quasar variability, what form did you take for the variations and what fraction of the total flux was assumed to vary?

Boldt: A source is considered variable when the time-average of (S^2) is large compared with $(<S>)^2$.

Peebles: If the N(S) slope is not steep enough, most of the variance of the X-ray flux comes from the nearest quasar. Does this happen?

Boldt: We assume that the slope of the log N-log S relation is steeper than that for an index of 2, such as the index 2.3 discussed for quasars by Schmidt (this symposium) and Setti and Woltjer (Astronomy and Astrophysics, 1979). In such a situation, the variance comes from sources close to the lowest S (i.e., the most distant sources).

THE OPTICAL AND X-RAY LUMINOSITY (MULTIPLICITY) FUNCTION OF GALAXY SYSTEMS.

Neta A. Bahcall
Princeton University Observatory
Princeton, New Jersey 08540

Abstract: The luminosity function of galaxy systems – from single galaxies and small groups to rich clusters - is determined. I first determine the rich-cluster luminosity function, then a "groupings" luminosity function, which includes all systems from small groups to rich clusters, and finally, I estimate a general luminosity function of all galaxy systems (AGS), from single galaxies to rich clusters. The catalogs of Abell (rich clusters) and Turner and Gott (groups, singles) are used. The AGS luminosity function is found to be a smooth and steeply decreasing function of number density with luminosity; it spans over twelve decades of number densities in the observed five-decade range of luminosities ($\sim 10^9$ to 10^{14} L$_\odot$). The function exhibits an exponential cutoff in the rich-cluster luminosity domain, a power law dependence over most of the intermediate luminosity range, and a flattening of slope below the characteristic galactic luminosity, L*. A smooth transition between the groups and rich-clusters occurs near $\sim 10^{13}$L$_\odot$. The groupings LF (i.e., groups to rich-clusters) can be represented well by a Schechter-type analytic form $\eta(L) = \eta_0 (L/L_0)^{-\alpha}$ exp $(-L/L_0)$, where $\eta(L)$ is number of systems per unit volume per unit luminosity. The best-fit parameters are $\alpha = 2.0 \pm 0.1$, $L_0 = 1.0 \pm 0.2 \times 10^{13}$ L$_\odot$, and a normalization $\eta_0 = 1.6 \times 10^{-7}$ systems Mpc^{-3} $(10^{12}$L$_\odot)^{-1}$. The "break" at $L_0 \simeq 1 \times 10^{13}L_\odot$ occurs at the typical luminosity of rich Abell-type clusters. The slope α is found to depend on the density enhancement factor with which the groups are selected, but the dependence is relatively weak. The exponential decay at the bright end (rich clusters) has an equivalent power-law slope of approximately -5, i.e., $\eta(L) \propto L^{-5}$. The functional form of the LF is consistent with the Press-Schechter (1974) model of galaxy formation, and with the recent Silk-White (1978) model (predicted slope of -2.0). When single galaxies are added to the groupings luminosity function, the luminosity function of all galaxy systems (AGS) can be roughly described as $\eta(L) = 1.6 \times 10^{-7}$ $(L/L_0)^{-2}e^{-L/L_0}$ $(1 + 1.6 \times 10^{10}/L)^{-0.75}Mpc^{-3}(10^{12}L_\odot)^{-1}$ with $L_0 = 1 \times 10^{13}$L$_\odot$. This function approaches Schechter's field galaxy luminosity function at low luminosities, and the groupings function at high luminosities. Comparisons of the LF with the LF of galaxies and X-ray clusters of galaxies are discussed, and an estimate of $\mathcal{L}(B(0)) \simeq 10^8L_\odot$(B)Mpc^{-3} is given for

G.O. Abell and P. J. E. Peebles (eds.), Objects of High Redshift, 229–230.

the observed luminosity density in the universe.

The above luminosity function of galaxy systems could be compared with high redshift cluster data, when available, and thus shed light on possible evolutionary processes.

The X-ray luminosity function of clusters of galaxies is calculated using the general LF of optical clusters above and a relation between X-ray and optical luminosity based on a thermal bremmstrahlung model. The derived X-ray LF applies to clusters whose X-ray emission is mostly due to thermal bremmstrahlung from a smooth, hot intracluster gas. The main features predicted by the LF are a change of slope in the log η_x -log L_x plane at $L_x \sim 10^{44}$ ergs/s, a moderately steep slope of -2.5 at high luminosities ($\sim 10^{44}$-10^{46} ergs/s), and a flatter slope of ~ -1.3 below 10^{44} ergs/s (down to $\sim 10^{42}$ or 10^{43} ergs/s, where the smooth thermal bremmstrahlung may not dominate the total X-ray emission). Simple analytic expressions of the LF are given for various X-ray energy bands. X-ray data from Uhuru and Ariel-5, available for the 10^{44}-10^{46} ergs/s luminosity range (including upper limits in the 10^{45}-10^{46} range), agree well with the LF calculated here. The present calculations suggest that at least to $L_x \sim 10^{46}$ ergs/s the X-ray LF does <u>not</u> decay faster than $\sim L_x^{-2.5}$ (for a wide enough energy band). It is also suggested that fewer faint X-ray clusters than previously expected may exist (and be detected by the Einstein satellite).

DISCUSSION

Jaffe: The breaks in the slope of your optical luminosity function occur at the points where you changed catalogs (Abell to Gott-Turner; Gott-Turner to single galaxies). Do you think that the different selection criteria in different catalogs cause the change in slope, or did the changes in slope provide natural limits where the catalog compilers stopped counting?

Bahcall: No. The flattening in the luminosity function near $L_0 \sim 1 - 2 \times 10^{13}$ $L_\odot$ occurs even when the Abell catalog is used by itself, independent of any group catalog. The flattening at the fainter end occurs without any change of catalogs; both the groups and the "single" galaxies used are the entire complete sample given by the same Turner-Gott catalog. The uncertainties at the faint end, however, are large, but the agreement with the standard field galaxy luminosity function is apparent.

Perrenod: Your optical cluster luminosity function has an exponential tail and your X-ray luminosity-optical luminosity relation is a power law. Shouldn't your X-ray luminosity function have an exponential to a weak power law in it?

Bahcall: Yes. If expressed as an exponential $\eta_x(L_x)$ would roughly satisfy $\sim e_x^{-1/3}$ at the bright end. This is a very weak exponential, roughly equivalent to the $\sim L_x^{-2.5}$ power law for $L_x \lesssim 5 \times 10^{45}$ ergs/s.

EVOLUTION OF THE CLUSTER X-RAY LUMINOSITY FUNCTION

Stephen C. Perrenod
Kitt Peak National Observatory

ABSTRACT

   I predict the evolution of the X-ray luminosity function of
clusters of galaxies. Predominantly, I treat the assumption that
galaxies form first, then cluster purely due to gravitation. I show
that the richness distribution of Abell clusters favors this scenario,
rather than the protocluster hypothesis. The luminosity function is
produced by combining a generalized (for all Ω) Press-Schechter
evolutionary mass function for clusters (derived herein) with a power
law X-ray luminosity-mass relation; a power law relation is supported
by observations of low-redshift clusters.

   I find very steep evolution in the luminosity function, and thus
in the source counts, for large Ω, and moderate evolution for small Ω.
For a variety of models for the gas supply rate to the intracluster
medium, the evolution of the luminosity function does not vary greatly.
Thus it appears that the Ω dependence will dominate and that number
counts of X-ray clusters will yield cosmological information. The
power of a test of Ω with an evolving luminosity function is considerably enhanced relative to a test which involves solely global
cosmological effects on a non-evolving population. This occurs
because of the well-known result that, at late times, clustering
tends to proceed slowly for universes of small Ω and rapidly for
large Ω.

Submitted to The Astrophysical Journal.

G.O. Abell and P. J. E. Peebles (eds.), Objects of High Redshift, 231.
Copyright © 1980 by the IAU.

DISCRETE SOURCE SYNTHESIS OF THE X-RAY BACKGROUND

A. Cavaliere, L. Danese, G. De Zotti
and A. Franceschini
Istituto di Astronomia - Padova

The recent deep X-ray surveys suggest that discrete sources comprise most, if not all, of the energy content of the background (cf. Giacconi, this volume). We have shown that, if sources have flat power law spectra and relatively sharp high-energy cutoffs, their combined emission can also mimic very accurately its extended (2 - 400 keV) spectral shape. A broad distribution of cutoff energies E_c is required, in this case, to model the high-energy ($E \gtrsim 20 - 40$ keV) part: a power law envelope of distribution functions of various types of sources can be envisaged; alternatively, the well-known fact that most of the 2 - 10 keV background should be produced by low flux, low E_c sources can suggest that $E_c \propto (1 + z)^{-\eta}$. The quality of the fit turns out to be not very sensitive to the amount of number/luminosity evolution assumed, though the minimum χ^2 test slightly favours strongly evolving sources. In the case of differential luminosity $\propto E^{-\gamma}$ and evolution $\propto (1 + z)^{\alpha}$ with $0 \lesssim \alpha \lesssim 6$, the best fits to pre HEAO-1 data are obtained for mean values of the spectral index $\bar{\gamma} \simeq 0.5 - 0.9$ and for dispersions $\Delta\gamma \simeq 0.5 - 0.7$. Rather wide ranges of values of $\bar{\gamma}$ and $\Delta\gamma$ are, however, still allowed; e.g., for $\alpha = 6$, the allowed intervals are $0.2 \lesssim \bar{\gamma} \lesssim 1.3$, $0 \lesssim \Delta\gamma \lesssim 0.7$.

On the other hand, we could not obtain a satisfactory fit using source spectra similar to that of the background; indeed, the smearing effect of source redshift distribution leads, at intermediate energies, to a combined spectrum smoother than observed even if extreme evolution and no dispersion of source spectral properties are assumed.

Preliminary checks indicate that the HEAO-1 data essentially confirm our conclusions: only minor changes of the best fit values of the parameters seem to be required. Of course, these much more precise measurements would strongly narrow the allowed ranges.

Complementary information is expected from deep source counts and from measurements of the energy spectrum of fluctuations at E > 40 keV.

SOME IMPLICATIONS OF ULTRAVIOLET OBSERVATIONS OF QUASARS AND ACTIVE
GALAXIES

Arthur F. Davidsen[*]
Department of Physics, The Johns Hopkins University,
Baltimore, Maryland 21218, U.S.A.

ABSTRACT

The problem of the order of magnitude discrepancy in the expected
and observed ratios of the Lyman and Balmer lines in quasars and active
galaxies is reviewed. Whereas early photoionization models for the
emission line regions predicted $F(L\alpha)/F(H\beta) \gtrsim 40$, the observations give
values for this ratio in the range 3-8. Attempts at explaining the
observations have involved dust, both external and internal to the
emission line regions, and improved treatments of the collisional processes and radiative transfer effects in dense ($N_e \sim 10^{10}$ cm^{-3}), optically thick clouds. None of the effects considered is, by itself, able
to explain all the observations, and a combination of several of them
is probably required.

The most significant result which has so far come from ultraviolet
observations of quasars and active nuclei is the realization that the
intensity ratios of ultraviolet and optical lines are not at all what
was expected a few years ago. In attempting to construct photoionization models for quasar emission line regions, most authors (e.g.
Davidson 1972, MacAlpine 1972) assumed $F(L\alpha)/F(H\beta) = 40$, based on
ordinary radiative recombination theory with a small additional contribution of collisionally excited $L\alpha$. However, a compilation of photographic data on equivalent widths by Chan and Burbidge (1975) suggested
$F(L\alpha)/F(H\beta) \sim 18$, while spectrophotometric data by Baldwin (1977)
yielded $F(L\alpha)/F(H\beta) \sim 3$ for a composite of 26 QSO's of various redshifts.
Reddening by dust was an obvious candidate to explain the order of magnitude discrepancy in the predicted and observed $L\alpha/H\beta$ ratio, but
Baldwin argued that the absence or weakness of the 2175 Å feature in
several quasars indicated the continua were not generally reddened.
Steep Balmer decrements in quasars and Seyfert nuclei had also often
been ascribed to reddening, but observations of the $P\alpha/H\beta$ ratio in
3C 273 by Grasdalen (1976) apparently ruled out reddening of the emission
line region for at least that one object.

[*]Alfred P. Sloan Foundation Research Fellow.

G.O. Abell and P. J. E. Peebles (eds.), Objects of High Redshift, 235–245.
Copyright © 1980 by the IAU.

Baldwin (1977) also pointed out that it was not at all clear that a depressed Lα flux was responsible for the discrepant Lα/Hβ ratio. Indeed, the general absence of Lyman continuum absorption in high redshift quasars was taken to indicate that the optically thick emission line clouds surrounding the central continuum source covered only a small fraction of the sky, of order 10 per cent, and the typical Lα equivalent widths observed were consistent with this only if Lα had not been substantially reduced by dust or any other means. Baldwin was lead to suggest that perhaps the Balmer lines have instead been substantially enhanced by collisions or radiative transfer effects. Alternatively it was possible that the composite QSO spectrum was meaningless because high and low redshift QSOs were intrinsically different. Baldwin concluded by recommending that simultaneous uv and optical observations of Lyman and Balmer lines in quasars and Seyfert galaxies should be made.

Preparations for such observations with a rocket-borne telescope had been underway in our group since Spring of 1976, and in April 1977 the Faint Object Telescope (FOT) obtained the first far-ultraviolet spectrum of the quasar 3C 273 (Davidsen, Hartig and Fastie 1977). The order of magnitude low Lα/Hβ ratio was independently discovered in this observation and also inferred at about the same time by Wu (1977), from broad band uv observations of 3C 273 with the ANS satellite. The ANS observations did not extend to wavelengths as short as Lα, but yielded an estimate of the C IV flux, from which F(Lα) could be inferred. Wu (1977) interpreted the ANS results in terms of reddening of the lines and continuum in 3C 273.

The absolute fluxes reported by Davidsen, Hartig, and Fastie (1977) have been revised, due mainly to the discovery of a previously unknown telemetry system deadtime correction during preparations for a second launch of the FOT (Hartig 1978). The corrected observed Lα flux is $F(Lα) = 12.3(\pm2.0) \times 10^{-12}$ erg cm^{-2} s^{-1}. The Lα flux in 3C 273 has also been measured with IUE, (Boksenberg *et al.* 1978, Boggess *et al.* 1979) with the result that $F(Lα) = 11.5 \times 10^{-12}$ erg cm^{-2} s^{-1} in the observed frame (Boggess *et al.*). Thus there is excellent agreement between the IUE and FOT observations of Lα in 3C 273. Using the recent optical spectrophotometry reported by Boggess *et al.*, which gives $F(Hβ) = 2.1 \times 10^{-12}$ erg cm^{-2} s^{-1} in the observed frame, we have $F(Lα)/F(Hβ) = 5.7$, uncertain by perhaps ±25 per cent. Note that differences among various measures of the Hβ flux are as large as those for Lα.

The continuum flux reported for 3C 273 by Davidsen, Hartig and Fastie (1977) must also be revised, for the same reason mentioned above, with the result that $F_ν(\lambda_{obs} = 1400$ Å$) = 1.2(\pm0.2) \times 10^{-25}$ erg cm^{-2} s^{-1} Hz^{-1}. This is also in excellent agreement with the results of Boggess *et al.* (1979) using IUE. Extrapolating to the Lyman limit as $ν^{-α}$, with $α = 0.43$ (Boggess *et al.*) and converting to the emitted frame ($H_0 = 50$ km s^{-1}, $q_0 = 1$), we find $L_ν = 1.0 \times 10^{31}$ erg s^{-1}Hz^{-1} for the luminosity at 1 Rydberg. This result includes no correction for any possible reddening of the continuum. Boggess *et al.* detected no 2175 Å feature, either in the rest frame of 3C 273 or of our Galaxy, with a limit $E_{B-V} \leq 0.06$ for the extinction curve of Code *et al.* (1976). The corresponding upper limit on the factor by which $F(L\alpha)/F(H\beta)$ could be reduced (assuming the same reddening limit applies to lines as well as continuum) is 1.4. Taken together with the $P\alpha$ observations (Grasdalen 1976, Puetter *et al.* 1978) it is clear that reddening by intervening dust cannot explain the uv/optical line ratios in 3C 273 unless the properties of the dust are quite different from that in our Galaxy.

Shortly after the first Lyman α observations of quasars in the ultraviolet were made, infrared astronomers were able to measure redshifted $H\alpha$ in the spectra of high redshift objects, where $L\alpha$ was observed in the optical. Ratios $F(L\alpha)/F(H\alpha) = 1.7 \pm 0.6$ for PKS 0237-23 (Hyland, Becklin and Neugebauer 1978), ~ 0.8 for B2 1225 + 317 (Soifer *et al.* 1979) and ~ 0.34 for the same object (Puetter, Smith and Willner 1979) were found. Using typical Balmer decrements for quasars, the corresponding values of $F(L\alpha)/F(H\beta)$ are ~ 3-8, consistent with that observed for 3C 273.

Subsequently the ultraviolet spectra of a large number of galaxies and quasars have been observed with IUE. Objects for which uv fluxes have appeared in the literature are listed in Table 1, where I have given, in addition to the $L\alpha/H\beta$ ratio, the ratios He II(1640)/He II (4686) and C IV(1550)/$H\beta$. Many additional objects will be added from continuing observations with IUE (see the papers by M. Penston and R. Green in

TABLE 1. UV/optical Line Ratios in Quasars and Active Galaxies

Object	$L\alpha/H\beta$	He II(1640)/He II(4686)	C IV/$H\beta$
3C 273	5.7	--	2.0
PG 0026+129	6	--	--
NGC 4151	2.7-4.4	1.3-2.0	2-4
NGC 1068	--	3-4	4.1
3C 120	4.8	2.3	9.7
Mrk 79	2.2	24[1]	2.1
3C 390.3 broad lines	3.1	--	1.9
narrow lines	35	$\lesssim 10$	14
Composite QSO[2]	2.8	1.5	1.2

notes: [1]Only a weak narrow He II $\lambda 4686$ feature was found by Oke and Lauer (1979), but the spectrum in this region is affected by strong Fe II blends.

[2]Compiled by Shuder and MacAlpine (1979).

this Symposium). With one exception (the narrow line component in
3C 390.3 [Ferland *et al.* 1979], discussed below) the observations all
indicate that the Lα/Hβ intensity ratio is about an order of magnitude
smaller than expected from simple models for the emission line regions
in quasars and active nuclei.

Efforts to understand the small uv to optical line intensity ratios
may be divided into two general classes: (1) those which have considered
various collisional and radiative transfer effects in a purely gaseous
emission line cloud, and (2) those which have considered the possible
effects of dust on the line ratios. The latter class may be further
divided into those where the dust is (a) internal, affecting the ionization structure and thermal balance of the cloud as well as attenuating
resonance lines which are multiply scattered before escaping, and
(b) external, reducing the observed intensity of all uv lines and continuum relative to the optical by an extinction law more or less similar
to that produced by dust in our own Galaxy.

The earliest extensive discussion of the Lα/Hβ problem is by Krolik
and McKee (1978), who considered models with $N_e \geq 10^{10}$ cm^{-3} and found
that collisional de-excitation of n = 2, together with large optical
depths, could potentially explain the Lα/Hβ ratio, the steep Balmer
decrement, and the Pα/Hβ ratio in 3C 273. A limitation of the calculation, however, was that it was essentially a homogeneous model, employ-
ing a mean escape probability calculation for the line strengths. Subsequently Ferland and Netzer (1979) employed Monte Carlo calculations,
which take into account the optical depth where each photon is created,
and found that for $N_e \leq 10^{10}$ cm^{-3} the overall destruction of Lα is not
substantial. They showed that much of the Lα is produced by recombina-
tions at relatively small optical depth where it can escape easily.
Only the collisionally excited Lα, created deeper in the cloud, was
effectively destroyed. Since observations of C III] λ1909 in quasars
indicate that $N_e \sim 10^{10}$ cm^{-3}, Ferland and Netzer concluded that collisional and radiative transfer effects in a purely gaseous nebula
cannot reduce the Lα/Hβ intensity ratio substantially below its recombination value of ∼35. Instead they suggest that dust must be invoked
to explain the low observed ratio. London (1979) also suggested that
dust was the most promising mechanism for reducing F(Lα)/F(Hβ).

The effect of internal dust, mixed with the emission line gas in
quasars, has been considered in some detail by Baldwin and Netzer (1978),
Ferland and Netzer (1979), and Shuder and MacAlpine (1979). Internal
dust can, of course, be very effective at destroying resonantly trapped
Lα (and to a lesser extent C IV λ1550), but if one tries to reduce
F(Lα)/F(Hβ) to the observed value in this way, other line ratios
present problems. For example, He II λ1640 becomes too strong relative
to Lα. Baldwin and Netzer (1978) cite the observed He II λ1640/Lα
ratio, which is about a factor 2 larger than expected in many quasars,
as evidence that a small amount of internal dust is attenuating Lα.
However, Ferland and Netzer and Shuder and MacAlpine demonstrate that
the amount of internal dust permitted by this and other line ratios is

too small (perhaps 10 per cent of the galactic dust-to-gas ratio) to
explain the observed Lα/Hβ ratio. They suggest that, in addition to
internal dust, there must be external dust, either surrounding the
quasar or somewhere along the line of sight.

The case for extinction in quasars and Seyfert nuclei by external
dust has been made by Netzer and Davidson (1979). They suggest that
the ratio of the He II recombination lines λλ1640, 4686 provides a good
measure of reddening. Since the theoretical ratio, F(1640)/F(4686) = 6.6
for n_e = 10^4 cm^{-3}, T_e = 10^4°K (Brocklehurst 1971), is not very sensitive
to density or temperature, and because neither line involves the ground
level so that the lines should be optically thin, this ratio should not
be affected much by internal destruction mechanisms. Since the observed
ratio in several cases (see Table 1) is ∿1.5-2.0, and that inferred for
a composite QSO is ∿1.5 (Shuder and MacAlpine 1979), differential extinction of a factor 3-4 between λλ1640 and 4686 may be inferred. Using
the reddening law of Code et al. (1976), this corresponds to E_{B-V} ∿
0.29-0.36. The same reddening would yield differential extinction by
a factor 5.5-8.3 between Lα and Hβ.

Netzer and Davidson have also suggested that the O I lines λλ1303,
8446 provide another good reddening indicator. Approximately equal
numbers of photons are expected in these two lines, so that
F(λ1303)/F(λ8446) ∿ 6.5. O I λ1303 is typically quite weak in high
redshift quasars and this ratio is therefore difficult to measure
accurately, but Netzer and Davidson conclude F(λ1303)/F(λ8446) ∿ 1.5
generally. The FOT observation of 3C 273 yielded a marginal detection
of O I λ1303 which gave F(λ1303)/F(Lα) = 0.06 (Hartig 1978). Oke and
Shields (1976) give F(λ8446)/F(Hβ) = 0.33, so that F(λ1303)/F(λ8446) ∿
1.0. The implied reddening is E_{B-V} ∿ 0.26. The FOT observation of
NGC 4151 (Davidsen and Hartig 1978) detected no O I λ1303 (see Table 2).
When combined with the O I λ8446 flux (Netzer and Penston 1976) the
data give F(λ1303)/F(λ8446) ≲ 1.

We see that the ratios Lα/Hβ, He II λλ1640/4686, and O I λλ1303/8446
all may be consistent with reddening by a normal extinction law with
E_{B-V} ∿ 0.3-0.4. As Netzer and Davidson point out, the extinction must
occur outside the broad emission line regions because internal dust
would affect the ratios differently. In some objects reddening of the
narrow line gas surrounding the nucleus had already been deduced from
the [S II] λλ4072, 10320 lines (Wampler 1968, 1971). In NGC 4151, for
example, this method gives E_{B-V} = 0.24 ± 0.08. Thus, perhaps there
is dust surrounding all of these objects. If this is the case, then
the continuua are also reddened, and one is lead to the conclusion that
the luminosities of quasars and Seyfert nuclei are intrinsically an
order of magnitude larger than previously thought. In many cases the
intrinsic continuum must then be rising toward higher frequencies in
the ultraviolet (Netzer and Davidson 1979). If this conclusion is
supported by further work it will surely represent a very significant
contribution of ultraviolet studies to our understanding of active
nuclei and quasars.

Let me return now to some of the more recent observations with IUE that are relevant to this problem. Lα was measured in the quasar PG 0026 + 129 by Baldwin *et al.* (1978), while Puetter *et al.* (1978) measured Hβ and Pα in the same object. The ratios are Lα/Hβ/Pα = 6:1:1.4 and are consistent with reddening by normal intervening material (Baldwin *et al.* 1978), contrary to the result for 3C 273.

Oke and Zimmerman (1979) observed the galaxies 3C 120 and Markarian 79 with IUE and found a depression at $\sim$2200 Å in both objects. Interpreting this as evidence for the 2175 Å interstellar extinction feature, they find $E_{B-V} = 0.22$ and 0.38 in Mrk 79 and 3C 120, respectively. The small observed Lα/Hβ ratios (see Table 1) are then at least partially or completely (for 3C 120) explained by the inferred reddening.

An extremely important observation relevant to this problem has recently been reported by Ferland *et al.* (1979), who observed the radio galaxy 3C 390.3 with IUE. They were able to measure separately the broad and narrow components of Lα (and some other lines) and calculate separately Lα/Hβ for the broad line gas and the more extended, lower density narrow line gas. The results are listed in Table 1. We see that the narrow line gas displays normal ratios, perhaps reddened slightly by dust within our Galaxy ($E_{B-V} \stackrel{\sim}{\sim} 0.1$), but the broad line component has the severely depressed uv to optical line ratios of the other objects which have been observed. (All the previous measurements referred to broad line gas). Thus Ferland *et al.* conclude that external reddening cannot generally explain the reduced Lα/Hβ ratios observed. They also use an energy budget analysis to argue convincingly that Lα has not been severely attenuated by internal dust, since it carries its expected fraction of the total cooling (>25 per cent) in both the broad and narrow line gas in 3C 390.3 and also in 3C 273. Ferland *et al.* conclude that it is not that Lα has not been reduced, but that the Balmer lines have been enhanced by collisional processes at $N_e \geq 10^{10}$ cm^{-3}.

The observation of NGC 4151 by the FOT (Davidsen and Hartig 1978) is also relevant to this problem. The spectrum obtained is shown in Figure 1 and fluxes are given in Table 2, where they are compared with the results of preliminary IUE observations (Boksenberg *et al.* 1978). Because of the small redshift of NGC 4151, Lα is contaminated by geocoronal emission, for which a substantial correction has been made in the fluxes quoted in Table 2. Lα may also be affected by interstellar absorption within our Galaxy. The Lα/Hβ ratio is 2.7 ± 0.8 (FOT) and 4.4 ± 2 (IUE, Boksenberg *et al.* 1978). The He II λ1640/λ4686 ratio is 1.3–2.0 when the total line strengths are compared, but the situation here is similar to that for 3C 390.3. The λ4686 profile consists mostly ($\sim$80 per cent) of a very broad component with FWHM $\sim$6000 km s^{-1} (Osterbrock and Koski 1976) along with a narrow component having FWHM $\sim$ 450 km s^{-1}. Figure 2 shows the observed profile of He II λ1640, which is much narrower than the broad component of λ4686. (The feature longward of λ1640 is identified as O III] λ1663, while the rise toward shorter λ is the beginning of the very broad wings of C IV.) The

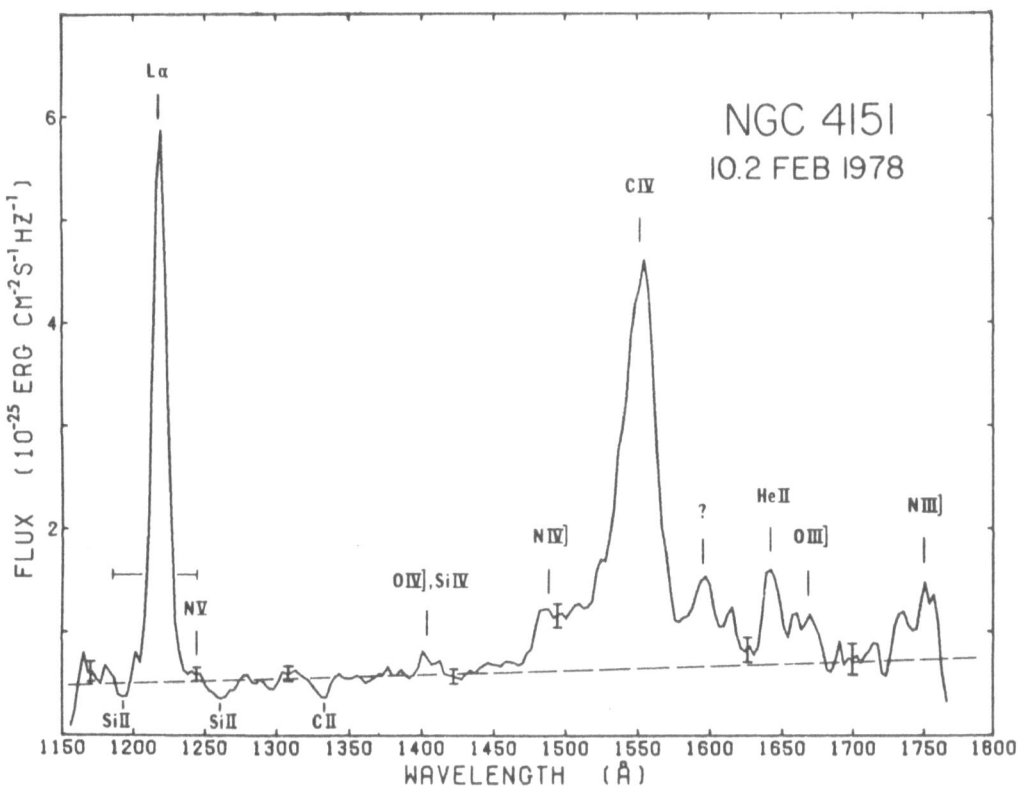

Figure 1.   Far-ultraviolet spectrum of NGC 4151 as observed with the
FOT.   The Lα profile has been corrected for geocoronal emission
extending across the width of the horizontal bar.   The dashed line
shows the assumed continuum level.

uv/optical ratios must therefore be very different for the broad and
narrow line gas, just as Ferland *et al*. (1979) found in 3C 390.3.   A
best fit for the He II λ1640 profile yields F(1640)/F(4686) $\approx$ 5 for
the narrow line gas, consistent with only a small amount of reddening
E_{B-V} $\lesssim$ 0.15 (Hartig 1978).   The broad line ratio is much smaller
but undetermined.   External reddening therefore does not appear capable
of explaining the uv/optical line ratios in NGC 4151.

Two very recent papers (Canfield and Puetter 1979, Kwan and
Krolik 1979) have reported improved photoionization calculations for
purely gaseous emission line clouds, incorporating excitation and
ionization from excited states of hydrogen.   In these calculations it
is found that at large optical depths within the cloud the n=2 level
is strongly populated, and both photoionizations and collisional
ionizations from this level are important.   The ionized fraction is
therefore maintained at a very significant level deep within the cloud,

and the Balmer lines become the major coolant. The resulting Lyman/
Balmer line ratios agree well with the observations, without recourse
to dust. They also show that Lα is not reduced in these models, but
rather the Balmer lines are substantially enhanced. The results are
consistent with the conclusions of Baldwin (1977), Puetter, Smith
and Willner (1979), and Ferland et al. (1979). It still remains to
be seen whether other uv/optical line ratios, e.g. those of O I and
He II, can also be explained without resorting to dust.

It is clear that ultraviolet spectrophotometry of quasars and
Seyfert galaxies is having a major impact on efforts to understand
their emission line regions. It is also clear that the problems are
still not resolved. On the one hand there is some direct evidence
(Oke and Zimmerman 1979) that reddening has significantly affected
the uv/optical flux ratio in two cases, while on the other hand there
is also direct evidence that reddening alone cannot explain the observed effects in two other cases (Ferland et al. 1979, Hartig 1978).
Also there is not yet agreement on whether the hydrogen line problem
is telling us that Lα is too weak or that the Balmer lines are too
strong. Several authors (e.g. Baldwin 1977, Peutter, Smith, and
Willner 1979) have argued that the covering factor ε (the per cent of
the sky which appears opaque to the central source at the Lyman limit)
in quasars is < 0.1 and that Lα therefore has the correct equivalent
width while the Balmer lines are enhanced. However, Boggess et al.
(1979) have argued that ε ∿ 1/3 and that the Balmer lines have the
correct equivalent width while Lα is severely attenuated. A much
better determination of the value of ε is certainly needed. For high

TABLE 2. Comparison of Observations of NGC 4151 by IUE and FOT

Identification	λ_o (Å)	Flux(10^{-12} erg cm^{-2} s^{-1})	
		FOT	IUE
Lα	1216	14(±3)	29(±13)
N V	1240	< 0.4	(abs)
O I	1303	< 0.4	(abs)
O IV]+Si IV	1400	0.5	(abs)
N IV	1486	$\stackrel{\scriptstyle\sim}{\scriptstyle\cdot}$ 1.8	0.7
C IV	1550	12–20	24
He II	1640	1.4	2.3
O III]	1663	0.7	0.9
F_ν (continuum) (10^{-26} erg cm^{-2} Hz^{-1})	1450	6.3(±1.6)	16

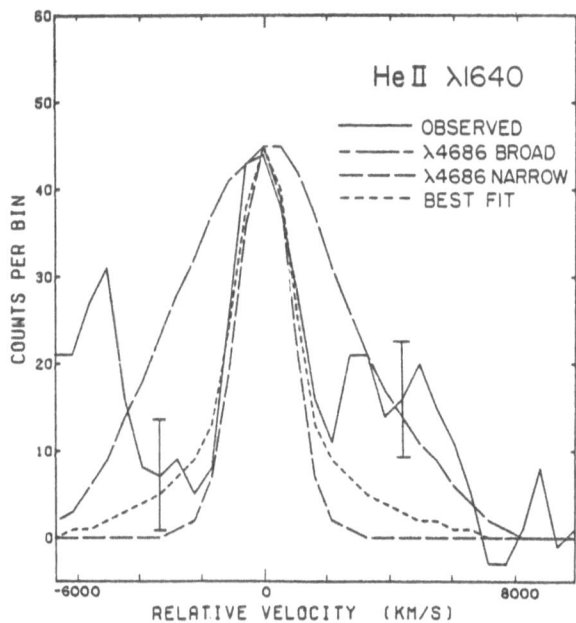

Figure 2. Line profile of He II λ1640 in NGC 4151, with continuum subtracted. The observed profile is much narrower than the broad He II λ 4686 profile, implying that F(1640)/F(4686) is larger for the narrow line gas than for the broad line gas.

redshift quasars observations at the Lyman limit can be done from the ground, but for low redshift quasars they require IUE and eventually the Space Telescope. For Seyfert galaxies the redshifts are too small, so that Lyman limit observations to determine ε cannot be performed with any existing or currently planned instrumentation.

Another interesting problem which the ultraviolet observations have raised concerns the absence of absorption or emission in the uv resonance lines of Fe II which were expected from optical observations of Fe II in the spectra of many Seyferts and some quasars (Phillips 1978, Boggess et al. 1979, Oke and Zimmerman 1979). This problem is closely related to that of the covering factor, since the most likely excitation mechanism for the Fe II lines is resonance fluoresence, requiring a very efficient conversion of uv continuum photons to Fe II line emission (Phillips 1978).

In conclusion, it seems likely that at least a small amount of reddening affects the spectra of many, if not all, active galaxies and quasars. Perhaps some internal dust is also involved in some cases. In addition, collisional processes and radiative transfer effects in the dense clouds in these objects produce significant modifications of the line intensities expected in simple models. In order to unravel all these effects accurate ultraviolet (and optical) spectrophotometry is required. Improvements in sensitivity are needed to measure weak lines, and improved resolution is needed to clearly separate the broad and narrow components of emission in objects with Seyfert 1 type spectra. The needed improvements will be available with the instrument complement planned for Space Telescope.

Research in ultraviolet astronomy at The Johns Hopkins University is supported by NASA grant NGR 21-001-001. I also acknowledge the support of the Alfred P. Sloan Foundation.

REFERENCES

Baldwin, J. A.: 1977, Mon. Not. R. astr. Soc. 178, pp. 67P-74P.
Baldwin, J., and Netzer, H.: 1978, Ap. J. 226, pp. 1-20.
Baldwin, J. A. et al.: 1978, Ap. J. 226, pp. L57-L59.
Boggess, A., et al.: 1979, Ap. J. (Letters) 230, pp. L131-L136.
Boksenberg, A., et al.: 1978, Nature 275, pp. 404-414.
Brocklehurst, M.: 1971, Mon. Not. R. astr. Soc. 153, pp. 471-490.
Canfield, R. C., and Puetter, R. C.: 1979, preprint.
Chan, Y.-W.T., and Burbidge, E. M.: 1975, Ap. J. 198, pp. 45-55.
Code, A. D. et al.: 1976, Ap. J. 203, pp. 417-434.
Davidsen, A. F., and Hartig, G. F.: 1978, Proc. COSPAR/IAU Symposium on
    X-Ray Astronomy, Eds. W. A. Baity and L. E. Peterson, Pergamon, N. Y.
Davidsen, A. F., Hartig, G. F., and Fastie, W. G.: 1977, Nature 269,
    pp. 203-206.
Davidson, K.: 1972, Ap. J. 171, pp. 213-231.
Ferland, G.. and Netzer, H.: 1979, Ap. J. 229, pp. 274-290.
Ferland, G. J. et al.: 1979, Mon. Not. R. astr. Soc., pp. 65P-71P.
Grasdalen, G. L.: 1976, Ap. J. (Letters) 208, pp. L11-L12.
Hartig, G. F.: 1978, Ph.D. thesis, The Johns Hopkins University, Balto.Md.
Hyland, A. R., Becklin, E. E., and Neugebauer, G.: 1978, Ap. J. (Letters)
    220, pp. L73-L75.
Krolik, J., and McKee, C.: 1978, Ap. J. Suppl. 37, pp. 459-483.
Kwan, J., and Krolik, J.: 1979, preprint.
London, R.: 1979, Ap. J. 228, pp. 8-12.
MacAlpine, G. M.: 1972, Ap. J. 175, pp. 11-30.
Netzer, H., and Davidson, K.: 1979, Mon. Not. R. astr. Soc. 187,pp.871-882.
Netzer, H., and Penston, M.: 1976, Mon. Not. R. astr. Soc. 174,pp. 319-325.
Oke, J. B. and Lauer, T. R.: 1979, Ap. J. 230, pp. 360-372.
Oke, J. B., and Zimmerman, B.: 1979, Ap. J. 231, pp. L13-L17.
Osterbrock, D. E., and Koski, A. T.: 1976, Mon. Not. R. astr. Soc. 176.
    pp. 61P-66P.
Peutter, R. C. et al.: 1978, Ap. J. (Letters) 226, pp. L53-L56.
Peutter, R. C., Smith, H. E., and Willner, S. P.: 1979, Ap. J. (Letters)
    227, pp. L5-L7.
Phillips, M. M.: 1978, Ap. J. 226, pp. 736-752.
Shuder, J. M., and MacAlpine, G. M.: 1979, Ap. J. 230, pp. 348-359.
Soiffer, B. T. et al.: 1979, Ap. J. (Letters) 227, pp. L1-L3.
Wampler, E. J.: 1968, Ap. J. (Letters) 154, pp. L53-L56.
Wampler, E. J.: 1971, Ap. J. 164, pp. 1-9.
Wu, C.-C.: 1977, Ap. J. (Letters) 217, pp. L117-L120.

DISCUSSION

Smith: An observation that may be relevant is the data on polarization in NGC 4151, which shows that the narrow emission line components are unpolarized, while the broad components are polarized, as if the broad lines may have been scattered by dust, while the narrow lines are not.

Rieke: Unpublished infrared spectroscopy of NGC 4151 by Lebofsky, Thompson, Tokunaga, and me indicates that the broad component of Bγ (2.16 μm) is somewhat stronger than predicted from the Balmer lines and recombination ratios, whereas the narrow component is undetected, requiring that its strength follow recombination ratios more closely than the broad component does. Thus, the same general trend seen for 3C390.3 in the ultraviolet is continued into the infrared for NGC 4151.

Oke: A large group of us have made IUE, visual and infrared observations of NGC 1068, which is a type II Seyfert where all the lines have comparable breadths. There are four pairs of lines which we have measured to derive the reddening; Lα/Hβ, $\lambda1640$/$\lambda4686$ of He II, Hα/Hβ, Pα/Hβ. There is available in the literature a Brackett line measure-which gives a Brackett/Balmer ratio. The [S II] lines also have been used by Wampler to derive the reddening. Within the accuracy of the measurements, all these ratios are consistent with E (B–V) = 0.40, provided simple recombination theory is used for the permitted lines. It is probably significant that NGC 1068 has no $N_e \approx 10^{10}$ cm^{-3} region, but a maximum N_e of the order of 10^6 cm^{-3}. The abnormal Lα/Hβ ratio found in quasars and type I Seyferts probably is confined to the 10^{10} cm^{-3} gas.

Rees: The model considered by Kwan and Krolik and by Canfield and Puetter requires that there must be a substantial energy input into parts of the cloud with very large optical depth in the Lyman continuum. Is photoionization by a power-law continuum sufficient, or does this require some additional energy input (e.g., Compton recoil from hard X-rays, or fast particles)?

Davidsen: Photoionization of helium and heavier elements by the unobserved extrapolation of the power-law continuum to soft X-ray energies is all that is required to provide the heating of the deep layers of the emission line clouds.

EXTRAGALACTIC WORK WITH IUE

M.V. Penston
Astronomy Division, ESTEC, Villafranca Satellite Tracking
Station, European Space Agency, Apartado 54065, Madrid, Spain

1. INTRODUCTION

The International Ultraviolet Explorer (IUE) is an 18 inch (45 cm)
space telescope for ultraviolet spectrophotometry in geosynchronous
orbit and is a joint project of the European Space Agency, the American
National Aeronautics and Space Administration and the British Science
Research Council. (Boggess et al. 1978 ab). It is a small instrument
to use for objects of truly high redshift and therefore IUE observations
relevant to extragalactic astronomy in general will be covered here.

2. NORMAL GALAXIES

The general form of the ultraviolet energy distributions from normal
galaxies had already been foreshadowed before the flight of IUE by the
OAO-2 data (Code & Welsh 1979). These are generally confirmed by IUE
(e.g. Boksenberg 1978, Johnson 1979, Bertola et al. 1979, Benvenuti
et al. 1979, Fosbury et al. 1979). In the wavelength range between
2000-3000 Å, galaxy spectra are basically those of the late-type stellar
content of the galaxy and form a natural and expected extension of the
optical spectrum. Various programmes are being pursued with IUE on
normal galaxies in this spectral region to establish better K-corrections and to reduce the degree of non-uniqueness inherent in population
synthesis work.

By contrast there is a less expected result shortward of 2000 Å
where a rising component (rising $F\lambda$ with decreasing λ) is seen. In
most cases this excess is probably due to a population of hot stars.
It is also seen in globular clusters with blue horizontal branches
(Dupree et al. 1979) and can very reasonably be attributed either to
the stars in these clusters on or above the horizontal branch. In
addition the excess from the nucleus of M87 has been shown to be
spatially extended in the same way as the late-type stars seen in the
optical (Bertola et al. 1979, Perola & Tarenghi 1979). On the other
hand more exotic explanations of the ultra-violet excess may be

G.O. Abell and P. J. E. Peebles (eds.), Objects of High Redshift, 247–255.
Copyright © 1980 by the IAU.

appropriate in other cases since IUE would be a sensitive way to detect
very underluminous Seyfert nuclei.

One should note of course that several of the "normal" galaxies
observed to date include objects like M87 and NGC 1052 which have some
exceptional qualities. Certainly establishing the systematics of the
ultraviolet properties of truly bog-standard galaxies is an important
project but it will be a long one given the speed of IUE.

As a footnote to this section, one should record the work of
Perola & Tarenghi (1979) on the jet of M87. The brightest knot is
detected to 1150 Å with a featureless spectrum similar to that of the
BL Lac objects.

3. BL LAC OBJECTS

Before the flight of IUE, there were two rival explanations for the
weakness of emission lines in BL Lac objects being the possible
absence of gas or alternatively the absence of an ionizing continuum.
IUE immediately demonstrated (Boksenberg et al. 1978) the extension
of the continuum in Mk421 to 1150 Å and the absence of ultraviolet
emission lines. New models soon appeared however (e.g. Krolik et al.
1978) which allowed the presence of hot gas with only OVI λ1032
emission. The only clear statement which can be made is that gas is
not present in the same physical and geometrical state as in quasars!

The IUE data show a flattening (in the f_ν/ν plane) of the spectrum
shortward of 2000 Å in the spectra of Mk 421 and Mk 501 (Boksenberg
et al. 1978, Snijders et al. 1979). This apparent flattening can be
removed by allowing a small amount of galactic reddening. In the case
of Mk 501 at least, the value of this reddening is consistent with the
galactic latitude and leaves a dereddened ultraviolet continuum of the
form $f_\nu \propto \nu^{-\alpha}$ which points back (under the optical flux distribution
due to stars) and joins up well with the high-frequency radio spectrum.

4. QUASARS

The IUE observations of the objects of largest redshift are those of
Wilson et al. (1979) who have since extended this work (Boggess et al.
1979) to detect two 17th magnitude quasars (one radio-noisy, one radio-
quiet) down to about 300 Å in the rest frame. The basic result is that
the observed ultraviolet flux fits well to the extrapolation of the
observed optical spectrum. There is a suggestion that He II λ304
emission may be seen in at least one of the quasars.

By contrast the brightest quasar, 3C273, has been well observed
in the ultraviolet both by rocket (Davidsen et al. 1977) and IUE
(Boksenberg et al. 1978, Boggess et al. 1979, Ulrich et al. 1979). The
last investigation has produced the best data since it used the

averages of several spectra at each wavelength. The most significant
result is the similarity to the spectra of higher redshift quasars.

This data is shown in Fig. 1.

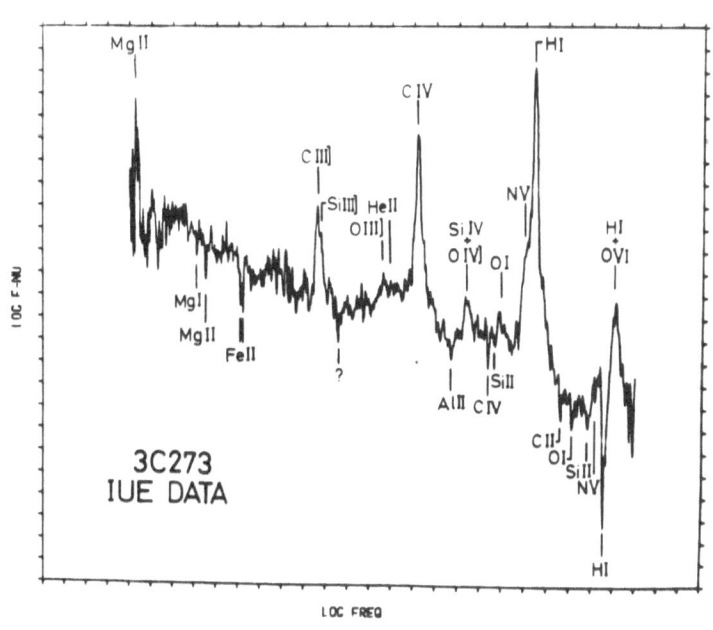

Figure 1. The mean of many IUE spectra of the quasar 3C273. Note the
emission lines shifted in frequency by the quasar's redshift, Z=0.158,
compared with the absorption lines which originate in the halo of our
own galaxy.

One of the most intriguing discoveries is the presence of absorption
lines which are at zero redshift and caused by the halo of our galaxy.
One interesting point is the excess strength of the CIV absorption
compared to its normal value in the disk of the galaxy. This is similar
to the discovery of strong CIV in the halos of our galaxy and the
Magellanic Clouds by Savage & de Boer (1979). In the strength of CIV
(and in the list of lines that are seen) the galactic halo resembles
the high-redshift absorption systems in quasars and therefore is
consistent with the origin of these features in intervening galaxies.

The previous paper (Davidsen 1979) discussed the Lα/Hβ problem
including the important IUE data of Ferland et al. (1979). The 3C273
data have been used by Ulrich et al. (1979) to undermine the conclusions
of Netzer & Davidsen (1979) that all the ultraviolet lines might be too
weak. Ulrich et al. favour a much lower continuum than earlier investigators (joining the continuum longward of CIII] to that shortward of
Lα) and this increases the measured intensities of He II λ1640 and

OI λ1304 on which Netzer & Davidson base their case. Of course such
a change in the continuum has a smaller effect on Lα and the original
Lα/Hβ anomaly remains.

Before the IUE observations of 3C273, there were two alternative
theories for the excitation of the optical lines of Fe II. Proponents
of resonance fluorescence predicted absorption in the ultraviolet lines
while those favouring collisional excitation predicted emission. In
the event (Boksenberg et al. 1978),both schools were confounded with no
apparent ultraviolet Fe II lines observed in either emission or absorption! The situation is now clearing as theoreticians (Jordan 1979,
Collin-Souffrin et al. 1979ab) postulate high optical depths in the
ultraviolet lines leading to line scattering and a "leak" of photons
from the optical lines.

A last result from 3C273 is the confirmation of the excess ultraviolet continuum first discovered by Davidsen et al. (1978). The
Ulrich et al. (1979) data suggest the presence of 2200 dust absorption
at the redshift of the quasar and lead to a stronger intrinsic excess
when reddening corrections are made. This excess is too big to be
explained by gaseous continuum emission processes like Balmer continuum
or two-photon emission. As discussed by Rees (1979) earlier in this
volume possibly thermal emission from accreting material can account
for this excess.

Other IUE observations of quasars have been reported by Baldwin
et al. (1978), Wu et al. (1979), Gondhalekar & Wilson (1979) and
Green (1979).

5. SEYFERT GALAXIES

Seyfert galaxies as observed in the ultraviolet also have spectra
which are basically similar to those of quasars as seen from the ground.
An important point however is that they seem to violate the correlation
between CIV equivalent width and luminosity proposed by Baldwin (1977)
for quasars. Basically the CIV equivalent widths for Seyferts are
similar to those for quasars but the luminosities are of course much
lower.

It is interesting to note that in some Seyfert galaxies some
ultraviolet FeII lines have now been seen. Blended emission from
ultraviolet multiplets 62 and 63 is identified in unpublished spectra
of II Zw 136, Mk 231 and NGC 1566. These decays are to low-lying
quartet states and it is still so that no resonance lines have been
detected in emission.

The brightest Type I Seyfert galaxy, NGC 4151, has been well
studied with IUE at both high (Penston et al. 1979) and low (Boksenberg
et al. 1978, Perola et al. 1979) dispersion. Prior to IUE (by one
day!) a spectrum of NGC 4151 was obtained by rocket (Davidsen & Fastie

1978). This data demonstrated both continuum and absorption line variability that have been confirmed by the later IUE data.

The high dispersion spectra of NGC 4151 are of low signal-to-noise in the continuum but the profiles of the three strong emission lines, $L\alpha$, CIV and CIII] are well defined and by contrast to the case of quasars each have clearly distinct profiles. The $L\alpha$ profile is narrow (FWZI $\sim$ 1200 km s^{-1}) and obliterated by absorption to the blue of the line centre. The $L\alpha/H\beta$ varies from 15-25 at the line centre to less than 1.5 elsewhere. The CIV is also cut by absorption in the velocity range -100 to -1100 km s^{-1} relative to the optical narrow emission lines and is broader (FWZI $\sim$ 8000 km s^{-1}). The absorption varies in depth but not velocity. The CIII] line is narrow (FWZI $\sim$ 2000 km s^{-1}) and is shifted as if 1907-to-1909 are in their low density ratio i.e. $n_e < 10^{4.5}$ cm^{-3}.

Other reported results on active galaxies include work by Oke & Zimmerman (1979) Elvius et al. (1979), Wu et al. (1979) and Clavel et al. (1979).

REFERENCES

Baldwin, J.A., 1977. *Astrophys. J.*, 214, 679.

Baldwin, J.A., Rees, M.J., Longair, M.S. & Perryman, M.A.C., 1978. *Astrophys. J.*, 226, L57.

Benvenuti, P., Casini, C. & Heidmann, J., 1979. Presented at conference "First Year of IUE", UCL, April 4-6th 1979.

Bertola, F., Capaccioli,M., Holm, A.V. & Oke, J.B., 1979. Presented at conference "First Year of IUE", UCL, April 4-6th 1979.

Boggess, A., Carr, F.A., Evans, D.C., Fischel, D., Freeman, H.R., Fueschel, C.F., Klinglesmith, D.A., Krueger, V.L., Longanecker, G.W., Moore, J.V., Pyle, E.J., Rebar, F., Sizemore, K.O., Sparks, W., Underhill, A.B., Vitagliano, H.D., West, D.K., Macchetto, F., Fitton, B., Barker, P.J., Dunford, E., Gondhalekar, P.M., Hall, J.E., Harrison, V.A.W., Oliver, M.B., Sandford, M.C.W., Vaughan, P.A., Ward, A.K., Anderson, B.E., Boksenberg, A., Coleman, C.I., Snijders, M.A.J. & Wilson, R., 1978a. *Nature (London)*, 275, 372.

Boggess, A., Bohlin, R.C., Evans, D.C., Freeman, H.R., Gull, T.R., Heap, S.R., Klinglesmith, D.A., Longanecker, G.R., Sparks, W., West, D.K., Holm, A.V., Perry, P.M., Schiffer, F.H. III, Turnrose, B.E., Wu, C.C., Lane, A.L., Linsky, J.L., Savage, B.D., Benvenuti, P., Cassatella, A., Clavel, J., Heck, A., Macchetto, F., Penston, M.V., Selvelli, P.L., Dunford, E., Gondhalekar, P., Oliver, M.B., Sandford, M.C.W., Stickland, D., Boksenberg, A.,

Coleman, C.I., Snijders, M.A.J. & Wilson, R., 1978b.
Nature (Lond.), 275, 377.

Boggess, A., Daltabuit, E., Torres-Peinbert, S., Estabrook, F.B.,
Wahlquist, H.D., Lane, A.L., Green, R., Oke, J.B., Schmidt, M.,
Zimmerman, B., Morton, D.C. & Roeder, R.C., 1979. *Astrophys. J.*,
230, L131.

Boggess, A., Gondhalekhar, P.M., Wilson, R. & Wu, C.C., 1979. Presented
at conference "First Year of IUE", UCL, April 4-6th 1979.

Boksenberg, A., Snijders, M.A.J., Wilson, R., Benvenuti, P., Clavel, J.,
Macchetto, F., Penston, M., Boggess, A., Gull, T.R., Gondhalekar,
P., Lane, A.L., Turnrose, B., Wu, C.C., Burton, W.M., Smith, A.,
Bertola, F., Capaccioli, M., Elvius, A.M., Fosbury, R., Tarenghi,
M., Ulrich, M.H., Hackney, R.L., Jordan, C., Perola, C.G., Roeder,
R.C. & Schmidt, M., 1978. *Nature (Lond.)*, 275, 404.

Clavel, J., Benvenuti, P., Cassatella, A., Heck, A., Macchetto, F.,
Penston, M.V. & Selvelli, P.L., 1979. (preprint).

Code, A.D. & Welsh, G.A., 1979. *Astrophys. J.*, 228, 95.

Collin-Souffrin, S., Joly, M., Heidmann, N. & Dumont, S., 1979a.
Astr. Astrophys., (in press).

Collin-Souffrin, S., Dumont, S., Heidmann, N. & Joly, M., 1979b.
Astr. Astrophys., (in press).

Davidsen, A.F., Hartig, G.F. & Fastie, W.G., 1977. *Nature (Lond.)*,
269, 203.

Davidsen, A.F. & Hartig, G.F., 1978. IAU/COSPAR Symposium, Innsbruck.

Davidsen, A.F., 1979. This volume, pp. 235.

Dupree, A.F., Black, J.H., Davis, R.J., Matilsky, T.A., Raymond, J.C.,
& Gursky, H., 1979. *Astrophys. J.*, 230, L89.

Elvius, A., Lind, J. & Lindegren, L., 1979. Presented at conference
"First Year of IUE", UCL, Apr. 4-6th 1979.

Ferland, G.F., Rees, M.J., Longair, M.S. & Perryman, M.A.C., 1979.
Mon. Not. R. astr. Soc., 187, 65P.

Fosbury, R.A.E., Snijders, M.A.J., Boksenberg, A. & Penston, M.V.,
1979. (preprint).

Gondhalekhar, P. & Wilson, R., 1979. Presented at conference "First
Year of IUE", UCL, April 4-6th 1979.

Green, R. 1979. This volume, pp.73.

Johnson, H.M., 1979. *Astrophys. J.*, 230, L137.

Jordan, C., 1979. *Prog. in Atomic Spectroscopy*, ed. Haule & Kleinpoppen (Plenum Publ. Corp.), p 1453.

Krolik, J.H., McKee, C.F. & Tarter, C.B., 1978. *Pittsburgh Conference on BL Lac Objects* (University of Pittsburgh), p 277.

Netzer, H. & Davidson, K., 1979. *Mon. Not. R. astr. Soc.*, 187, 871.

Oke, J.B. & Zimmerman, B., 1979. *Astrophys. J.*, 231, L13.

Penston, M.V., Clavel, J., Snijders, M.A.J., Boksenberg, A. & Fosbury, R.A.E., 1979. *Mon. Not. R. astr. Soc.*, (in press).

Perola, G.C. & Tarenghi, M., 1979. Presented at conference "First year of IUE",UCL, April 4-6th 1979.

Perola, G.C., Boksenberg, A., Bromage, G., Carswell, R., Elvius, A., Gabriel, A., Gondhalekhar, P.M., Jordan, C., Lind, J., Lindegren, L., Longair, M.S., Penston, M.V., Perryman, M.A.C., Pettini, M., Rees, M., Snijders, M.A.J., Tanzi, E.G., Tarenghi, M., Ulrich, M.H. & Wilson, R., 1979. Presented at conference "First year of IUE", UCL, April 4-6th 1979.

Rees, M., 1979. This volume, pp. 207.

Savage, B.D. & de Boer, K., 1979. *Astrophys. J.*, 230, L77.

Snijders, M.A.J., Boksenberg, A., Barr, P., Sanford, P.W. & Penston, M.V., 1979, *Mon. Not. R. astr. Soc.*, (in press).

Ulrich, M.H., Boksenberg, A., Bromage, G., Carswell, R., Elvius, A., Gabriel, A., Gondhalekhar, P.M., Lind, J., Lindegren, L., Longair, M.S., Penston, M.V., Perryman, M.A.C., Pettini, M., Perola, G.C., Rees, M., Sciama, D., Snijders, M.A.J., Tanzi, E., Tarenghi, M. & Wilson, R., 1979. Presented at conference "First year of IUE", UCL, April 4-6th 1979.

Wilson, R., Carnochan, D.J. & Gondhalekhar, P.M., 1979. *Nature (Lond.)* 277, 457.

Wu, C.C., Boggess, A. & Gull, T.R., 1979. Presented at conference "First year of IUE", UCL, April 4-6th 1979.

DISCUSSION

Gaskell: I would like to make a couple of remarks. First, I don't
think that the mere strength of the 2200 Å absorption feature
can be taken as a quantitative indicator of reddening. There are cases
in our own galaxy where $\lambda 2200$ is very much weaker or very much stronger
than expected for a given E(B - V).

Second, in a survey of the ultraviolet continua of a large
number of flat radio spectrum QSOs that I have almost finished, the continuum to the blue of Lyman-alpha is usually well-represented by an
extrapolation of the continuum of 2100 to 1300 Å, if one ignores absorption. Also, in this survey I find that most of the continua have a
power-law index (corrected for local extinction) of about 0.8, with
almost no indication of intrinsic reddening greater than E(B - V) $\sim$ 0.06

Penston: Well, to take the second point first, I must say that my
feeling on looking at the published data was that there were
several cases which did have the same kind of jump in the continuum
across Lyα as is seen in 3C 273.

On the first point, there are, of course, some special cases
where $\lambda 2200$ is anomalous, but in the vast majority of cases
in our galaxy it conforms to a standard law.

Wolfe: I was very interested in the galactic MgII absorption in
3C 273. It seems to me that you could calculate the Mg^+
column densities from a curve-of-growth study, using the 21-cm profiles.
You could then calculate the HI column density from the 21-cm emission
fluxes. In this way you can calculate the Mg/H abundance ratio in the
galactic halo. Would you comment on this?

Penston: What we have done is use the 21-cm profile and the assumption
of normal abundances and ionization to compute the equivalent
widths. These agree with those observed except in the case of the CIV,
which is observed to be too strong.

P. Veron: Are your results affected by the calibration problem recently
discovered?

Penston: Of course, there has been a problem with the intensity transfer function for the IUE short wavelength cameras, but it
does not seriously affect the data presented here.

Marscher: I'm not convinced that the presence of ultraviolet continuum
combined with the absence of detectable emission lines neces-

sarily rules out the presence of dense gas in BL Lac objects. Blandford
and Rees have suggested that the continuum radiation might be beamed
toward the observer in these sources. If the gas lies in a disk with
an axis parallel to the beam, it sees a continuum which is greatly
reddened at a very low flux level.

Penston: I am sure that there are many alternative explanations!

INFRARED OBSERVATIONS OF ELLIPTICAL GALAXIES

M. J. Lebofsky
Steward Observatory, University of Arizona, Tucson, AZ 85721

Abstract

A program to search for luminosity evolution of giant elliptical galaxies at high redshift has been begun at 2.2μm. Observing at infrared wavelengths offers the possibility of avoiding large and uncertain K corrections at redshifts near 1. First results of this program are in agreement with conclusions drawn from much larger optical studies, and demonstrate that luminosity evolution may be present.

I. INTRODUCTION

Giant elliptical galaxies have served as the standard candles in studies aimed at measuring cosmological parameters such as q_0, but whether the luminosity of these galaxies is actually constant or whether galaxies were more luminous in the past is not known. Various theoretical considerations (e. g. Tinsley 1977) indicate that galaxies should have been brighter in the past, and the observation of strong CO bands in nearby ellipticals (Frogel et al. 1978) supports this contention (Tinsley 1978); Sandage (1961) was the first to consider the effect of luminosity evolution on attempts to determine q_0. His conclusion then was very similar to that of Kristian et al. (1978): that either $q_0 \sim 1$ and galaxies do not show any luminosity evolution or $q_0 \sim 0$ and galaxies are about ~ 0.5 mag dimmer now than at 5×10^9 years ago. Many corrections are required to derive these results from optical data; some of the same corrections are required for an analysis of the infrared data but some such as the K correction (very uncertain for optical magnitudes as $z \rightarrow 1$) are more secure in the infrared.

Because of the possibility of detecting galaxy evolution and checking optical measurements, a program of measuring the brightest cluster ellipticals has been started with observations made principally at K (2.2μm). This wavelength has the advantage that the portion of the galactic energy distribution which is redshifted into

G.O. Abell and P. J. E. Peebles (eds.), Objects of High Redshift, 257–262.
Copyright © 1980 by the IAU.

the filter pass band is well-measured in nearby galaxies for redshifts
even larger than 1. Other useful features are the relative flatness
of the near-infrared energy distribution of ellipticals, the uniformity
of the near-infrared colors, and the lack of aperture dependence on
the near-infrared colors, all of which make the K corrections small
and well-determined. This paper presents the first observations at
2.2μm of 6 giant elliptical galaxies ranging in z from 0.136 to 0.947
with a preliminary analysis of the results.

II. DATA

All measurements were obtained with the Steward Observatory 2.25
meter (90 inch) telescope equipped with a liquid helium-cooled InSb
photometer. The data were obtained at K (2.2μm) and also at H (1.6μm)
for most galaxies; Table I presents the data and the apertures used.
Standard infrared chopping techniques were used but no corrections
for galaxy flux in the reference beam were made because such corrections were estimated to be less than 4% in all cases. All measurements
were made using offset guiding using position measurements derived from
Palomar plates for the galaxies and nearby stars. The V band data
used in deriving the V-K colors in Table II are from Gunn and Oke
(1975) or Kristian et al. (1978). The K corrections for the V data
were taken from Whitford (1975) through z = 0.46 and from the plots
in Code and Welch (1977) for the companion to 3C330.

Before the infrared data can be used for absolute magnitude
estimates or for cosmological analyses, several corrections need to be
considered. Since this paper is only a first examination of a small
number of measurements, only preliminary values have been derived for
some of these corrections. In particular, the energy-dependent portion
of the K correction has been derived using a smoothed galaxy energy
distribution and rectangular approximations to the filter functions.
This portion of the K correction at 2.2μm is never larger than 0.28
mag so the uncertainty introduced by using this approximate K correction is small.

Table I. Photometry of Giant Ellipticals

	Z	H (1.6μm)	K (2.2μm)	Aperture (")
UMa II, Cl.1	0.136	14.79±0.09	14.56±0.06	5.9
Hydra, G8+G9	0.202	15.43±0.08	15.21±0.08	7.8
1318.1+3157	0.270	15.37±0.08	14.60±0.06	7.8
1613+31	0.415	16.25±0.13	15.38±0.09	7.8
3C330, companion	0.532		16.22±0.13	7.8
1305+2952	0.947		$17.77^{+0.82}_{-0.46}$	7.8

The aperture correction was derived by assuming a fixed diameter
of 86 Kpc and using the 2.2μm curve-of-grow in Frogel et al. (1978)
and following a procedure similar to that outlined in Sandage (1972a).
This choice of diameter was made for ease in comparison of absolute
magnitudes with those derived by Sandage and others at V where
$< M_V > \sim -23.3$ which yields $< M_K > \sim -26.6$ (V-K $\sim$ 3.3 [Frogel et al.
1978]) for 86 Kpc, $q_0 = 1$, and $H_0 = 50$. As discussed by Kristian et
al. (1978), the aperture correction has only a small dependence on
q_0; however the conversion of linear diameter to aperture size
depends linearly on the value of H_0 (see equations in Sandage 1972a)
so the data have been reduced using two choices (50 or 100 Km/sec/Mpc)
for H_0 and two choices for q_0 (0 or 1). Since the main objective at
this time is to see whether galaxies are brighter at high redshift,
this method will provide acceptable aperture corrections, but zero-
point shifts will exist between the different choices of H_0 and q_0.
The most appropriate galaxy diameter to use will be considered more
carefully in a later paper.

The corrections usually applied for interstellar absorption
within our galaxy will not be used since $A_K \sim .1 A_V$. The corrections
for Bautz-Morgan contrast class and richness class usually applied
by Sandage and his co-workers (Sandage and Hardy 1973; Kristian et al.
1978) have not been used because it is not yet clear whether such
corrections are applicable at 2.2μm.

III. DISCUSSION

The small number of sources measured so far in this program
precludes any definitive discussion of whether luminosity evolution
has been detected. However, a brief consideration of these results
will show the potential of 2-μm measurements for studying properties

Table II. Colors and Absolute Magnitudes

	z	$(V-K)_c^*$	$(H-K)_c$	M_K q=0 H=50	q=0 H=100	q=1 H=50	q=1 H=100
UMa II, Cl.1	0.136	2.89	0.03	-25.9	-25.0	-25.8	-24.9
Hydra, G8+G9	0.202	2.37	-0.09	-25.4	-24.7	-25.5	-24.5
1318.1+3157	0.270	3.89	0.40	-26.8	-25.8	-26.6	-25.6
1613+31	0.415	3.26	0.33	-27.0	-25.9	-26.7	-25.6
3C330, companion	0.532	3.05		-26.9	-25.7	-26.4	-25.2
1305+2952	0.947			-27.3	-25.9	-26.5	-25.3

* V magnitudes from references in text.

of galaxies at high redshift. The 2-µm Hubble diagram is shown in
Figure 1 fit with a line appropriate to the q_0 = 1, H_0 = 50 case
which was chosen because the choice of 86 Kpc as the physical diameter
used in computing the aperture correction is correct only for these
choices of q_0 and H_0. The constant was calculated by minimizing
the residuals. The observed points are well-represented by a linear
fit demonstrating that the infrared data follows the same general
behavior as optical data--there are not infrared excesses or other
unexpected properties destroying the usefulness of these measurements.

The absolute magnitudes at K(2.2µm) and $(V-K)_c$ and $(H-K)_c$ colors
derived from the data are presented in Table II. The H-K colors have
an uncertainty of 0.10 mag while the V-K colors are uncertain by
0.15 mag. Since the main purpose of this initial investigation is to
determine what infrared measurements will reveal, the values in
Table II will be discussed in terms of what problems or general trends
the data show.

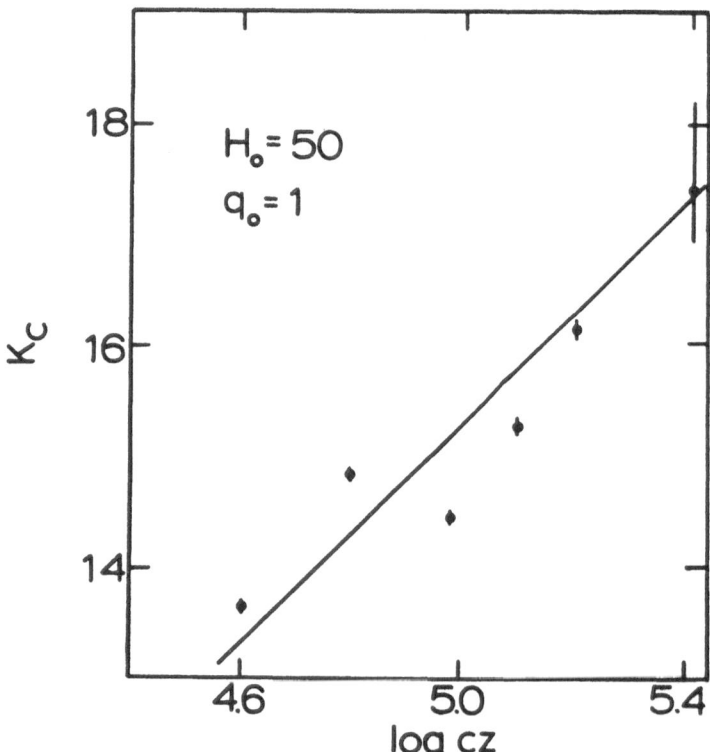

Figure 1. The infrared Hubble diagram with the
data fit by the line K_c = 5 log cz + C, C = -9.72,
(Sandage 1972b).

One problem evident from Table II is that the measurements of Hydra are anomolous--the colors are too blue and the total K flux is too small. More observation will be required to learn if this galaxy is peculiar. The other abnormal color in Table II is the very red V-K color of 1318.1+3157. The H-K color of this galaxy is redder than the normal H-K ~ 0.22 for ellipticals, but it is only 2σ away from the nominal color. One possible explanation for the red colors is absorption by intergalactic dust; this explanation is supported by the $2.2\mu m$ absolute magnitude which is close to the nominal value of -26.6 in the $H_0 = 50$ $q_0 = 1$ case. Another possibility is that the galaxy has an infrared excess although the absolute magnitude does not support this.

The surveys at optical wavelengths of cluster ellipticals (e. g. Kristian et al. 1978) have come to the conclusion that either q_0 is near 0 and galaxies were brighter in the past or that $q_0 > 1$. The absolute magnitudes in Table II support a similar conclusion from the infrared data, but a larger sample will have to be studied before any stronger statements can be made. A sample large enough to circumvent the intrinsic scatter in absolute magnitudes and colors will be needed as will a resolution of why some galaxies such as 1318.1+3157 have such red colors.

REFERENCES

Code, A. D., and Welch, G. H. 1977, preprint.

Frogel, J. A., Persson, S. E., Aaronson, M., and Matthews, K. 1978, Ap. J., 220, 75.

Gunn, J. E., and Oke, J. B. 1975, Ap. J., 195, 255.

Kristiàn, J., Sandage, A., and Westphal, J. A. 1978, Ap. J., 221, 383.

Sandage, A. 1961, Ap. J., 133, 916.

Sandage, A. 1972a, Ap. J., 173, 485.

Sandage, A. 1972b, Ap. J., 178, 1.

Sandage, A. 1975, Ap. J., 202, 563.

Sandage, A., and Hardy, E. 1973, Ap. J., 183, 743.

Tinsley, B. M. 1977, IAU Coll. No. 37, Decalages vers le Rouge et Expansion de l'Univers, p. 223.

Tinsley, B. M. 1978, Ap. J., 222, 14.

Whitford, A. E. 1975, Galaxies and the Universe, ed. A. Sandage, M. Sandage, and J. Kristian, p. 159.

DISCUSSION

Ellis: What is the scatter in the infrared colours for ellipticals
                at zero redshift?

Lebofsky: The scatter in the V–K colors at zero redshift is ~ 0.3 mag-
                nitude for giant ellipticals. The dispersion in the H–K
colors is only about 0.05 magnitude. The larger scatter in the V–K
color may not reflect a cosmic dispersion but may only reflect scatter
introduced by calculating colors from data taken with different apertures, etc.

Peebles: How long does it take you to measure the apparent magnitude
                at 2μ and redshift ~ 1? Is it feasible to count galaxies in
blank fields at 2μ at this depth?

                I might note that the absolute magnitudes you compute at
                H = 50 and 100 km s^{-1} Mpc^{-1} should differ by just 5 log 2
because H is just a scale factor.

Lebofsky: The 2.2 μ magnitude at z = 1 takes about three hours to
                measure with a 90-inch telescope. Since 1305+2952 may
be unusually blue, other galaxies at z = 1 may be measured more accurately or in less time, and the availability of larger telescopes such
as the MMT will also speed the observations. A carefully designed
experiment might be able to count galaxies at z = 1 at 2.2 μ over limited
areas of the sky.

                The computed absolute magnitudes do not differ by 5 log 2
                because the product of H$_O$ and linear diameter used in the
aperture correction was not held constant; the linear diameter was held
constant independent of H$_O$. Holding the diameter constant then produced
aperture corrections which have an unphysical dependence on H$_O$ and produced the differences you noticed in Table II. A more careful consideration of the aperture correction will use a linear diameter appropriate
to 2.2 μ observations and will not depend on H$_O$.

Schild: Can you tell us what the sky brightness is in magnitudes per
                square arc sec at 2.2 μ?

Lebofsky: The sky brightness at 2.2 μ is produced by thermal radiation
                from the sky and telescope and depends on the temperature
and the emissivity of both the sky and telescope. For a typical temperature and an emissivity appropriate to a low background telescope,
K ≈ +17 per square arc sec. For the fields of view used in this work,
this corresponds to a detector N.E.P. of about 10^{-16} watts Hz$^{-1/2}$ at
2.2 μ.

IDENTIFICATION OF INFRARED SOURCES IN "EMPTY" FIELDS

G. H. Rieke
M. J. Lebofsky
Steward Observatory, University of Arizona, Tucson, AZ 85721

Abstract

From infrared observation of a radio selected sample of QSOs, it is shown that optical identification programs produce samples biased against very red sources, even if identifications are made without regard to color.

I. INTRODUCTION

The optical identification of flat-spectrum radio sources has been pursued very actively because it appears possible to obtain a complete set of identifications for a statistically complete sample of radio sources (see, e.g., Condon, Jauncey, and Wright 1978). This possibility is suggested by the high positional accuracy that can be achieved for these objects in the radio, by the correlation between optical and radio brightnesses, by the exact positional coincidence between radio and optical counterparts, and by the average brightness of the optical identifications which is well above the Palomar sky survey limit. Nonetheless, 10 to 15 per cent of these sources cannot be identified to the limit of the Palomar survey (Condon, Hicks, and Jauncey 1977); it is frequently assumed that the unidentified objects have optical counterparts similar to the identified ones except for being relatively faint, at least at the time of the Palomar survey.

The following is an interim report on a study of the infrared properties of these sources. We have found that at least half of the unidentified sources have optical-infrared spectra distinctly different from the optically identified portion of the sample. Any identification program which works only to the limit of the Palomar survey will produce a biased sample, even if it ignores color in making identifications. By combining optical and infrared observations, it appears possible to identify at least 95% of the brighter flat-spectrum radio sources.

G.O. Abell and P. J. E. Peebles (eds.), Objects of High Redshift, 263–268.
Copyright © 1980 by the IAU.

It is likely that the identifications of steeper-spectrum radio
sources suffer from similar biases, although it will be difficult to
prove this hypothesis because the radio and optical counterparts do
not always coincide.

II. OBSERVATIONS

Measurements were obtained with a high-performance near infrared
photometer on the University of Arizona 1.54 m (61 inch) and 2.25 m
(90 inch) telescopes. The infrared beam, of diameter 8", was centered
on the radio sources by offset guiding from nearby field stars. The
positions of these stars had been measured to an accuracy of $\sim$ 0".7; an
additional $\sim$ 1" of error was introduced by uncertainties in the offset guiding. The radio sources themselves had positional errors of
$\sim$ 1" or less. Measurements were begun at K(2.2μm) and extended to
H(1.6μm) and J(1.25μm) if a detection was achieved.

Initially, we concentrated on empty fields (to the limit of the
Palomar survey). When it was clear that a substantial number of
these objects could be detected and had spectra much steeper and
redder than other types of extragalactic source (Rieke, Lebofsky, and
Kinman 1979), the survey was extended to include infrared measurements
of a complete radio sample. This sample is the sources listed by
Condon, Hicks, and Jauncey (1977) brighter than 0.6 Jy at 8 GHz and
lying in the declination range 4° < δ < 25°. To be included in the
sample, the sources must have spectral indices α < 0.5 between 2.7
and 5 GHz (where the spectrum is represented by S = $\nu^{-\alpha}$). Of a total
of 69 sources in the sample, 23 have been observed in the infrared, or
1/3. Roughly 1/3 of each class of identification (EF, BSO, NSO, RSO)
has been included. The K magnitudes, spectral indices (B to K), and
identification class are listed in Table 1.

III. DISCUSSION

The spectra of the sources have been characterized by their
spectral indices, α_{KB}, assuming they are power laws between K and
B(0.44μm). The B magnitudes given by Condon, Hicks, and Jauncey (1977)
were assumed, although for about half of the identifications it was
verified with a TV acquisition system that the objects had not varied
significantly (more than $\sim$ 0.5 mag.). For the empty fields, the B
magnitudes were estimated from Rieke, Lebofsky, and Kinman (1979) or
by extrapolating from the J-K color.

The distribution of α_{KB} for this sample is compared with that for
other QSOs in figure 1. The optically brighter objects, including
those that are part of the radio sample, have α near 1, whereas the
fainter objects have α ranging fairly uniformly from 0 to 3, with a
greatly increased number of objects with large α and possibly an
increased number with very small α.

Table 1.   New Measurements of Flat Spectrum Radio Sources

Name	ID	K	α_{KB}
0722+145	BSO	15.08±0.15	1.1
0745+241	NSO*	14.08±0.06	1.7
0748+126	BSO	13.85±0.09	1.2
0754+100	NSO	11.86	1.5
0759+183	BSO	>16.7	<0.5
0829+046	BSO	12.27	1.6
0952+179	BSO	15.05±0.13	0.2
1014+208	BSO	>16.1	<0.8
1155+169	BSO	13.91±0.07	1.2
1257+145	BSO	>15.8	<0.1
1402+044	R(G)[+]	>16.3	<0.1
1427+109	BSO	16.7±0.4	∿−0.1
1434+235	BSO	16.4±0.4	∿0.1
1614+051	BSO	16.3±0.4	∿0.7
1656+053	BSO	14.04±0.08	1.1
1756+237	BSO	14.90±0.08	0.6
2149+056	EF	15.51±0.16	>2.0

* Listed by Condon, Hicks, and Jauncey (1977) as BSO;
  remeasurement of Palomar plate indicates NSO

[+] Z = 3.2

The increase of α with increasing m_B is not a selection effect.
We have observed six more objects with $m_B \geq 20$ that meet all the
criteria for inclusion in the radio sample except that they fall
outside the 4-25° declination range. From the total of 11 sources
with appropriate radio properties, 5 have been detected and shown to
have large values of α. It is therefore clear that a large percentage
of the optically very faint objects have steep, red spectra.

Two lines of evidence indicate a general similarity between the
very red sources and the rest of the sample. First, in common with
many of the optically brighter sources, at least two of the red
sources are variable on timescales of a month or less (Rieke, Lebofsky,
and Kinman 1979). Second, the ratio of 2-μm flux to radio flux is
similar for these objects to the average ratio for the whole sample,
corresponding to a spectral index of $\alpha = 0.64$. In comparison, the
average radio-optical spectral index is $\alpha = 0.71$ (Condon, Jauncey,
and Wright 1978). The red sources do not appear to have any
distinguishing properties in the radio.

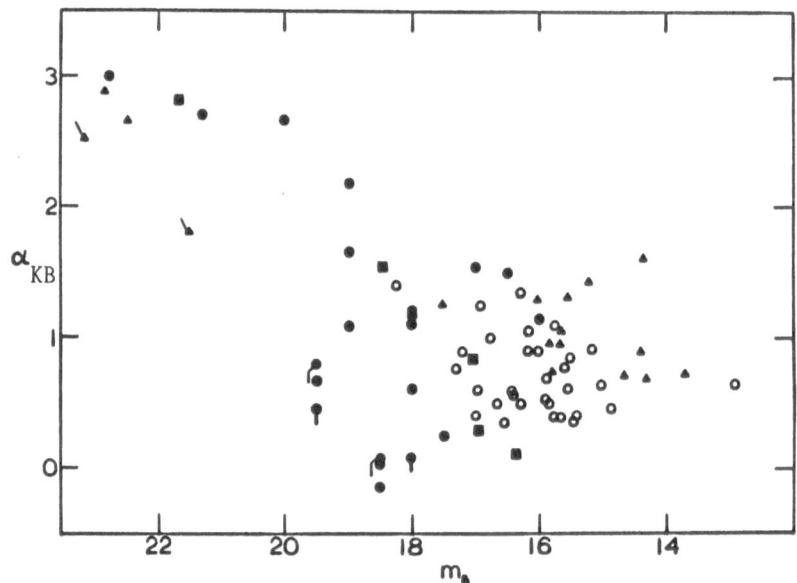

Figure 1. α_{KB} (2μm-to-blue power law slope) vs. m_B.
Filled dots represent radio-selected QSOs from this
study. Open dots are optically selected QSOs from
Neugebauer et al. (1979); squares are from the same
reference, but with Z > 1.0. Triangles are from other
work (see references summarized by Rieke and Lebofsky
1979).

The explanation for the trend of α_{KB} with m_B is probably that the
observed correlation between radio and optical fluxes (Condon, Jauncey,
and Wright 1978) is a manifestation of a proportionality of the luminosities in the radio and the ultraviolet-optical-infrared spectral
regions. For sources with UV-O-IR spectral slopes near 1, m_B is a
reliable indicator of the luminosity. Sources with slopes near B
substantially different from one will have larger UV-O-IR luminosities
than indicated by m_B alone and therefore will be discriminated against
in optically selected samples. The possible increase in the percentage
of sources with very small α_{KB} near m_B = 18 is also consistent with
this hypothesis. Some of these sources have extremely blue colors,
with B-R ∿ -0.4 to -0.8 estimated from the Palomar plates.

Of the radio sources listed by Condon, Hicks, and Jauncey (1977)
and Condon, Jauncey, and Wright (1978) that meet all the criteria for
inclusion in our sample except for declination range, 89% have optical
identifications. With allowance for sources missed because of
variability and sensitivity limitations in the infrared, our success
rate in detecting the empty fields indicates that optical-infrared

counterparts can be found for 95% or more of these objects. The goal of a virtually complete set of identifications can therefore be met, but until it is, it must be recognized that the optical identifications represent a biased sample.

REFERENCES

Condon, J. J., Hicks, P. D., and Jauncey, D. L. 1977, A. J., 82, 692.

Condon, J. J., Jauncey, D. L., and Wright, A. E. 1978, A. J., 83, 1036.

Neugebauer, G., Oke, J. B., Becklin, E. E., and Matthews, K. 1979, Ap. J., 230, 79.

Rieke, G. H., Lebofsky, M. J. 1979, Ann. Rev. Astron. Ap., in press.

Rieke, G. H., Lebofsky, M. J., and Kinman, T. D. 1979, Ap. J. (Letters), in press.

DISCUSSION

Schild: At Mt. Hopkins we have been observing the Einstein Observatory deep fields with our CCD camera. In two fields which were blank on the Palomar Sky Survey and on a deep IIIa-J plate taken with the Palomar 48-inch Schmidt by Sargent, we found faint stellar sources with very red (R-I) color indices. If these are M stars they would have an unexpectedly high ratio of X-ray to optical luminosity. We think it is more likely that they are the same kind of very red quasar that Dr. Rieke has found. This suggests that X-ray as well as radar surveys turn up more red quasars than do the optical searches.

Smith: Perhaps you need not be so pessimistic about spectroscopy: Spinrad and I have been observing QSO identifications at the faint end. About 10% of 3C QSOs are quire red ($1.5 \leq \alpha_{opt} \leq 4.5$). Although the ionizing continuum at CIV may be down by a factor of 10 (or more) with respect to hydrogen, the line strengths of these objects are relatively normal.

Epstein: You said that the 2-micron magnitudes vary by a factor of as large as 2 or 3 on a time scale of about one month. Do any other objects have 2-micron variability this large?

Rieke: BL Lac-type sources have been observed to vary in the infrared by large amounts in a few days.

Murdoch: We can add one more steep optical spectrum QSO (with a flat spectrum) to the Smith and Spinrad sample. The optical spectral index is 3.5, and with a redshift of 1.71, Lyα is in the observation window, but was only marginally detected due to dispersion of the UV image. The continuum slope of 3.5 corrected for dispersion extends to Lyα.

Rieke: That is undoubtedly an interesting QSO. However, before
            classifying it with the ones found in the infrared, we need
to measure it there. All of the QSOs discovered so far in the optical
that have red optical continua have a change in spectral slope near
one micron, and a much flatter continuum in the infrared. These sources
were discovered only because their continua allow their energy to emerge
predominantly near the optical despite their red colors, so that the
selection effects I have discussed are reduced.

Wolfe: Are these objects too faint to measure infrared polarization?
            If not, the detection of significant polarization would put
them in the BL Lac class, and thus help you to classify these objects.

Rieke: The sources are a little too faint for easy infrared polarime-
            try, although if we find one just slightly brighter we cer-
tainly plan to try.

INFRARED OBSERVATIONS OF GIANT ELLIPTICAL GALAXIES: (V-K) COLORS AND THE INFRARED HUBBLE DIAGRAM

Gary L. Grasdalen
Kitt Peak National Observatory and Department of Physics and
Astronomy, University of Wyoming

ABSTRACT

The (V-K) colors of giant elliptical galaxies as a function of
redshift are discussed. Present data are consistent with mild color
evolution at $z \sim 0.45$. An infrared Hubble (redshift-magnitude) diagram
is given. Cosmological models with $q_0 = 0$ and no luminosity evolution
are clearly excluded by the present data. A wide variety of models
including those with $q_0 = 0$ are permissible if luminosity evolution is
included. Instrumental and programmatic implications of these results
are summarized.

I. INTRODUCTION

In this contribution I will summarize the results of an infrared
program aimed at directly observing the color and luminosity evolution
predicted by big bang cosmological models for the stellar content of
galaxies. In searches for cosmological tests an enormous amount of
effort has already been expended in making optical observations of the
first ranked elliptical galaxies in rich clusters of galaxies (e.g.
Sandage 1975; Kristian, Sandage and Westphal 1978). This work has
clearly demonstrated that these galaxies are highly uniform in their
absolute visible luminosity as well as their energy distributions.
Furthermore these early type galaxies contain little evidence for
recent star formation, a fact which leads to the expectation that luminosity and color evolutionary effects will be both predictable and
small for these systems. Since these galaxies are also the most
luminous stellar systems in the universe they are logical candidates
for use in delineating the relation between time and velocity that
applies for the universe. The time scale of course is measured backwards from the present epoch by estimating the distance or light travel
time to the galaxy from its apparent brightness. In applying this
cosmological test it is crucial to accurately assess the past luminosity
of the selected class of galaxies.

G.O. Abell and P. J. E. Peebles (eds.), Objects of High Redshift, 269–277.
Copyright © 1980 by the IAU.

II.   (V-K) COLORS OF GALAXIES

The most direct test of the correctness of the predicted luminosity
evolution due to stellar evolution is the expected change in the color
of the galaxy.   In Figure 1 we have plotted the energy distribution for
a giant elliptical galaxy derived by Whitford (1971) for the optical
spectrum and derived from the broadband infrared colors given by Frogel
et al. (1978).   Examination of this figure reveals that as a function of
increasing redshift the observed $V(0.55\mu)$ or $R(0.70\mu)$ magnitudes refer
to successively fainter portions of the energy distribution, while the
$K(2.2\mu)$ magnitude will measure brighter portions of the energy distribution.   Thus in attempting to observe color changes in elliptical
galaxies it does not seem very profitable to use observed wavelengths
shortward of the V magnitude band.   The effect of the blanketing break
near 0.4μ is catastrophic on the observed brightness, making observations shortward of the break exceedingly difficult.   The (V-K) or (R-K)
colors appear far more attractive choices.   They also have a long
baseline in wavelength which should enhance the evolutionary effects
on the color.

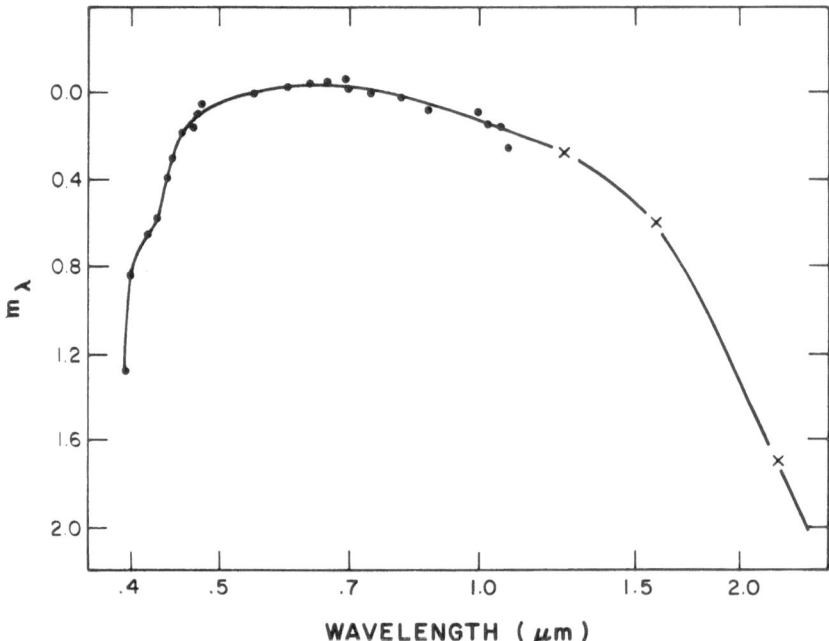

Figure 1.   Energy distribution for a giant elliptical
galaxy.

A program to measure the (V-K) colors of giant elliptical galaxies
was undertaken at Kitt Peak National Observatory in 1974.   Preliminary
and erroneous results from that program have been previously reported

(Grasdalen 1975, Frogel et al. 1975). The difficulty in those results
arose from inadequate Fabry action in the implementation of the new
InSb detectors and hence erroneous values for the beam size. (Rieke
and Lebofsky 1979). The observations reported here were either made
with more sophisticated systems or were rereduced with the corrected
beam sizes. Except for a few nearby galaxies the V magnitudes were not
remeasured but were taken from the photometry given by Sandage (1972b
& c, 1973b). The V magnitudes were generally interpolated, occasionally
extrapolated to the diaphragm size used to make the K observations using
the growth curve given by Sandage (1972a). Galactic extinction was
removed following the prescription given by Sandage (1973a) for A_V as a
function of galactic latitude and then assuming that $E(V-K) = 0.9\ A_V$
(Sneden et al. 1978).

     The results from this program are plotted in Figure 2. The solid
curve in this figure is the expected (V-K) color as a function of redshift based on the energy distribution in Figure 1. For redshifts below
0.2 the data are in moderately good agreement with expectations.
Unfortunately there are no galaxies in the redshift range 0.2 to 0.4.
The two galaxies between $z = 0.4$ and 0.5, 3C 295 and 0024+16 have (V-K)
colors slightly bluer than would be expected on the basis of the
observed local energy distribution. The effect is not particularly
large, only 0.25 and 0.5 mag. This is however the appropriate sign and
approximate size for the expected color change (Tinsley and Gunn 1976).

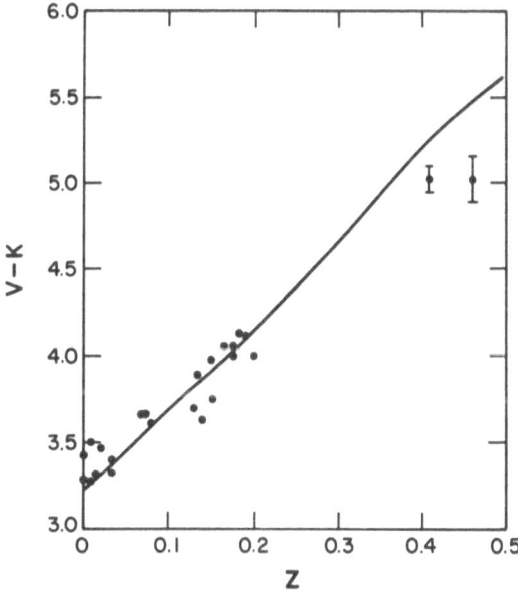

Figure 2. Observed and predicted (V-K) colors of elliptical
galaxies.

An obvious worry is that some spurious effect is mimicking the effect expected on evolutionary grounds. At redshifts beyond 0.4 the V magnitude refers to an emitted wavelength less than 0.4μ. This makes the observed V magnitude very sensitive to the metal content. The rather rare blue stars can also begin to influence this spectral region; especially worrisome in this regard is the possible influence of differences in horizontal branch morphology.

III. THE INFRARED HUBBLE DIAGRAM

The difficulties besetting the interpretation of the (V-K) colors illustrate why the optical portion of the spectrum as well as being observationally challenging is not a very attractive wavelength to observe early type galaxies at high redshift. The most directly interpretable data will be obtained at rest wavelengths beyond 0.4μ. Ground based observers are of course plagued by the very strong OH bands present in the night sky beyond 0.7μ. The severe effect of these bands can be seen by examining the spectrophotometric scans presented by Spinrad et al. (1976). Because of the enormous improvements in infrared detectors in recent years, the wide and dark atmospheric window at 2.2μ is the most attractive wavelength for ground based observations of distant galaxies.

To this end, as redshifts become available for more distant first ranked cluster galaxies (Spinrad 1975, Spinrad et al. 1976, Smith et al. 1979), infrared (2.2μ) observations were obtained. The relative ease with which photometric observations could be made prompted the thought that infrared photometry could be carried out on galaxies that are not visible at optical wavelengths at all. To test this idea I chose three 3CR radio sources which 1) had good radio interferometric observations that showed them to be classical double lobed radio sources (Pooley and Henbest 1974, Longair 1975), and 2) had been searched optically with deep photographs (Kristian, Sandage and Katem 1974, Longair and Gunn 1975). The 4-meter telescope was then used to make photometric observations at 2.2μ at a point midway between the two radio lobes. The beam size for these observations was 6.5". The linear intensity results and the final 2.2μ magnitudes for 3CR 427.1, 65 and 68.2 are given in Table 1. Clearly 2.2μ light was detected for 3CR 427.1 and 65. Simultaneously, H. Spinrad (this symposium) had succeeded in obtaining photographs of 3CR 427.1 and after many nights of spectroscopic integration has proposed a redshift of 1.11. Currently the galaxy associated with 3CR 65 has not been detected optically (Smith, Burbidge and Spinrad 1976).

In Figure 3 the available infrared data has been plotted as a Hubble diagram. The mode of plotting the data has been somewhat modified from past diagrams; only corrections for aperture size and galactic reddening have been applied to the data to produce K_{SM} the 2.2μ magnitude at a standard metric diameter. (q_o = 0 has been assumed in making that correction.) The position of 3CR 65 has been indicated

as a vertical line and 3CR 68.2 as an upper limit. In comparing the
data with the cosmological and evolutionary models, corrections for the
variable rest wavelength of the observations have been incorporated
into the predicted relations.

Table 1.   2.2μ Results for Double Lobed Sources

Source	Linear Reading σ	Integration Time 10^3 sec	2.2μ mag
3CR 427.1	11.6 ± 1.8	6.4	
	16.8 ± 2.6	3.2	
	10.7 ± 2.3	3.2	
mean	12.7 ± 1.3	12.8	16.4 ± 0.1
3CR 65	6.5 ± 2.8	6.4	
	7.7 ± 1.9	6.4	
	8.0 ± 2.4	6.4	
mean	7.4 ± 1.4	19.2	17.0 ± 0.2
3CR 68.2	-2.0 ± 4.2	3.2	
	2.2 ± 1.9	6.4	
	0.6 ± 2.0	6.4	
mean	0.6 ± 1.4	16.0	>17.5 (3σ)

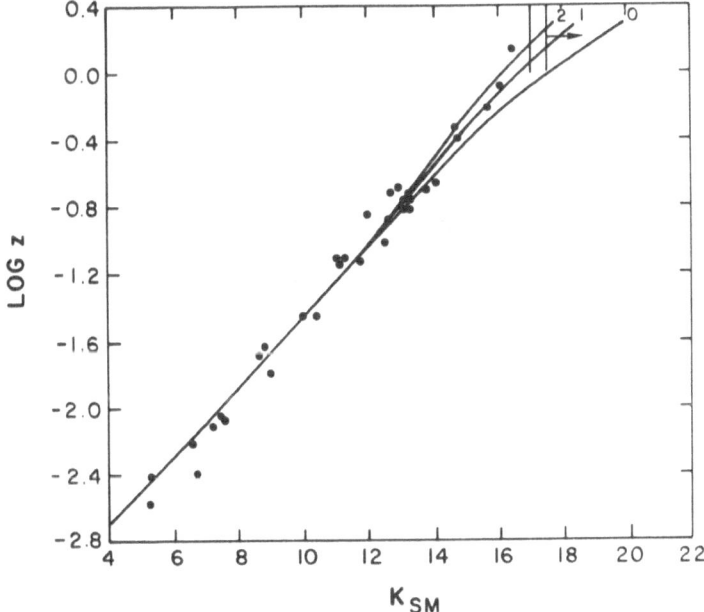

Figure 3.   The observed infrared Hubble diagram. Cosmological models with q_o = 0,1,2 and no evolutionary corrections
have been superposed.

The predictions, also plotted in Figure 3, were made from cosmological models without any evolutionary corrections. The curves
correspond to three values of the deceleration parameter $q_0 = 0, 1, 2$.
Examination of Figure 3 leaves no doubt that models with $q_0 = 0$ and
no evolution are ruled out by the present data. If evolutionary
corrections are admitted the number of adjustable parameters becomes
large enough to fit the data with any value of q_0 larger than or equal
to zero. The evolutionary luminosity calculations were taken from
Tinsley and Gunn (1976). In Figure 4 we show the Hubble diagram with
a particular set of $q_0 = 0$, evolutionary predictions superposed. Here
the parameter, X, used by Gunn and Tinsley to describe the main sequence
mass function has been set equal to zero. This produces the fastest
luminosity evolution, but is still consistent with the observed properties of nearby elliptical galaxies.

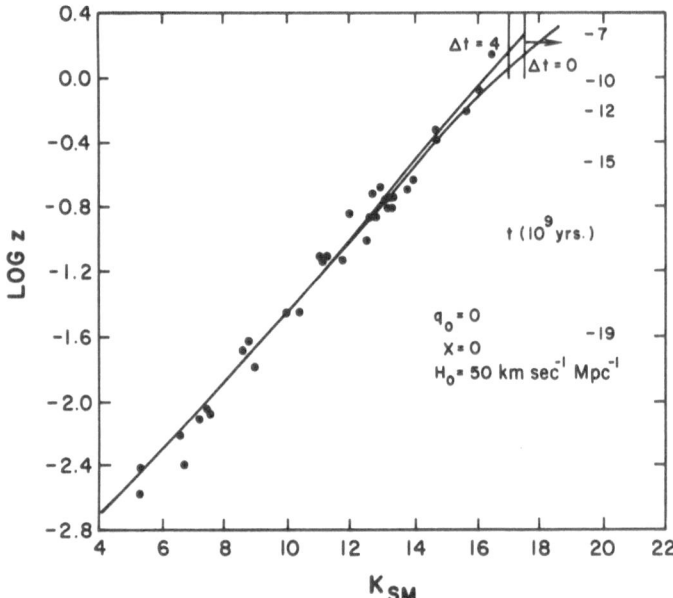

Figure 4.   The observed infrared diagram: The parameter Δt
is the time, in units of 10^9 years, between the big bang and
the epoch of galaxy formation.

The curves have been drawn for two assumptions; $\Delta t = 0$, galaxies
were formed immediately after the beginning of the expansion, and
$\Delta t = 4$, galaxies were not formed until 4×10^9 years had elapsed since
the beginning of the expansion. If we accept that quasars are at the
nuclei of galaxies we cannot postpone the formation of galaxies beyond
this epoch since quasars are already observed at that epoch. Clearly
however, evolutionary models can be constructed that are compatible
with the data.

IV. CONCLUSIONS

For redshifts greater than ~0.7 it appears unlikely that ground based observations can succeed in making the sample adequately large. This fact arises from the faintness of galaxies in the ultra-violet portion of the spectrum. The distances involved are not especially large. If new redshifts are to be obtained they must come from space telescope observations of the blanketing break redshifted into the photoelectric infrared or from the rest frame ultra-violet features. Space telescope will be required to avoid the strong OH emission in the night sky in the first case and is required in the second case in order that very small angular apertures can be employed.

Our expectation that photoelectric and near infrared photometry give the most directly interpretable information on the luminosity evolution of elliptical galaxies is directly confirmed by the present data.

These considerations lead me to conclude that the development of high efficiency-low noise detectors in the wavelength range 0.7-2.4μ is a necessary requirement for the further progress of galaxian observational cosmology.

Our observations demonstrate that the remaining optically blank 3CR fields correspond to distant radio galaxies.

The presently available infrared redshift-magnitude diagram is consistent with cosmological models for which $q_o = 0$ and only the effects of stellar evolution are included in the luminosity evolution of elliptical galaxies.

Finally we must caution that in order to accurately separate the effects of geometry and evolution it may be necessary to observe more rapidly evolving systems (i.e. late type galaxies), another task that will require the spatial resolution of Space Telescope.

REFERENCES

Frogel, J. A., Persson, S. E., Aaronson, M., Becklin, E. E., Matthews, K. and Neugebauer, G.: 1975, Ap. J. 200, p. L123.
Frogel, J. A., Persson, S. E., Aaronson, M. and Matthews, K.: 1978, Ap. J. 220, p. 75.
Grasdalen, G. L.: 1975, Ap. J. 195, p. 605.
Kristian, J., Sandage, A. and Katem, B. N.: 1974, Ap. J. 191, p. 43.
Kristian, J., Sandage, A. and Westphal, J. A.: 1978, Ap. J. 221, p. 383.
Longair, M. S.: 1975, M.N.R.A.S. 173, p. 309.
Longair, M. S. and Gunn, J. E.: 1975, M.N.R.A.S. 170, p. 121.
Pooley, G. G. and Henbest, S. N.: 1974, M.N.R.A.S. 169, p. 477.
Rieke, G. H. and Lebofsky, M. J.: 1980, Ann. Rev. Ast. and Astrophys. 17, p. 477.

Sandage, A.: 1972a, Ap. J. 173, p. 485.
Sandage, A.: 1972b, Ap. J. 178, p. 1.
Sandage, A.: 1972c, Ap. J. 178, p. 25.
Sandage, A.: 1973a, Ap. J. 183, p. 711.
Sandage, A.: 1973b, Ap. J. 183, p. 731.
Sandage, A.: 1975, Ap. J. 202, p. 563.
Smith, H. E., Burbidge, E. M. and Spinrad, H.: 1976, Ap. J. 210, p. 627.
Smith, H. E., Junkkarinen, V. T., Spinrad, H., Grueff, G. and Vigotti,
    M.: 1979, Ap. J. 231, p. 307.
Sneden, C., Gehrz, R. D., Hackwell, J. A., York, D. G. and Snow, T. P.:
    1978, Ap. J. 223, p. 168.
Spinrad, H.: 1975, Ap. J. 199, p. L3.
Spinrad, H., Liebert, J., Smith, H. E. and Hunstead, R.: 1976, Ap. J.
    206, p. L79.
Tinsley, B. M. and Gunn, J. E.: 1976, Ap. J. 203, p. 52.
Whitford, A. E.: 1971, Ap. J. 169, p. 209.

DISCUSSION

P. Veron: Why do you believe that the objects you have detected at
                the positions of 3CR radio sources are galaxies when
Rieke calls objects with similar infrared properties found at the positions of flat spectrum radiosources QSOs?

Grasdalen: The radio sources are quite different. These are double-
                lobed sources without a central source. In the case of
427.1 the observed color from 1.25 μ to 2.2 μ is consistent with a redshifted giant elliptical galaxy.

M. Burbidge: If one has to wait for the second generation of instru-
                ments on the Space Telescope to get deep red or near
infrared response, perhaps one should pin hopes on forthcoming large
ground-based telescopes (10-m class, or the Next Generation Telescope
in the 25-m class).

Grasdalen: Yes, for the wavelengths around 2 microns where the air
                glow is rather small. However, for the photoelectric
infrared (0.7 - 1.2 μ), the strong OH emission lines in the atmosphere
make the Space Telescope much more attractive than the large ground-
based telescopes of the future.

Peebles: Which are the stars that contribute most of the light at
                1 to 2 μ? Are the evolution corrections as touchy in the
infrared as the visible?

Grasdalen: The late giants dominate the infrared light. This has
                been demonstrated by Frogel and his collaborators, by
showing that the CO band at 2.4 μ is quite strong in the spectra of
giant ellipticals. Thus, the evolutionary corrections are as sensitive
as the rest frame V-band corrections.

Scheuer: Could you and Dr. Rieke and Dr. Longair between you say
                 how many of Longair's complete sample of 166 3CR sources
still remain unidentified after the IR work you have just described?

Grasdalen: This question was answered by Longair.

Longair: Of the four 3CR radio sources which lie in empty fields
                 in the statistical sample of 166 sources, two are 3C 65
and 68.1.   3C 65 has been studied with excellent plate material on at
least three occasions and no identification has been found.   If we
accept Dr. Grasdalen's infrared object as an "optical" identification
(and I see no reason not to), there are only three blank fields in the
sample of 166.

LOW RESOLUTION INFRARED SPECTRA OF QUASARS

B. T. Soifer, G. Neugebauer, J. B. Oke and K. Matthews
Hale Observatories* and California Institute of Technology

Abstract

Low resolution spectra of a significant sample of quasars show
that the Paschen α and Balmer line ratios do not agree with the radiative recombination case B result and vary widely within the quasars
sampled. The range in Pα:Hβ ratios is a factor of ∼ 6, while the range
in Lyα:Hα ratios is a factor of ∼ 5. For the Pα:Balmer series, the
deviations from case B recombination are not consistent with reddening,
but appear, within large dispersions, to be consistent with optical
depth effects in the Balmer lines affecting the line ratios. The
Lyα:Hα ratio is, however, correlated with the continuum spectral index,
and can be explained as due to reddening affecting both the lines and
continuum.

* * *

In this paper we summarize recent observational results based on
a joint infrared/optical survey of the hydrogen line spectra of a
significant number of the brightest low and high redshift quasars
(Soifer et al., 1980). This survey includes 12 quasars in the redshift range $0.07 \leq z < 0.30$, where the Paschen α line of hydrogen is
redshifted into the 2.2 μm atmospheric window, and seven quasars with
$z > 1.5$, where Hα and/or Hβ is redshifted into the 1.65 μm or 2.2 μm
atmospheric windows.

For the low redshift quasars, the equivalent width of Pα in the
quasar rest frame varies from ∼ 200 Å to less than 30 Å. The large
variation in Pα flux is reflected in a similar large variation in the
ratio Pα/Hβ which is found to vary from less than 0.15 to ∼ 0.9. The
intensity ratios Pα:Hβ:Hα are found to deviate significantly from the
nominal case B recombination theory result; the scatter between quasars
is large. Although in the case of the Seyfert 2 galaxy NGC 1068
the case B recombination ratios combined with a normal extinction
law can explain the observed line ratios including Lyα (Neugebauer et
al., 1980), the observations of quasars do not appear to be consistent

---

*Operated jointly by the Carnegie Institution of Washington and the
California Institute of Technology.

G.O. Abell and P. J. E. Peebles (eds.), Objects of High Redshift, 279–282.

with this simple result. Rather the results for quasars appear to be more consistent with the deviations from case B predicted due to effects of large optical depth in the Balmer lines (Netzer, 1975; Krolik and McKee, 1978).

In addition, there appears to be no correlation in the low red-shift quasars between the line ratio Pα:Hβ and the continuum spectral index measured at these wavelengths. This results seems to rule out external reddening as a dominant factor in determining the line ratios and is consistent with the result found previously by Neugebauer et al. (1979) from considerations of the lack of correlation between Hα:Hβ, and the continuum slope.

Hα and/or Hβ has been observed at infrared wavelengths in a total of seven high redshift quasars by the Caltech group. The range of equivalent width of Hα is $\sim$ 250 Å to $\sim$ 650 Å in the quasar rest frame, with an average of 420 Å; no attempt has been made to remove any contribution from the blended N II lines 6548+6584 (the results of Baldwin (1975) suggest these are a negligible contribution compared to the uncertainties in the measurements). This agrees remarkably well with the range and average equivalent widths (440 Å) found in the low red-shift quasars.

The Lyα/Hα line ratio has been calculated for each of these quasars and is found to vary from 1 for Ton 490 to 5 in the case of 1623+26. The average value is Lyα/Hα $\sim$ 2. This result is consistent with the previously reported deviation of the Lyα/Hα ratio by a factor of $\sim$ 10 from that expected for simple quasar models. No account was taken of the contribution of N V to the Ly + N V total flux, but this is expected to be only $\sim$ 20% of the total measured flux (Baldwin, 1977).

The ratio of Lyα:Hα flux may correlate with the continuum spectral index in high redshift quasars. The apparent correlation is in the sense that the decreasing Lyα/Hα ratio implies a more negative spectral index α (where $F_\nu \propto \nu^\alpha$). The low redshift quasars 3C273 and PG 0026+129 appear to obey this same relation. The slope of this correlation agrees remarkably well with the predictions for a simple model of external reddening affecting both the lines and the continuum in the same amount. Explanations of the discordant quasar line ratios as due to this effect have been put forth by Netzer and Davidsen (1979), and Shuder and MacAlpine (1979), but many problems may exist with these models. To name just two obvious ones: (1) Why do the Balmer and Paschen series lines not show any evidence for reddening? and (2) Where is all of the energy coming out? In the case of 3C273 there is no evidence of any excess energy emergant, yet the simple model predicts 3C273 is at least 10 times more luminous than observed. Clearly the apparently conflicting observations indicate that the resolution of this problem is not a simple one.

This work was supported by NSF grant AST77-20516 and NASA grant NGL 05-002-207.

References

Baldwin, J. A.: 1975, Astrophys. J. 201, p. 26.
Baldwin, J. A.: 1977, Mon. Not. R. astr. Soc. 178, p. 67p.
Krolik, J., and McKee, C.: 1978, Astrophys. J. Suppl. 37, p. 459.
Netzer, H.: 1975, Mon. Not. R. astr. Soc. 171, p. 395.
Neugebauer, G., Oke, J. B., Becklin, E. E., and Matthews, K.: 1979,
    Astrophys. J. 230, p. 79.
Neugebauer, G., et al.: 1980, in preparation.
Shuder, J. M., and MacAlpine, G. M.: 1979, Astrophys. J. 230, p. 348.
Soifer, B. T., Neugebauer, G., Oke, J. B., and Matthews, K.: 1980,
    in preparation.

DISCUSSION

Tyson: Why couldn't the dust be local to us?

Soifer: The agreement in the Lyα/Hα ratios in low and high redshift
              quasars means that if due to extinction, the extinction
must be either local to the quasars or local to us. To cite a specific
example, for the high redshift (z = 2.20) quasar B2 1225+31, Lyα/Hα ˜ 1,
so that if due to galactic extinction then A (4000 Å) ˜ 3 mag. This
quasar is very near the north galactic pole, where visual extinctions
are likely to be a few tenths of a magnitude at most. Similar arguments
can be made based on other high redshift quasars, which leads me to rule
out galactic extinction as a possibility.

B.J. Wills: Soifer has suggested that the Netzer and Davidson (1979,
              M.N.R.A.S., 187, 871) photoionization models, including
dust, may be untenable because no excess infrared emission is observed.
However, the above authors have suggested that the missing energy may be
radiated (perhaps by some electronic transitions on dust grains) as at
least part of the 3000 Å "bump." Evidence that this "bump" is associated with emission from a large volume (e.g., the broad line emitting
region) comes from spectrum time variability arguments (Netzer et al.,
1979, Ap. J. (Letters), in press). If the 3000 Å bump is due to the
dust causing the "Lyα/Hβ problem," then it must co-exist with or lie
beyond the broad line emitting clouds; this is inconsistent with the
QSO model presented by Rees (this symposium), where the "bump" originates between the continuum source and the line-emitting region.

Soifer: The problem with this explanation is the energetics of the
              sources. The Netzer-Davidson model predicts that quasars
are intrinsically a factor of ˜ 10 more luminous than their observed
ultraviolet luminosities. This energy should be detectable somewhere.
It does not emerge in the 3000 Å band, and in the case of 3C 273, at
least, there is no observational evidence for this energy emerging at
any longer wavelengths.

Smith: I have the (unquantitative) impression that the Lyα/Hα ratio
              is relatively stable, as if there were some regulation

mechanism, while there is a larger dispersion in the Pα/Hα ratio. Do you see this in your data?

Soifer: I have not specifically looked at the Pα/Hα ratio for our sample of low redshift quasars; however the Pα/Hβ ratio varies by a factor of ~ 10. On the other hand, for the high redshift quasars, we find the Lyα/Hα ratio varies by a factor of 5. However, if the QSO 1623+26 is eliminated from our sample, the variation in this ratio is only a factor of ~ 2.

THE SPECTRUM OF THE MICROWAVE BACKGROUND

P. L. Richards and D. P. Woody[†]
Department of Physics, University of California,
and Materials and Molecular Research Division,
Lawrence Berkeley Laboratory, Berkeley, California 94720

ABSTRACT

The spectrum of the microwave background radiation closely resembles that of a blackbody with a peak at 6 cm^{-1}. Many experimenters have contributed to the determination of this simple but important fact. In this review we describe the experimental difficulties which have plagued background measurements in the high frequency Wien region. An evaluation of the present status of our knowledge of the entire spectrum will be given and prospects for future experiments will be mentioned.

MEASUREMENTS IN THE RAYLEIGH-JEANS REGION

Spectral measurements at frequencies $\lesssim$ 3 cm^{-1} have made use of ground based microwave antennas with well controlled sidelobe response and conventional heterodyne receivers. Atmospheric contributions were removed by measuring the sky temperature as a function of zenith angle. Calibration was accomplished with an ambient temperature (hot) load and a liquid He temperature (cold) load. The addition of the cold load to a relatively conventional radio telescope system made possible the original detection of the background radiation by Penzias and Wilson (1965). A second generation apparatus developed by the Princeton Group (Wilkenson 1967) was used for several experiments in the frequency range $0.3 < \nu < 3$ cm^{-1}. It employed a stationary downward looking receiver with an external cold load and a tiltable mirror to scan the zenith angle.

The data obtained using these microwave techniques have not changed appreciably in nearly 10 years. They provide a reasonably consistant picture in the Rayleigh-Jeans region over the frequency range from 0.01 to 3 cm^{-1}. The accuracy of the data was limited at the low frequency end by galactic synchrotron radiation and at the high frequency end by atmospheric emission. This review will not list all measurements of the spectrum of the microwave background. Such a list has been given recently by Danese and De Zotti (1977).

[†]California Institute of Technology, Pasadena, California 91125

G.O. Abell and P. J. E. Peebles (eds.), Objects of High Redshift, 283–291.
Copyright © 1980 by the IAU.

MEASUREMENTS AT AND BEYOND THE PEAK

Direct measurements of the spectrum of the microwave background radiation at and beyond the peak have proved troublesome for three important reasons:

(i) The background radiation must be measured in the presence of emission from nearby sources, including the atmosphere, the earth, and the apparatus. The ratio of the brightness of a 300 K blackbody to that of a 3 K blackbody is 10^2 at low frequencies. Because of the exponential cutoff of the 3 K curve, however, this ratio is 5×10^2 at 6 cm^{-1} and 1.5×10^5 at 20 cm^{-1}. Avoidance of the emission from ambient temperature objects thus becomes extremely difficult. In practice, high frequency experiments must use liquid He temperature apparatus.

(ii) The density and strength of emission lines in the earth's atmosphere increases rapidly with frequency. The combination of (i) and (ii) means that ground-based measurements can be made at 1 cm^{-1} with a relatively small atmospheric correction. By 12 cm^{-1}, however, large atmospheric corrections are required for measurements made from the highest available balloon altitudes.

(iii) The technology for high frequency cosmic background measurements was and is relatively undeveloped. In the Rayleigh-Jeans spectral region where adequate microwave technology was generally available, any investigator who seriously attacked the measurement problem had, in principle, an opportunity to do a measurement of good quality. At higher frequencies, however, the early workers simply did not have the technology base required. Despite intensive and imaginative efforts, misleading results were sometimes announced. In retrospect, the value of many of these early experiments lay in their contributions to measurement technology. Our present knowledge of the microwave background spectrum at and beyond the peak comes from the most recent experiments which could not have been done without antennas, spectrometers, and detectors that were invented during the past ten years.

It seems useful to summarize the development of the technology on which current direct spectral measurements rely. A great variety of approaches has been used for the high frequency measurements. The early rocket experiments of the Cornell, Naval Research Laboratory, and Los Alamos groups using far infrared detectors and band-limiting filters were pioneering efforts in a very real sense. It was very difficult to make any background measurements from sounding rockets at that time because the sensitivity of the available detectors was not sufficient to permit detailed diagnostic studies of instrumental performance to be made during the flight. There were also serious size and complexity limitations for rocket payloads. The conical antenna introduced by the Los Alamos group was of great value in later experiments. References to this early work are given by Danese and De Zotti (1977).

The field was advanced considerably by the balloon experiments of
the MIT group (Muehlner and Weiss 1973a, 1973b). Their experiments
showed that balloon platforms did permit the required complexity of
payload and long observing times, but with problems from the residual
atmosphere at balloon altitude. Since the MIT experiment used fairly
broad spectral bands, the atmospheric signal contained contributions
from both saturated and unsaturated emission lines. As a consequence,
a multiparameter fit to the zenith-angle dependence of their data was
required which limited the accuracy of the experimental results. This
group explicitly recognized the seriousness of diffraction of 300 K
radiation from the horizon into the apparatus and developed the apodized
antenna and ground shield to reduce this problem. They demonstrated
that balloon experiments could be operated open port, thus avoiding the
use of a warm emissive window. The MIT group also introduced the procedure of fitting their data to detailed atmospheric calculations.

The most recent generation of balloon measurements by the groups
at Queen Mary College (Robson et al., 1974 and Robson 1976) and
Berkeley (Woody, et al., 1975, Woody and Richards 1979)have made use of
Fourier transform infrared spectrometers to obtain the spectral resolution required to separate atmospheric lines from the background
radiation. The Martin-Puplett polarizing interferometer has nearly
ideal properties for measurements of this type and was used in both
experiments.

The Berkeley experiments were accompanied by an extensive program
of technology development. Innovations for the first Berkeley flight
included the use of an unobstructed conical antenna to define the beam
on the sky, the introduction of antenna pattern calculations using the
geometrical theory of diffraction, and the invention of a method for
measuring the antenna pattern over a broad range of angles and spectral
frequencies (Mather, et al., 1974). Innovations introduced in the
second Berkeley experiment included the use of a Winston cone (Winston
1970) for the primary antenna, and the use of a ^{3}He cooled composite
bolometric detector (Nishioka, et al., 1978).

Because of the continuous technical improvements in the higher
frequency measurements only the results from the most recent Berkeley
balloon experiment will be discussed. The primary conclusions of this
experiment are shown as a plot of flux versus wavenumber at a spectral
resolution of 1 cm^{-1} in Fig. 1. The two thin lines above and below the
crosshatched region indicate the ± 1 standard deviation limits of the
measurement assuming that all known errors are random and can be added
in quadrature. There are gaps in the data at the frequencies of strong
atmospheric emission lines where the error limits become very large.
The data clearly indicate that the spectrum of the background radiation
peaks in the neighborhood of 6 cm^{-1} and falls rapidly at higher frequencies. The integrated flux is equal to that from a 2.96 K Planck
curve which is shown for comparison. These new data are consistent with
the first Berkeley experiment which was best fit by a 2.99 K Planck curve
and also with the MIT results.

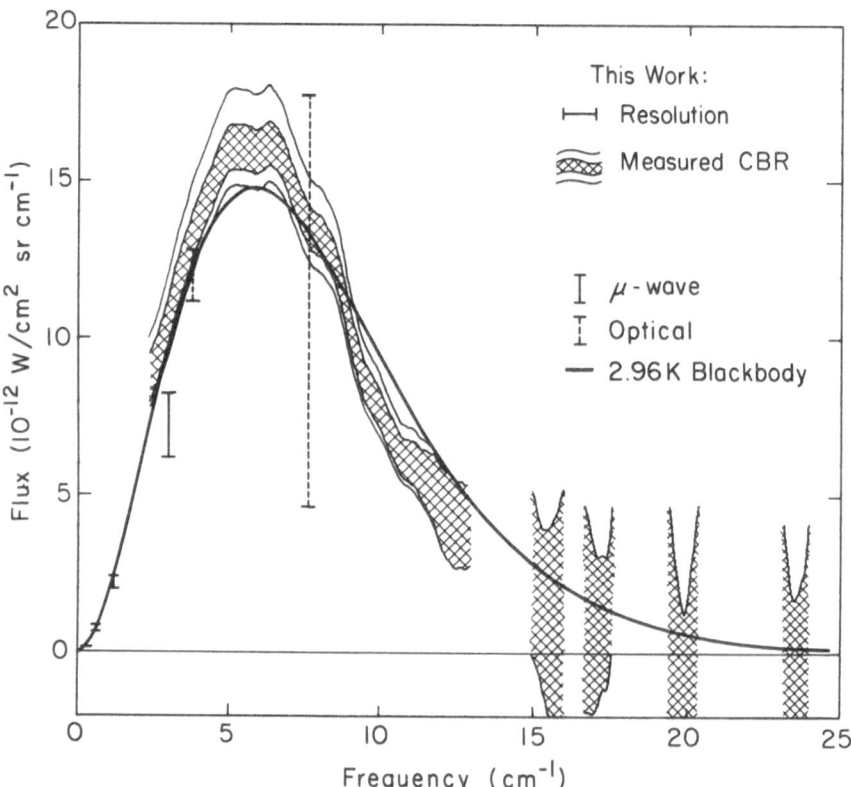

Figure 1.   Second Berkeley spectrum plotted as ± 1 σ error limits with
a distinction made between two types of experimental error.   The crosshatched region gives the error limits considering only those sources of
error which are essentially statistically independent between one
spectral resolution element and the next.   When sources of error are
included which affect the overall scale factor the crosshatched region
can be shifted up or down within the limits indicated by the thin solid
lines.   For comparison we also show the spectrum of the 2.96 K blackbody
which fits the measured integrated flux, as well as selected microwave
and optical measurements of the cosmic background radiation.

Because of the interest in possible deviations from a Planck curve
the errors in this experiment have been analyzed carefully.   One type of
error, which arises from sources such as amplifier gain, helium in the
antenna, etc., has the effect of expanding or contracting the scale
factor for the overall curve.   Another type of error, which arises from

sources such as detector noise and uncertainties in atmospheric line
parameters, is essentially statistically uncorrelated from one spectral
resolution element to the next. In Fig. 1 the $\pm$ 1 standard deviation
limits of these latter errors are plotted as a crosshatched region.
This region can be shifted up or down within the limits set by the
thin solid lines in order to include the effects of uncertainties in
the overall scale factor.

A statistical analysis which separates the effects of errors which
are correlated across the spectrum from those which are uncorrelated
between neighboring resolution elements shows that the data and the
Planck curve with the same included flux are 5 standard deviations
apart. Possible spurious sources for these discrepancies have been
carefully explored. They are unlikely to arise from the atmospheric
correction which is small for frequencies below 12 cm^{-1}. A large
(~25%) reduction in the overall scale factor would bring the data into
agreement with a lower temperature Planck curve, but no cause for such
an error has been identified. If it had occurred, the measured atmospheric spectrum would no longer be in agreement with the known mixing
ratio of atmospheric oxygen.

COLLECTED RESULTS

A comparison between the Berkeley data (full 1 σ error limits) and
the microwave data is given in Fig. 2 in the form of a plot of temperature versus frequency. Measurements of the rotational temperature of
interstellar CN molecules deduced from optical spectra are also shown
at 3.8 and 7.6 cm^{-1}. The temperatures obtained in this way are relatively insensitive to systematic errors and should be thought of as
hard upper limits. Corrections have been applied to the 3.8 cm^{-1} data
to account for local excitation of the CN molecules. When this is done
the temperature is reduced by about 0.1 K below the plotted point.
(Thaddeus 1972).

There is a hint of a discrepancy in the data in the 3 cm^{-1} region
where the highest frequency microwave points fall below the Berkeley
data. It is interesting to note in this regard that the atmospheric
contribution to these ground based measurements was more than 10
times the background signal. The atmospheric contribution to the
balloon experiment by contrast, was small for frequencies $\leq$ 12 cm^{-1}.

In the presence of possible systematic errors the assignment of
error limits is a subjective process. It is difficult therefore to
compare error limits set by different investigators. Two rather
extreme points of view will illustrate the problem:

(i) We can assume that differences in the errors assigned by
different investigators are significant and weight each data point in
accordance with its quoted error. If this point of view is accepted,
then the data set is dominated by a few measurements which have narrow
error limits.

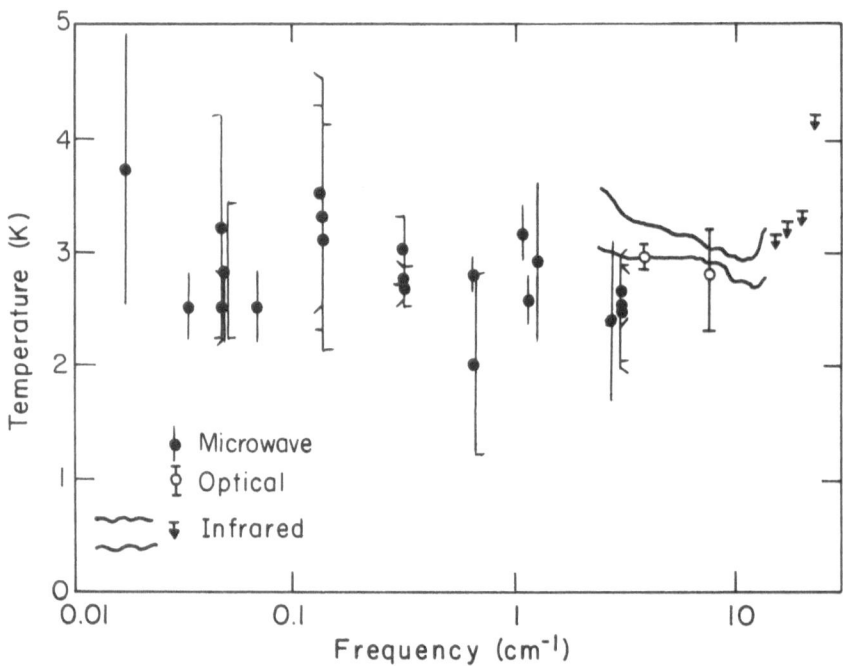

Figure 2. Collected microwave, interstellar CN and Berkeley results plotted as a frequency dependent thermodynamic temperature. The error limits are intended to be ± 1 σ.

(ii) We can assume that the differences in assigned error are primarily subjective and average only the measured values. From this point of view, a weighted average introduces a high degree of subjectivity into the conclusions. Because of the differences between these possible interpretations it is not clear to us how to correctly analyze the entire data set.

We will restrict our comments to the Berkeley data which are a consistent data set and which are not in agreement with a thermal spectrum. Some of the mechanisms which could lead to deviations from a blackbody spectrum are Compton scattering of the background photons by "hot" electrons, radiative damping of turbulence, and annihilation of matter and antimatter (Zeldovich, et al., 1972). The net result of these mechanisms is to scatter low-energy photons to higher energy and hence to shift the peak in the spectrum to higher frequencies. These models, however, do not fit the data as well as a simple Planck curve. The fit is degraded by 1 σ for an energy exchange between the

photons and an optically thin hot plasma equal to 3% of the energy
in the microwave background at the time of interaction.

One idea which appears to account adequately for the shape of the
Berkeley spectrum has been discussed by Rowan-Robinson, et al., (1979).
They suggest that the microwave background could have been re-heated
by interaction with dust which was created during an early period of
star formation. If this occurred at z = 200 then a silicate feature
in the emissivity of the dust could account for the observed fall in
thermodynamic temperatures with frequency beyond the peak.

If the deviation is correct, it will require some revisions in our
thinking about the early universe. In order to understand the degree
to which such revisions are imposed upon us by these experimental
results it is necessary to consider the nature of physical measurements.
In any difficult experiment it is impossible to be absolutely certain
that all systematic errors have been identified and either eliminated
or corrected. The degree of confidence in any one experiment obviously
increases with evidence that the experiment has been carefully done. A
high degree of confidence, however, is obtained only when several experiments give the same conclusion. It is preferable that these experiments
be done using different techniques.

NEW EXPERIMENTS

A consensus appears to be developing in the interested community
that improved measurements are both possible and desirable over the
whole microwave frequency range. It has been suggested by Wilkinson
(1979) that measurements of the rate of change of flux with frequency
could have better accuracy than absolute flux measurements. Better
experiments for frequencies above ≤ 1 cm^{-1} would have to be done above
the surface of the earth. At low frequencies large antennas would be
required to accurately sift out the effects of galactic synchrotron
radiation.

A measurement of the spectrum at and beyond the peak is being
planned as part of the COBE satellite. The design is related to the
Berkeley experiment reported above in that it makes use of bolometric
detectors and a polarizing Fourier spectrometer. An accurate independent measurement should be available from this experiment in about
1984.

Also, a new balloon experiment is being developed at Berkeley
which is designed to be as different as possible from the previous
Berkeley measurements so as to provide relatively independent information. The Fourier transform spectrometer will be replaced by a set
of 5 band-pass filters over the frequency range from 3 to 10.3 cm^{-1}.
The center frequencies of the filters have been selected to avoid
strong atmospheric lines. The residual atmospheric contribution in
the filter bands at balloon altitude will range from 0.5% to 47% of

the background signal. This contribution will be removed by measuring
signals as a function of zenith angle and extrapolating to zero atmosphere. The instrument will be calibrated in flight with an ambient
temperature calibrator that fills the entire antenna beam. The results
of this new experiment should be available in about one year.

ACKNOWLEDGMENTS

The authors are greatly indebted to their former colleagues,
Dr. J. C. Mather and Mr. N. S. Nishioka who participated actively in
the Berkeley experiments. The third Berkeley experiment is being prepared with the collaboration of Professor T. Timusk and Mr. J. Bonomo.

REFERENCES

Danese, L., and De Zotti, G.: 1977, Rivista del Nuovo Cimento 7,
   pp. 277-362.
Mather, J. C., Richards, P. L., and Woody, D. P.: 1974, IEEE Trans.
   MTT-22, pp. 1046-1048.
Muehlner, D., and Weiss, R.: 1973a, Phys. Rev. D7, pp. 326-344.
Muehlner, D., and Weiss, R.: 1973b, Phys. Rev. Letters 30, pp. 757-760.
Nishioka, N. S., Richards, P. L., and Woody, D. P.: 1978, Appl. Optics
   17, pp. 1562-1567.
Penzias, A. A., and Wilson, R. W.: 1965, Ap. J. 142, pp. 419-421.
Robson, E. I., Vickers, D. G., Huizinga, J. S., Beckman, J. E., and
   Clegg, P. E.: 1974, Nature 251, pp. 591-592.
Robson, E. I.: 1976, in M. Rowan-Robinson, (ed.), "Far Infrared
   Astronomy", Pergamon Press, London, pp. 115-124.
Rowan-Robinson, M., Negroponte, J., and Silk, J.: 1979 (to be published).
Thaddeus, P.: 1972, Ann. Rev. Astron. Astrophys. 10, pp. 305-309.
Wilkinson, D. T.: 1967, Phys. Rev. Letters, 19, pp. 1195-1198.
Wilkinson, D. T.: 1979, Proceedings of the Symposium "The Universe at
   Large Redshifts," Copenhagen, June 25-29, (to be published).
Winston, R.: 1970, J. Opt. Soc. Am. 60, pp. 245-247.
Woody, D. P., Mather, J. C., Nishioka, N. S., and Richards, P. L.:
   1975, Phys. Rev. Letters 34, pp. 1036-1039; A more complete des-
   cription of the first Berkeley experiment is found in Ph.D. Thesis
   by Mather and Woody which can be obtained by writing to the authors.
Woody, D.P., and Richards, P. L.: 1979, Phys. Rev. Letters 42, pp.
   925-929, (and to be published).
Zeldovich, Y. B., Illarionov, A. Z., and Syunyaev, R. A.: 1972, Zh.
   Eksp. Teor. Fiz. 62, pp. 1217-1224 [Sov. Phys. JETP 35, pp. 643-648.]

DISCUSSION

Baum: Could you get a better fit to the observed microwave back-
                   ground radiation spectrum by assuming a <u>spread</u> of blackbody
temperatures (and associated emissivities) instead of a single blackbody
temperature?

Woody: The Compton scattering from hot electrons and many other
                   reheating mechanisms can be analyzed as a superposition
of different temperature blackbodies with emissivity less than one.
Such models do not improve the fit to our data. An improvement can
only be obtained if emissivities are allowed to exceed unity.

Epstein: Some of your graphs seem to show that the atmospheric
                   emission is quite large compared to the deviations from
the pure blackbody spectrum. I think some of us are confused and/or
skeptical as to how you convinced yourselves that you had correctly
removed the effects of the atmosphere. Would you please review this
point for us?

Woody: Although the atmospheric emission is large at frequencies
                   above 15 cm^{-1}, the atmospheric correction in the region
where deviations are reported is small. At the peak in the CBR the
atmospheric contribution is less than 1% of the observed night sky
emission. It increases to ~ 40% at 11 cm^{-1}. We have found no reasonable alteration in the atmospheric emission model that would account
for the deviations.

Segal: I wonder whether you know of any special reason to antici-
                   pate that the CBR does not have a substantial isotropic
angular momentum.

Woody: We have no special information to contribute on this
                   question.

J. Roberts: The atmospheric contribution which you subtract is negative
                   in just the frequency range where you find an excess. Why
is this?

Woody: Our instrument measures the differences between the sky
                   temperature and the instrumental reference temperature of
1.7 K. The plot referred to is a theoretical prediction of the instrumental response to the <u>atmospheric</u> emission. It goes negative at low
frequencies where the atmospheric temperature becomes less than 1.7 K.

OBJECTS AT THE HIGHEST REDSHIFT

P. E. BOYNTON
Department of Astronomy, University of Washington
Seattle, Washington 98195

Small angular scale fluctuations in the temperature of the relict radiation may provide crucial information regarding the evolution of large scale structure in the universe. Some aspects of the interface between theory and observation in this developing study are considered.

I. Introduction

A central, unresolved issue in modern cosmology is how to understand the large scale structuring process which has transformed a relatively homogeneous early universe (on scales of galaxies, clusters, and super-clusters) into the highly inhomogeneous world of the current epoch. Although several alternatives have been pursued (see for example Ozernoi 1974, Press and Schechter 1974), the classical gravitational instability picture whose linearized theory was first discussed by Lifschitz (1946) is generally regarded as the most appealing framework for the discussion of such a process (cf. Weinberg 1972).

This picture provides for the growth of structure from incipient density perturbations in the early universe (but does not address the question of their origin). More specifically, within the context of the Standard Big Bang cosmology, the growth of these "seed" perturbations is restricted in time to the era following hydrogen recombination, and initially (i.e. in the linear regime) growth proceeds only as a power law in time, and not exponentially. Such painfully slow development might be viewed as a difficulty in using this process to explain the striking density contrast currently found on large scales.

But I prefer to argue the converse. Based on a Friedmann cosmology (specified by H_o and Ω_o) the locally observed distribution of matter ($z < 1$) implies a distinct evolutionary history of structure, and in particular specifies the amplitude of density perturbations which must have been present at the time of recombination. The elements of this argument may be summarized as follows:

G.O. Abell and P. J. E. Peebles (eds.), Objects of High Redshift, 293–303.

(1) From studies of the spatial correlations of galaxies,
     (e.g. Davis et al. 1977) one finds that density
     perturbations with masses larger than $3 \times 10^{14} \, \Omega_o h^{-1} M_\odot$
     are just now entering the non-linear regime,
     $(\delta\rho/\rho)_{z=0} \simeq 1$.

(2) Density perturbations are free to grow from the time the
     Jeans mass falls below the perturbation mass. This
     transition takes place during recombination as a con-
     sequence of the decoupling of matter and radiation.

(3) In an open universe ($\Omega_o < 1$), growth in the linear
     regime continues as $\delta\rho/\rho \propto (1+z)^{-1}$ until the universe
     approaches "free" expansion; that is, until $\Omega(z)$
     begins to depart from unity, which occurs nominally
     at $1 + z_f \simeq \Omega_o^{-1}$. Growth is thereby limited to the interval
     $z_f < z < z_{r(ecombination)}$, and the resulting growth
     factor $(1+z_r)/(1+z_f)$ ranges from 10^2 to 10^3 for
     $0.1 < \Omega_o < 1.0$. Thus, for those mass scales just now
     approaching non-linear condensation, the fractional
     density fluctuation back at z_r must have been in the
     range 10^{-2} to 10^{-3}, depending on Ω_o.

(4) Because of the tight coupling between matter and radiation
     up to recombination, density perturbations of this amplitude
     and angular scale ($\sim$ minutes of arc) might produce *obser-
     vable* fine scale temperature anisotropy in the cosmic
     microwave background radiation.

Consequently, a natural by-product of this evolutionary scenario is
a "fossil record" of the texture of the early universe which may be
available to us for direct inspection. Such a record was necessarily
imprinted on the radiation field as the universe became transparent
(in the *last-scattering* process). So, by definition, the fractional
anisotropies, $\delta T/T$, produced in this manner are the signatures of the
most distant, highest redshift objects observable; hence the apparent
hyperbole expressed in the title of this contribution.

If we could understand in detail the coupling between $\delta\rho$ and δT,
this series of four concepts in the evolutionary scenario taken together
would provide a test of the gravitational instability hypothesis through
measuring or even through placing upper limits on temperature anisotropy
for angular scales corresponding to progenitors of structure which clearly
does exist at the current epoch. The motivation for fine-scale anisotropy
observations is therefore clear.

II. Status of Observations

With one exception, little has changed since IAU Symposium #79 and
there seems no need to repeat the summary of observational efforts I

presented at that time, especially in view of the recent, comprehensive
discussion of fine-scale observations by Partridge (1979a). Moreover,
the announcement by Partridge (1979b) of a 95% confidence upper limit
of roughly 10^{-4} for an rms fractional temperature fluctuation observed
on an angular scale of ~ 10 arc minutes, continues a well-established
trend: that each improvement in technique fails to provide a detection
of anisotropy, and instead places still more stringent upper limits on
the amplitude of "primordial" temperature fluctuations.

In the past, these descending observational limits may have been
paralleled by informal downward revisions in the expected fluctuation
amplitude based on gravitational instability theory. In any case, an
actual conflict between theory and observation on this topic has yet
to occur. Motivated at least partly by the resilient ingenuity of
theoreticians as a group, Harry van der Laan (1977) asked publicly at
Tallinn if a *firm* lower limit to expected temperature anisotropy could
be established. There are several considerations which could provide
lower bound estimates. An estimate of minimal anisotropy present at
recombination $(\Delta T/T)_{min}$ in the context of gravitational instability
theory can be made, and one approach is discussed in the following
section. But there is no lower bound to *observable* fluctuations
$(\Delta T/T)_{obs}$ because of the unsettled issue of the history of the optical
depth of the Universe following initial recombination. Reionization of
the intergalactic medium may have occurred during the collapse of an
early generation of high density contrast systems. The consequent
increase in optical depth to Thomson scattering subsequent to z_r could
erase all traces of information about structure present at recombination.
In this situation, the lower bound $(\Delta T/T)_{min}$ provides a critical goal
in planning the sensitivity of future observations. We may yet discover
small angular scale temperature anisotropies at some level between
$(\Delta T/T)_{min}$ and the current observational upper bound, and if successful
we will have clearly learned something. On the other hand, if $(\Delta T/T)_{obs}$
is forced below $(\Delta T/T)_{min}$ the situation is ambiguous: either the
universe is opaque back to recombination, or gravitational instability
theory is incorrect. Aside from questions of practicality, there seems
little motivation for improving observational sensitivity to search
for fluctuations below $(\Delta T/T)_{min}$.

III. Defining the Critical Measurement

Van der Laan's question, rephrased in the spirit of that viewpoint
developed in the Introduction, becomes: what is the smallest value of
$(\delta T/T)_{z_r}$ consistent with the currently perceived value of $(\delta \rho/\rho)_{z=0}$
on interesting angular scales and which are also accessible to
measurement? The question itself provides a prescription for formula-
ting a reply: characterize the current mass spectrum of inhomogeneities,
extrapolate that distribution back to the recombination era, define a
minimal coupling to the radiation field thereby determining a lower
bound for anisotropy consistent with gravitational instability. The
details of this procedure have been spelled out by Davis and Boynton
(1979); a selective review of that work follows.

What constitutes minimal coupling? An arbitrary perturbation in the pre-recombination matter-radiation fluid can be decomposed into familiar isothermal and adiabatic components, or alternatively considered in terms of matter and radiation perturbations. In the case of an adiabatic disturbance, matter and radiation are strongly correlated, $\delta T/T = 1/3\ \delta\rho/\rho$, prior to the onset of recombination. Although an isothermal disturbance may be thought of as a pure matter perturbation before passing through the horizon, subsequent gravitational coupling gives rise to a "spontaneous" radiation component. Therefore we have chosen to define *minimal* coupling through a disturbance which is artificially constrained as a *pure matter perturbation* throughout the recombination process. The only coupling to the radiation field is through Thomson scattering, and anisotropy is produced solely through large scale mass motion via the familiar Doppler effect, $\delta T/T \propto (v/c)$ $\cos\phi$, where the scattering element is traveling at speed v at an angle ϕ relative to the direction to the observer. Even this minimal coupling mechanism is fairly effective because of the necessarily simultaneous reduction in optical depth, τ, and Jeans mass brought about by recombination. Consequently, photons are last scattered in material just developing a turbulent velocity field through the condensation process. To carry out our prescription, this velocity field must be related to the "seed" density perturbation spectrum. Such a relation follows directly from mass conservation, $\partial/\partial t(\delta\rho/\rho) = -\nabla\cdot v$, yielding fourier amplitudes $v_k \propto \delta_k\ 1/k$. Then the temperature fluctuation is given by

$$\delta T/T = \int (v/c)\ \cos\ \phi\ e^{-\tau(z)}(d\tau/dz)dz, \qquad (1)$$

where $e^{-\tau(z)}d\tau/dz$ defines a shell of last scatter with mean redshift $z_r \simeq 1100$, and width Δz of ~ 150 (Zel'dovich and Sunyaev 1972).

It is customary to examine the power spectrum of fluctuations, and in this case

$$|a_k|^2 \equiv |(\delta T/T)_k|^2 \propto (|\delta_k|^2/k^2)e^{-k^2\cos^2\phi}\ \sigma_x^2$$

where σ_x is the coordinate width of the last scattering shell. This spectral density clearly exhibits the primary features of the minimal coupling process. First, the "velocity derived" temperature perturbations exhibit a considerably steeper spectrum (by a factor of k^{-2}) for small k compared to the "density derived" case: that is, compared to the adiabatic radiation component fluctuations for which $|a_k|^2 \propto |\delta_k|^2$.

This steepening implies a correspondingly broader autocovariance function, hence a more extended angular correlation for the minimal coupling perturbations. Second, the exponential cut-off at small scales arises directly from the finite width of the last scattering shell. For perturbations with linear size smaller than σ_x, the fluctuation

is reduced by averaging over the several scattering elements which may
contribute along the sight line. Third, not so obvious is the remarkable coincidence that the smallest mass scale which remains unaffected
by this exponential cut-off is also the minimum mass which still lies
in the linear growth regime at the present epoch[1]. This fact enables
a trivial extrapolation (both normalization and spectral shape) of the
current density spectrum back to z_r.

Although the power spectrum of temperature fluctuations adequately
illustrates these points, it does not provide a convenient interface
between theory and observation. The customary beam-switching observing
technique involves the measurement of an ensemble of *temperature
differences* between adjacent "patches" of sky. The "patch" angular size
is determined by the width of the antenna beam pattern, θ_B, and the
separation of patch pairs is just the antenna beam throw, $\Delta\theta$. These
temperature differences are then employed to compute a mean-square
measure of temperature fluctuation, $(\Delta T/T)^2_{obs}$. Calculating a comparable
measure from a theoretical power spectrum $|a_k|_T^2$ involves computing
the fourier transform of this single-difference sampling function
including the effects of beam shape. The square of this transform,
$|f_k|^2$, defines a power density filter whose product with the power
density spectrum, when integrated, yields the desired mean square
temperature difference:

$$\int |a_k|_T^2 \cdot |f_k|^2 \, d^3k = (\frac{\Delta T}{T})_T^2 \qquad (2)$$

An equivalent, yet computationally simpler and more transparent
procedure follows from recognizing the variance given by this equation
as an autocovariance $C(\Delta\theta)$ evaluated at zero lag:

$$\int e^{ik\cdot\Delta\theta} |a_k|_T^2 \cdot |f_k|^2 \, d^3k = C(\Delta\theta=0) = (\delta T/T)^2 \qquad (3)$$

By application of the convolution theorem, the fourier transform
of the product $|a_k|_T^2$ can be expressed as the convolution of the
corresponding autocovariance functions: $C = C_T * C_{sampling}$

$$\text{where} \quad C_T = \int e^{ik\cdot\theta} |a_k|_T^2 \, d^3k \quad \text{and} \quad C_{samp} = \int e^{ik\cdot\theta} |f_k|^2 \, d^3k, \qquad (4)$$

$$\text{and} \quad (\frac{\Delta T}{T})_T^2 = C(0) = \int 2\pi\theta \, C_{samp}(\theta) \cdot C_T(\theta) d\theta \qquad (5)$$

For the minimal coupling model, we characterize the density
fluctuation power spectrum as a power law $|\delta_k|^2 = |\delta_0|^2 \, k^n$. The result-
ing theoretical autocovariance function C_T is parametrized by this
power law index and Ω_0. A typical example (n = -0.8, Ω_0 = 1.0) is
shown in figure 1. Also depicted is an arbitrarily scaled covariance
$C_m(\theta)$ describing the matter distribution, or alternatively the
radiation component covariance for the adiabatic case prior to
decoupling. Note the striking contrast between C_T and C_m. For the
minimal coupling model, we define $(\Delta T/T)^2_{min} \equiv (\Delta T/T)_T^2$.

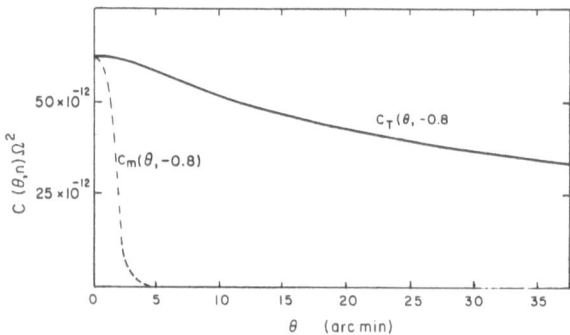

Figure 1. Theoretical
autocovariance for
minimal coupling model.

Comparison between this or any theory and a particular observing
program can be carried out as follows: the observation provides a value
of $(\Delta T/T)^2_{obs}$ corresponding to a particular sampling function $f(\theta)$
defined by the beam shape and switching amplitude. A one-dimensional
representation of a typical sampling function (the customary Gaussian
approximation to an antenna response function) is shown in figure 2a.
According to eq. 5, the autocovariance of this same sampling function
(fig. 2b) times $2\pi\theta$, when multiplied (at zero lag) by the theoretical
autocovariance of temperature fluctuations (fig. 2c) and integrated over
θ yields a value $(\Delta T/T)^2_{min}$ expected from the theory which is then

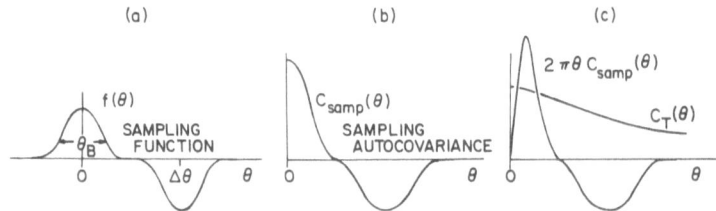

Figure 2. One dimensional representation of sampling function
(see note 2).

directly comparable to $(\Delta T/T)^2_{obs}$. Concise comparison between any
theory and observation is therefore possible if the theoretical
covariance function for sky fluctuations and the observational sky
sampling function are both made available.

The integral implied by figure 2c is particularly simple for
$\theta_B \ll \Delta\theta$ and $\theta_B < \theta_{corr}$ where θ_{corr} is the correlation angle
characteristic of $C_T(\theta)$. In that case, the sampling has a δ-function
character which yields $(\Delta T/T)^2_{min} = 2(C_T(0) - C_T(\Delta\theta))$. This result is
equivalent to that computed from a simple two-point difference spanning
an angle $\Delta\theta$, $(\Delta T/T)^2_{min} = var(\delta T/T(\theta) - \delta T/T(\theta + \theta\Delta))$, but from
elementary error propagation considerations $var(\delta T/T(\theta) - \delta T/T(\theta + \Delta\theta))$
$= var(\delta T/T) + var(\delta T/T) - 2 cov(\delta T/T(\theta), \delta T/T(\theta + \Delta\theta)) = 2C_T(0) - 2C_T(\Delta\theta)$

No such simple expression is possible in dealing directly with the
power spectrum representation.

For the large θ_{corr} encountered in the minimal coupling case,
typical millimeter wave beam switching geometries roughly satisfy the
point sampling approximation introduced above. The resulting $(\Delta T/T)_{min}$
calculated for the sky autocovariance function of figure 1 and also
for functions appropriate to $n = 0.0$ and $+0.8$ are shown in figure 3
for $\Omega_0 = 1.0$. From this figure, it is clear that $(\Delta T/T)_{min}$ is
insensitive (unfortunately) to the power law index n; therefore,
choosing $n = 0$ as representative allows the display of $(\Delta T/T)_{min}$
for various values of Ω_0 shown in figure 4.

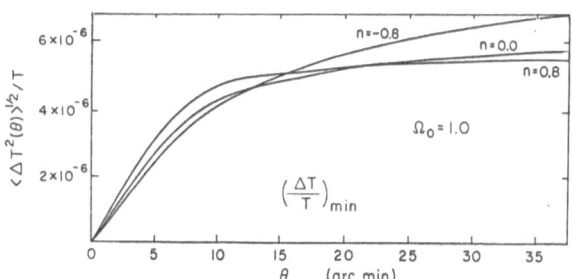

Figure 3. Angular dependence
of differentially measured
temperature fluctuations.

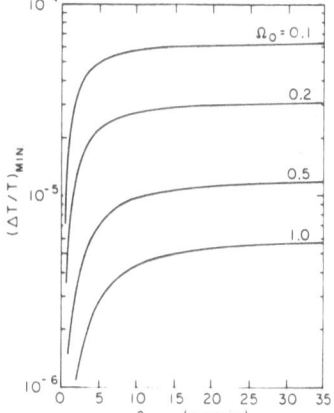

Figure 4. $(\Delta T/T)_{min}$ for
various values of Ω_0.

Being based on minimal coupling, these curves provide lower limit
values for $(\Delta T/T)_{obs}$. In a sense no definitive measurement is implied
if Ω_0 is not predetermined. If upper limits on $(\Delta T/T)_{obs}$ continue
to be reduced, larger values of Ω_0 must be accepted if the gravitation
instability hypothesis is to be retained. In this sense, from figure 4,
Partridge's upper limit implies $\Omega_0 > 0.05$. Although Ω_0 is still
poorly determined, current observations suggest $\Omega_0 < 0.3$, thus
$(\Delta T/T)_{min} \sim 2 \times 10^{-5}$ might be considered a tentative critical value in
designing future observing programs and instrumentation.

As already mentioned, it is always possible to reduce $(\Delta T/T)_{obs}$
to arbitrarily low values through postulating sufficiently early
reionization of the intergalactic medium following initial recombination.
Although information about fluctuations present at z_r is lost in that
case, the same process of scattering in the presence of large scale
mass motion produces temperature fluctuations at an epoch for which the
optical depth to Thomson scattering again becomes small. However,
scattering at lower z yields structure on larger angular scales,
degrees rather than minutes of arc (Davis 1979).

As emphasized by Rees (1979), if seed perturbations are purely
adiabatic, reionization is less likely. Perturbations on mass scales
which might collapse and release gravitational energy at early epochs
are strongly damped during recombination in this case. At the same time,
the expected temperature fluctuations at recombination are much larger
and consequently existing upper limits on $(\Delta T/T)_{obs}$ constrain Ω_0
to much larger values. Recent calculations of $(\Delta T/T)_T$ by Silk and
Wilson (1979) for initially adiabatic perturbations, when compared to
Partridge's 95% upper limit, require $\Omega_0 > 1$ to avoid rejecting the
gravitational instability hypothesis. Given the current uneasiness about
such large values of Ω_0, this comparison of theory and experiment
provides a fairly compelling argument against the presence of purely
adiabatic disturbances in the early universe.

It would be interesting to be forced to contemplate our non-existence
by observing sufficiently small upper limits for the amplitude of fine
scale anisotropy. But because reionization can always be invoked to
explain the absence of fluctuations observed from the current epoch, this
irony is easily avoided. Until it is possible to infer independently
the z-dependence of the optical depth of the universe for $z < 1000$
(Basko and Polnarev 1979), the utility of the minimal coupling limits
presented in figure 4 is primarily as that gauge of experimental effort
requested by van der Laan. That is, until observational limits are
forced below $\sim 10^{-5}$ there remains the possibility of discovering the
texture of the early universe, studying conditions in the shell of last-
scattering, and confirming a clear view back to the initial decoupling
of matter and radiation. However, if no cosmological anisotropy is
detected on small scales even at this level, our sense of *horizon*
may have to shrink a little.

FOOTNOTES

1. This smallest mass is also well above the "Silk damping" mass, and
   therefore the "velocity derived" component of temperature fluctuations
   is also correct for initially adiabatic perturbations.

2. The simple integral of the product of the two functions shown in
   figure 2c is a good approximation to $(\Delta T/T)^2$ for $\theta_B < \Delta\theta$. Note
   in figure 2c that the positive portion of the sampling autocovariance
   is multiplied by $2\pi\theta$ whereas the negative portion is not, this

choice is appropriate to an approximate one-dimensional representation of the two-dimensional convolution implicit in equation (5).
We stress this one-dimensional representation only because it
provides a simple visualization of this linkage between theory and
observation.

REFERENCES

Ozernoi, L.M., I.A.U. Symposium 63 (ed. M.S. Longair) Reidel Publ. Co.,
    Dordrecht, Holland (1974).

Press, W.H., and Schechter, P., Astrophys. J. 187, 425 (1974).

Lifschitz, E., Journal Phys. U.S.S.R. 10, 116 (1946).

Weinberg, S., Gravitation and Cosmology, John Wiley Co., New York (1972).

Partridge, R.B., to be published in Physica Scripta (1979).

Partridge, R.B., preprint (1979).

Van der Laan, H., 1977, I.A.U. Symposium 63 (ed. M.S. Longair) Reidel
    Publ. Co., Dordrecht, Holland, p. 403 (1974).

Davis, M., and Boynton, P.E., preprint, 1979.

Zel'dovich, Ya. B. and Sunyaev, R.A., Comments on Astrophys. and Space
    Sci. 4, 1973 (1972).

Davis, M. to be published in Physica Scripta (1979).

Rees, M.J., Private communication, see also discussion at end of this
    contribution (1979).

J. Silk and M.L. Wilson, preprint (1979).

Basko, M.M. and Polnarev, A.G., preprint (1979).

Davis, M., Peebles, P.J.W., and Groth, E.J., Ap.J. (Letters), 212,
    L107 (1977).

DISCUSSION

Turner: What influence on your calculations and conclusions would
                   changes in the assumed value of the amplitude and large-
scale cut-off or break in the present-day matter covariance function
have? Both of these quantities are significantly uncertain.

Boynton: Uncertainty in the amplitude of the matter covariance
                   function (on mass scales which would be expected to con-
tribute to observable temperature fluctuations) translates directly to
the same fractional uncertainty in the estimated mean square temperature fluctuation level. However, these estimates are rather insensitive
to the shape of the matter covariance function in the restricted sense
that they do not depend strongly on the local power law index (cf.
Fig. 3).

Gaskell: Why are you choosing to constrain the adiabatic fluctua-
                   tions rather than, say, that Ω_0 could be greater than
unity?

Boynton: Primarily because the conventional wisdom regarding the
                   value of Ω_0 allows one to pose this rather interesting
constraint. Heresy would appeal to me only if I had something new to
bring to bear on the closure issue, and I don't.

Birkinshaw: As a by-product of the microwave background observations
                   described later in this volume, we have a limit to the
value of $\Delta T/T$ on blank sky for a beam-throw of 15 arcmin; this limit is
~ 2×10^{-4}.

Rees: If there is an isothermal component to the initial inhomo-
                   geneities, the first generation of gravitationally-bound
systems (of masses 10^7 to 10^9 $M_\odot$) will condense out immediately after
recombination. The ensuing heat input may well be sufficient to reio-
nize a fraction of the remaining diffuse matter; one only needs $n_e/n \gtrsim$
$10^{-3} \Omega^{-1/2}$ in order to smear out and delay the "last scattering epoch"
and thereby reduce the Boynton-Davis fluctuations still further. (On
the other hand, your exclusion of a "pure adiabatic" model for the
fluctuations is on firmer ground, because in this model no bound systems
form until clusters of the Silk mass turn around at $z \lesssim 10$, which is an
epoch too recent to produce an opaque intercluster medium.)

                   One further point: if there is early production of
                   grains or molecules, the "last scattering" of the micro-
wave background may be due not to free electrons, but to some wavelength-
dependent opacity. This means that the last scattering surface must be
located at a wavelength-dependent redshift. There must then be no
detailed correlations between the $\Delta T/T$ observed at two different wavelengths from the same piece of sky, contrary to the expectations of the
standard model.

Boynton: I certainly agree with you about the possible impact of reionization on this question, but the point about excluding adiabatic fluctuations has been made by Silk and Wilson, not me. If wavelength-dependent opacity can be identified, it will make both the observing and interpretation tasks more difficult; but on the plus side, it might be exploited to provide a three-dimensional view of the pattern of early inhomogeneities. The angular scale of observed fluctuations will still be a crude index to the last-scattering epoch.

Silk: Wilson and I have also predicted that the minimal level of small-scale anisotropy expected from primordial isothermal density fluctuations is of the order 10^{-5} if there is no appreciable rescattering after decoupling. However, I believe that rescattering is probably inevitable for the following reason: in the hierarchical clustering model of galaxy and cluster formation, one begins with primordial isothermal fluctuations that systematically merge into deeper and deeper potential wells. These fluctuations must remain at least partly gaseous until clustering has developed. One can calculate the release of gravitational binding energy required in this model, and it must result in substantial amounts of ionizing radiation at very early epochs ($z > 30$ or more). The intervening medium will therefore be reionized at an early epoch if an isothermal fluctuation spectrum is the principal source of inhomogeneity in the early universe.

G. Burbidge: If no fluctuations can be found we have no direct evidence at all that galaxies are formed at early epochs through gravitational instability. That is, the current view of galaxy formation would be a purely theoretical concept.

Boynton: Lamentably so, but I think, rather, that galaxy formation would *remain* a purely theoretical concept.

SCATTERING OF THE MICROWAVE BACKGROUND IN CLUSTERS OF GALAXIES

George Lake
Astronomy Department, University of California
Berkeley, California 94720

I.   INTRODUCTION

Evidence from X-ray observations of clusters of galaxies suggests that they contain intergalactic plasma at a temperature of approximately 10^8 K which emits thermal bremsstrahlung (cf. Lea et al. 1973). Although numerous models for the origin and distribution of the plasma have been presented (cf. Gould and Raphaeli 1978), the amount and distribution of the plasma is still not certain. Several years ago, Sunyaev and Zel-dovich (1972) pointed out another observable consequence of hot gas in clusters of galaxies: inverse Compton scattering by electrons in the plasma would increase the energy of photons of the microwave background as these photons pass through the cluster. The fractional change in intensity of the microwave background is:

$$\frac{\Delta I}{I} = \frac{xe^x}{e^x-1}\left[\frac{x}{\tanh\frac{x}{2}} - 4\right]\int_{o.}^{\tau} \frac{kT_e}{m_e c^2} \, d\tau \tag{1}$$

where τ is the optical depth for Thomson scattering through the cluster, T_e is the electron temperature of the plasma, m_e is the electron mass, $x = h\nu/kT_r$, and T_r is the temperature of the microwave background radiation, which we take to be 2.8°K. It is worth noting explicitly that this effect lowers the intensity (and therefore the antenna temperature) of the microwave background in the Rayleigh-Jeans region. This follows from the fact that the inverse Compton process conserves photon number while increasing the photon energy. Since estimates of the optical depth for Thomson scattering are uncertain, $\Delta I/I$ cannot be precisely predicted. Estimates varying from 10^{-5} to 3×10^{-4} have been published, based on different models for the distribution of the plasma (see Gull and Northover, 1975; Sarazin and Bahcall, 1977; or Gould and Rephaeli, 1978).

The detection of this effect in clusters of galaxies would offer several benefits:

(1)   It would confirm the existence of plasma in the clusters thereby
      strengthening the hypothesis that it is thermal bremsstrahlung
      which produces the observed X-ray flux from clusters;

(2)   since $\Delta I/I$ is proportional to the electron number density, n_e,
      whereas X-ray flux is proportional to n_e^2, observations of both
      would permit n_e and the so-called "clumping factor", $\langle n_e^2 \rangle / \langle n_e \rangle^2$,
      to be found, and this in turn would help discriminate among the
      models for the distribution of the plasma;

(3)   combined X-ray and microwave observations can provide independent
      measures of H_0 and q_0 (Birkinshaw, 1979; Boynton and Murray,
      1978; Cavaliere, Danese and De Zotti, 1978; Gunn, 1978; Silk and
      White, 1978); and finally

(4)   detection of the effect would confirm the non-local origin of the
      microwave background.

II.   OBSERVATIONAL CONSIDERATIONS

i)   Observations by other groups

      Several other groups have sought or are seeking to detect the in-
verse Compton cooling of the microwave background produced by cluster
plasma. Pariiskii, working at a wavelength of 4 cm reported a positive
detection of approximately -10^{-3} K or -1 mK for the Coma cluster, a result in agreement with the early observations of Gull and Northover
(1976) at a wavelength of 3 cm, but not with the subsequent analysis of
Birkinshaw et al. (1978, hereafter BGN). Birkinshaw will be describing
their observations of several other Abell clusters, most of them X-ray
sources and their detection of the effect in two or possibly three of
them. However, Rudnick (1978) observing at 2 cm, reported no statistically significant evidence for cooling of the microwave background by
the clusters he has observed (at roughly comparable sensitivity), though
he now believes he sees cooling in Abell Cluster 2218 (private communication, 1979). Perrenod and Lada (1979, hereafter PL) claim roughly
2σ results in three clusters. The apparent disagreement reflects the
difficulty of these observations (see Rudnick, 1978). The expected signals are small. Most clusters contain radio sources which may produce
spurious signals (these could be of either sign in an observation which
employs beam switching). Finally, the plasma itself will emit bremsstrahlung radiation even at centimeter wavelengths, emission which may
mask the signal sought.

ii)   Our work at 9 mm

      In view of these problems, we decided to make our observations at
a shorter wavelength than used by others; $\lambda = 9$ mm, using the 11 meter
NRAO telescope in Tucson. Using a short wavelength offered the three
advantages summarized in Table 1 from Rudnick (1978). First, most radio

TABLE 1[1]

Possible Contributions to Sky Brightness (mK)

Source	Antenna Temperature	Brightness Temperature		
	Cluster Radio Galaxies	Diffuse Cluster Sources	Galactic Structure	Confusion
BGN (λ = 2.8 cm, D = 26 m)	$\gtrsim$ 4.	0.6	0.06	0.3
Rudnick (λ = 2 cm, D = 42 m)	$\gtrsim$ 5(2') – $\leqslant$ 0.4(24')	0.2	0.02	0.1
Lake and Partridge				
PL (λ = 0.9 cm, D = 11 m)	$\leqslant$ 0.2	0.02	0.002	0.01

[1]This table is reproduced from Rudnick (1978) with his permission.

sources are less intense at short wavelengths, so that our observations were less troubled by radio emission from sources within or near the clusters. Second, confusion is less a problem. The final advantage of shorter wavelength measurements is that thermal bremsstrahlung can be neglected.

Observations at short wavelengths present two problems, however. For a 2.8°K blackbody spectrum, λ = 9 mm is not truly in the Rayleigh-Jeans region. The antenna temperature at 31.4 GHz is only 2.1°K, and it is only for antenna temperature that $\Delta T/T = \Delta I/I$. This loss in sensitivity, however, is partially compensated by the increase in magnitude of the function f(x) in equation 1. A more fundamental difficulty is that receivers are noisier and the atmosphere is more emissive at 9 mm than at centimeter wavelengths. These are expected to be sources of statistical error, and we preferred to accept the possibility of larger statistical errors to avoid sources of possible systematic errors. The details of our observational technique and subsequent analysis are presented in a paper recently submitted to the Astrophysical Journal (Lake and Partridge, 1979).

Part way through our runs, we discovered two instrumental effects which produced spurious differences of a few millidegrees (mK) in antenna temperature. The first was a sudden jump in measured values of ΔT as the telescope crossed meridian if the transit occurred at high elevations. We have therefore excluded all data taken at elevations above 65°. The second effect was an elevation-dependent offset in the switched signal ("on" minus "off") which was present for sources in some, but not all, ranges of declination. Since we were not able to determine the origin of these offsets, we made extensive observations of nominally blank sky at the same declinations and over the same hour angles as our cluster observations. Finally we note that the elevation-dependent effect was particularly strong at $\delta \approx 66°$ and is responsible for the large, and probably erroneous, negative temperature offsets reported by Lake and Partridge (1977) for Abell clusters 2125 and 2218, both of which lie near δ = 66°. The magnitude of the effect is quite large, but data taken during different days and runs yield the same results, giving us confidence that we have been able to remove this systematic effect.

One must further make the various corrections for telescope efficiency, thermal bremsstrahlung and the geometry of the beam. This last correction is due to the non-filling of the main beam and the attenuation due to beam-switching in the nearby clusters which have an extent comparable to the size of the beam throw. Since this correction clearly depends on a model for the spatial extent and temperature run of the gas, it has not been made to the data appearing in Table 2.

The major conclusion of this work is that we find no reliable and significant evidence for the Sunyaev-Zel-dovich cooling in any of the clusters we observed, except for Abell 576.

TABLE 2

Brightness Temperature in mK for the Clusters Observed to Date

Source	Lake and Partridge	Perrenod and Lada	Birkenshaw *et al.*	Rudnick
376	1.86 ± 0.78		1.61 ± 0.33	
401				-0.4 ± 0.4
426	3.67 ± 1.12			
478			0.54 ± 0.57	
506		0.63 ± 0.76		
518		-1.56 ± 0.83		
545	1.68 ± 0.45			
576 .	-1.27 ± 0.28		-0.70 ± 0.19	
665	-1.03 ± 0.69	-1.30 ± 0.59	0.32 ± 0.35	
777	-0.22 ± 0.45			
910	0.22 ± 0.54			
1472A		1.24 ± 0.52		
1472B		-1.26 ± 1.02		
1656	0.19 ± 0.22		0.57 ± 0.53	0.4 ± 0.6
1656 (offset)	0.07 ± 0.43			
1689	-1.14 ± 0.88			
2079	-0.04 ± 0.24			
2125	0.73 ± 0.45		0.06 ± 0.30	
2142	-0.47 ± 0.78		0.41 ± 0.23	
2199				0.6 ± 0.5
2218 Center	0.80 ± 0.39	-1.04 ± 0.48	-1.09 ± 0.28	
East		-0.56 ± 0.73		
West		-0.17 ± 0.68		
2255				0.7 ± 1.2
2319A	-0.06 ± 0.20	1.37 ± 0.94	0.25 ± 0.28	0.8 ± 0.9
B		3.09 ± 1.61	0.98 ± 0.32	
2645	2.35 ± 0.71			
2666	0.63 ± 0.32		0.33 ± 0.29	

III. COMPARISONS AND CRITICISMS

Examining Table 2 we find a larger number of significant positive signals than negative ones. This should not destroy one's faith in the reality of the negative signals. Any source located at the center of the cluster produces a positive signal. However, since our telescope beam-switches in azimuth, the reference beam traces arcs about the cluster center. As a result, any source in the reference beam is only seen for a fraction of the total observing time and our method of analysis should detect its effect. This is also true for PL and BGN, but not for Rudnick who uses a telescope with a beam-switch in hour angle.

If we look across the columns of Table 2, we note a discrepancy in the case of A2218. This may be due to a difference in the source position used by us and the other groups. We used the position given by Abell (1958), whereas BGN have redetermined the position and obsere 1.5 to the North (roughly our halfpower beam width). PL also used the position of BGN.

If we look down the columns of Table 2, we see that half of the measurements of PL are roughly 2 σ from zero (3 negative and 2 positive out of 10). For BGN, 5 out of 11 are also ≳ 2 σ from zero (2 negative and 3 positive). In our results, 6 out of 17 are ≳ 2 σ from zero (5 positive and 1 negative). In our observations, I think that 545 and 2645 might well be spurious. These sources are both at $\delta \approx -10°$ and show roughly the same signal. We did not have good blank sky coverage here and because of their low declination the elevation changes very little during our observations and the systematic is more difficult to determine. This is being explored in more detail. I have no reason to suspect the other three significant results we claim, but it is certainly reasonable to suspect that some non-gaussion process may be at work.

IV. OUR GENERAL CONCLUSIONS

Only four of the 16 clusters we observed has a measured value of ΔT which can be more negative than -1 mK (sky or brightness temperature) at the 2 σ level. Since the clusters we surveyed have a wide range of X-ray luminosity, velocity dispersion, richness and cluster core radius, our limits on the microwave decrement may be helpful in constraining models for the origin and distribution of intergalactic plasma in clusters. The constraints on some well observed clusters, such as A777, A1656, A2079, A2125 and A2319, are particularly tight.

In general, we will leave to others detailed comparisons of our results with the theoretical models. We would, however, like to make three general points concerning our results.

First, it is clear that the apparent correlation we noted in 1977 (Lake and Partridge) between cluster richness and the size of ΔT is not supported by these further studies. The apparent correlation we reported was based on observations of three richness class 4 clusters; we now know that the values for ΔT reported for two of them were erroneous, because of the systematic effect mentioned in section II.ii. Our present work shows no significant correlation between ΔT and richness. This is in accord with the work of Pravdo, et al. (1979) who find no positive correlation between the richness of a cluster and its thermal bremsstrahlung X-ray emission.

Next, we note that of the roughly 20 clusters observed by us or by other groups, only two - numbers A576 and A2218 - are reported to have a significant decrement. Why have we (and others) not detected

the Sunyaev-Zel-dovich effect in other clusters? Our failure to detect a microwave decrement, especially in clusters known to be X-ray
sources such as A2079 and A2319, may indicate that bremsstrahlung
emission from cooler gas is masking the inverse Compton cooling (Tarter
1978). Such emission could help explain some of our positive values of
ΔT. It may also be that the so-called clumping factor $<n_e^2>/<n_e>^2$ is
substantially greater than unity. If the latter is the case, our results favor models where the hot gas is physically clumped; or more
probably models with a high central concentration (which raises
$<n_e^2>/<n_e>^2$ averaged over the cluster), rather than those with a more
uniform distribution of gas. High resolution X-ray studies by HEAO-2
may help resolve this issue.

Further, since the X-ray flux varies as $T^{1/2}$, whereas the microwave decrement varies as T, decreasing the temperature with a fixed
X-ray flux results in a smaller expected diminution. In the case of
A576 White and Silk (private communication, 1979) are finding a tempe-
rature of $\sim 10^{7}{}^{\circ}$K rather than $10^{8}{}^{\circ}$K. This particular case is rather
embarrassing, since it is the one cluster for which we claim to detect
the effect. Studies of silicon and magnesium lines with the crystal
spectrometer aboard HEAO-2 will provide much more reliable temperatures.

One is still left with the question: why does A576, and possibly
A2218 (Birkinshaw, et al. 1978b), show a microwave decrement? Since
many other comparable clusters do not appear to produce a significant
cooling, it is reasonable to ask what is special about these two?
Neither is an especially luminous X-ray source. Neither differs in
richness or morpohology from other clusters we have observed.

The questions raised in the previous sections brings us to our
final point: caution should be used when these or other microwave
measurements are used to constrain theoretical models, or to determine
the Hubble constant.

I am indebted to J. Negroponte, P. Nulson, R. B. Partridge, S.
Perrenod, L. Rudnick and S. White for valuable discussions, and to NSF
for research support.

REFERENCES

Abell, G. O. 1958, Ap. J. Suppl., 3, 211.
Bahcall, J. N., and Sarazin, C. L. 1977, Ap. J. Letters, 213, L99.
Birkinshaw, M. 1979, M. N. R. A. S., in press.
_____, Gull, S. F., and Northover, K. J. E. 1978a,
    M. N. R. A. S., 185, 245.
_____ 1978b, Nature, 275, 40.
Boynton, P. E., and Murray, S. S. 1978, privately communicated HEAO-B
    proposal.
Cavalieri, A., Danese, L., and DeZotti, G. 1978, preprint.
Gould, R. J., and Rephaeli, Y. 1978, Ap. J., 219, 12.
Gull, S. F., and Northover, K. J. E. 1975, M. N. R. A. S., 173, 585.
_____ 1976, Nature, 263, 572.
Gunn, J. E. 1978, in Observational Cosmology, eds. L. Maeder and G.
    Tammann, Geneva Observatory.
Lake, G. and Partridge, R. B. 1977, Nature, 270, 502.
Lake, G. and Partridge, R. B. 1979, Ap. J., submitted.
Lea, S., Silk, J., Kellogg, E., and Murray, S. 1973, Ap. J. Letters,
    184, L105.
Pariiskii, Yu. N. 1973, Sov. A. J., 16, 1048.
Perrenod, S. C. and Lada, C. J. 1979, preprint.
Pravdo, S. H., Boldt, E. A., Marshall, F. E., McKee, J., Mushotzky,
    R. F., Smith, B. W., and Reichert, C., Ap. J., in press.
Rudnick, L. 1978, Ap. J., 233, 37.
Sarazin, C. L., and Bahcall, J. N. 1977, Ap. J. Suppl., 34, 451.
Silk, J., and White, S. D. M. 1978, Ap. J. Letters, 226, L103.
Sunyaev, R. A., and Zel'dovich, Ya. B. 1972, Comments Ap. and Space
    Physics, 4, 173.
Tarter, J. C. 1978, Ap. J., 220, 749.

DISCUSSION

Jaffe: Could some of the scatter in your data be due to the first-
             order Doppler effect from a large amount of cool gas surround-
ing the cluster and moving with some peculiar velocity?

Lake: My first thought is that bremmstrahlung would be stronger.
             Second, how is such a cool gas supported?

Peebles: Have you computed the correlation coefficients among the
             independent sets of measurements of $\delta T/T$ for each cluster?

Lake: No, I have not. Part of the problem with such a comparison
             can be seen in Table 1. The contribution of radio sources
is quite different at the three frequencies that have been observed.

MEASUREMENTS OF THE SCATTERING OF THE MICROWAVE BACKGROUND RADIATION IN CLUSTERS OF GALAXIES

M. Birkinshaw
Department of Radio Astronomy, University of California
Berkeley, California 94720

Results for the decrements in the microwave background radiation towards the centres of 13 clusters of galaxies are presented. It is shown that these data imply central gas densities of about 2×10^{-24} kg m^{-3} and cluster masses of about 5×10^{45} kg.

INTRODUCTION

The prediction that a detectable diminution in the brightness of the microwave background radiation towards an X-ray cluster of galaxies should exist if the X-radiation is of thermal bremsstrahlung origin (Sunyaev and Zeldovich 1972) lead a number of workers to attempt to measure such an effect (Parijskij 1973a,b,c; Gull and Northover 1976; Lake and Partridge 1977; Birkinshaw, Gull and Northover 1978a,b; Rudnick 1978; Perrenod and Lada 1979). The overall lack of agreement between these workers may be ascribed to (1) the extreme difficulty of measuring brightness temperature differences of order 1 mK in the presence of much larger atmospheric effects that vary on timescales from hours to weeks, and (2) the possiblity of very small systematic errors entering the data from non-idealities in the beamshape or side-lobe structure of the telescope used.

In order to check against these effects, many hours of data using a carefully-tested instrument must be collected. This paper presents the latest results from about 4500 hours of observation towards, or almost towards, the centres of 13 clusters of galaxies. Full details of the experimental technique and other results will appear elsewhere (Birkinshaw, Gull and Northover 1980).

METHODS

The telescope used for these observations was the 25-m paraboloid of the SRC Appleton Laboratory at Chilbolton, Hampshire. At the observing frequency chosen (10.6 GHz), the telescope has a beamwidth

G.O. Abell and P. J. E. Peebles (eds.), Objects of High Redshift, 313–319.
Copyright © 1980 by the IAU.

of about 4.5 arcmin and an efficiency of 0.55. The receiver system
had a noise temperature of 180 K and a bandwidth of 250 MHz.

A twin-beam system in the configuration described by Birkinshaw
et al. (1978b) was used to eliminate the strongest atmospheric effects.
The 1-minute temperature difference measurements recorded by this
technique were then tested in a complex, but statistically-unbiased,
manner to remove interference spikes (caused by pigeons, helicopters
or Hampshire thunderstorms), to detect biased data-segments and to
find interfering sources in the reference arcs around each point
observed. It was found that subtle, long-term, atmospheric noise
fluctuations severely degraded the sensitivity of the system, so that
very long integration times were necessary to achieve measurement
accuracies of 0.3 mK or better. The final results from this analysis
were slightly adjusted for temperature contributions from the (small
angular size) sources detected in interferometric surveys at other
frequencies than that used to search for a decrement in the microwave
background.

RESULTS

The results at (or near, if the observing position was changed to
avoid strong radio sources) the centres of the well-observed clusters
are shown in Table 1. It can be seen that for A 576, 2218, and 2319

Table 1.   Results

Cluster	Observing Time (hr)	Measured Microwave Background Decrement, ΔT, corrected for radio sources (mK)
A 71	20	+0.29 ± 0.54
A 347	35	+0.34 ± 0.29
A 376	79	+1.22 ± 0.35
A 478	106	−0.71 ± 0.47
A 576	190	−1.12 ± 0.17
A 665	107	−0.53 ± 0.22
A 1656*	63	−0.41 ± 0.35
A 1904	24	+0.55 ± 0.40
A 2125	124	−0.39 ± 0.22
A 2142	40	−1.4 ± 1.0
A 2218	137	−1.05 ± 0.21
A 2319*	50	−0.77 ± 0.28
A 2666	175	+0.34 ± 0.29
Blank sky	114	+0.03 ± 0.21

* Near, not at, cluster centre

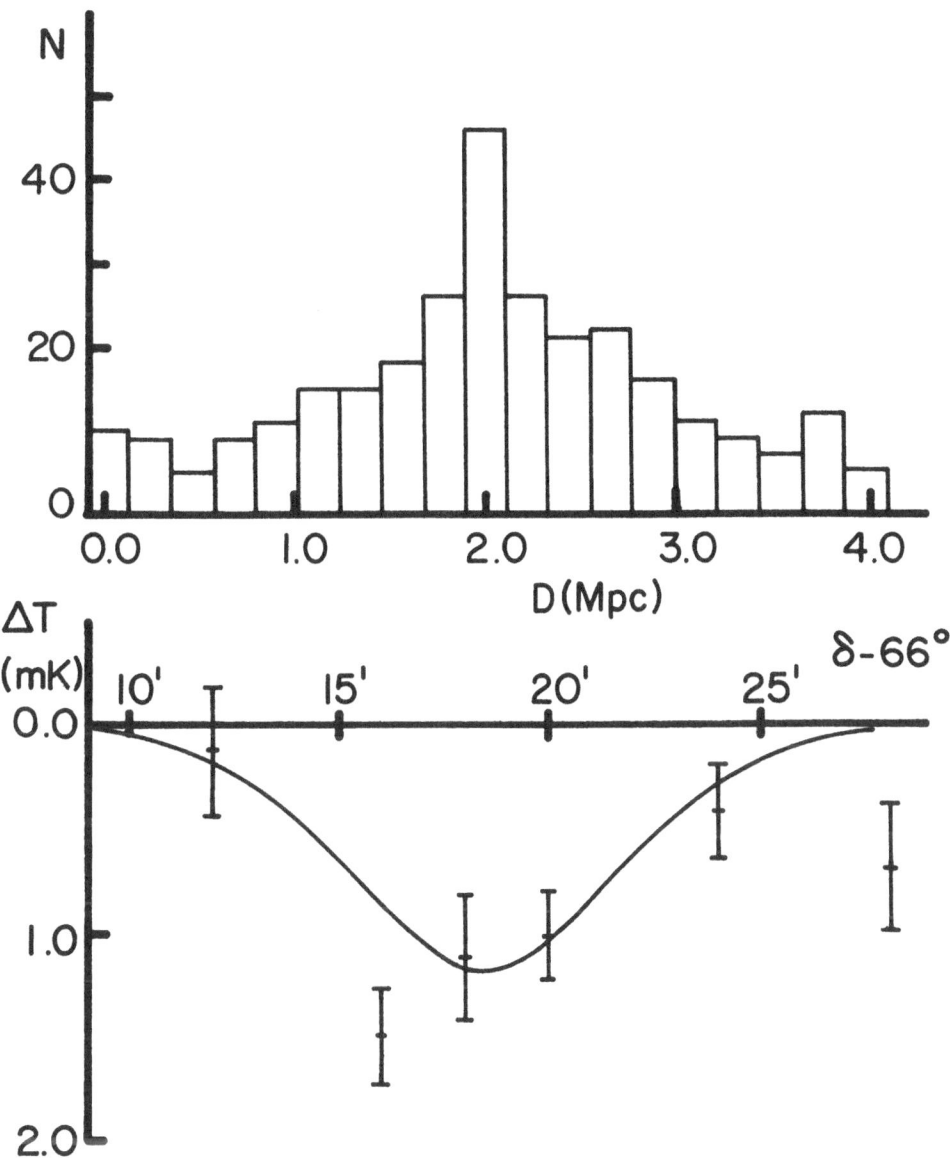

Figure 1. The galaxy strip-count, N, in arcminute bins and the observed temperature decrement at each point, ΔT, plotted as functions of declination (1950.0 coordinates). The solid line is the ΔT profile deduced from a reasonable model atmosphere lying in the potential well defined by the galaxy counts and convolved with the telescope beamshape. The linear scale, D, at the distance of the cluster is also indicated ($z = 0.168$, $H_0 = 50$ km s^{-1} Mpc^{-1}, $q_0 = 0.5$).

a cooling effect at a significance greater than 2.5σ has been found.
The other results may be consistent with zero decrement plus a confusion distribution from flat-spectrum background sources. A strong
check on systematic error is provided by the blank-sky observation,
made at the same declination as the A 576 observation, but 4 hours
later in RA, so that an observation of A 576 could be followed immediately by one on the blank sky.

In addition, several scans have been made. Figure 1 shows one
of these, the NS scan for A 2218, where the measurements are shown
plotted with the galaxy count across the cluster and the predictions
of a reasonable model for the gas distribution [described in Birkinshaw
et al. (1980)]. The close positional agreement between the large background decrements and the peak galaxy count comprises the best evidence
for a detection of a real background decrement yet available.

GAS PARAMETERS

From the cluster-centre results, and a reasonable model for the
cluster gas and galaxy distributions, model-dependent estimates of the
central density of the gas responsible for the X-ray emission and the
cluster mass may be found. These results are shown in Table 2 -- for
some clusters only limits to the parameters are derivable due to the
lack of significant X-ray or microwave-decrement detections. The
errors in these parameters are large and not marked in the Table --
but the scatter of results is fairly small about values

central gas density $\sim 2 \times 10^{-24}$ kg m^{-3}

total cluster mass $\sim 5 \times 10^{45}$ kg

Table 2. Model-dependent derived parameters of the clusters

Cluster	Central gas density 10^{-24} kg m^{-3}	Cluster mass 10^{46} kg
A 71	<2.7	--
A 347	<0.8	--
A 478	4.3	0.3
A 576	1.0	2.5
A 665	<11.0	>0.1
A 1656	2.2	0.8
A 1904	<2.1	<0.2
A 2125	<14.3	>0.1
A 2142	3.2	0.9
A 2218	2.1	1.7
A 2319	3.3	0.5
A 2666	<1.1	<0.6

If the gas is highly clumped these results scale with the clumping parameter to the powers 0.4 and -0.4 respectively.

CONCLUSIONS

The main conclusions of this work are:

1. the observed temperature decrements agree qualitatively with those expected on the basis of the *new* X-ray data on clusters of galaxies; and

2. the central gas densities in the clusters detected here are deduced to be about 2×10^{-24} kg m^{-3}, and the cluster masses are about 5×10^{45} kg, agreeing with their virial masses.

In any future continuation of this work, the following points should be noted:

1. These observations were atmosphere-limited, and long integration times were needed to eliminate all timescales and amplitudes of atmospheric signal. Even working at the comparatively low frequency of 10.6 GHz, the use of a better site than Chilbolton would be a good idea;

2. On the other hand, observations with other telescopes are often affected by telescope-generated effects -- which must be checked against carefully before attempting to use any telescope for this work;

3. For observing distant clusters, a narrower telescope beam than that used here would be desirable -- a factor of at least 2 increase in the peak decrement for A 2218 should be seen if a beamwidth less than about 2 arcmin were to be used; and

4. Spectral studies of the decrement would be interesting, and would limit theoretical models for the cluster environment.

ACKNOWLEDGEMENTS

The measurements reported in this work were made in collaboration with Drs. S.F. Gull and K.J.E. Northover -- without whom this work would never have started.

REFERENCES

Birkinshaw, M., Gull, S.F., and Northover, K.J.E.: 1978a. *Nature*, 275, 40.

Birkinshaw, M., Gull, S.F., and Northover, K.J.E.: 1978b. *Mon. Not. R. astr. Soc.*, 185, 245.

Birkinshaw, M., Gull, S.F., and Northover, K.J.E.: 1980. *Mon. Not. R. astr. Soc.*, in preparation.

Gull, S.F., and Northover, K.J.E.: 1976. *Nature*, 263, 572.

Lake, G., and Partridge, R.B.: 1977. *Nature*, 270, 502.

Parijskij, Yu. N.: 1973a. *Astr. Zhurn.*, 49, 1322.

Parijskij, Yu. N.: 1973b. *Astrophys. J.*, 180, L47. See also erratum in *Astrophys. J.*, 188, L113.

Parijskij, Yu. N.: 1973c. *Astr. Zhurn.*, 50, 453.

Perrenod, S.C., and Lada, C.J.: 1979. Preprint.

Rudnick, L.: 1978. *Astrophys. J.*, 223, 37.

Sunyaev, R.A., and Zeldovich, Ya. B.: 1972. *Comm. Astrophys. Sp. Phys.* 4, 173.

DISCUSSION

Ford: What is the dependence of ΔT on z if the cluster X-ray source does not evolve, and what is the optimum z for measuring microwave diminution?

Birkinshaw: For any given telescope beamwidth and beamthrow, the optimum z is determined principally by the criterion that the beamwidth should be less than ~ 2 cluster core radii. Of less importance is that the beamthrow should exceed ~ 5 core radii. Thus, the configuration described above works most efficiently at z ~ 0.07.

Peebles: Would you indicate which of the clusters you observe are observed to be X-ray sources? Is there a correlation between X-ray detection and negative δT?

Birkinshaw: All the clusters with the exceptions of A 665 and 2125 have at some stage been said to be detectable in the X-ray waveband. An investigation of any correlation between ΔT and the X-ray flux density (or, more correctly, the X-ray luminosity) must await complete, and accurate, X-ray flux measurements. However, the constancy of values of $\rho(0)$ and M in Table 2 indicates the existence of such a correlation.

Boynton: With regard to the status of A2218 as an X-ray source, Steve Murray, Bob Schommer, and I have made a 4 x 10^4-sec exposure with the high resolution imager of the Einstein Observatory. We find the flux level is within a factor of 2 of the Ariel V upper limit. This cluster appears symmetric and centrally condensed. Model fits suggest that the temperature is rather high, not inconsistent with the values required by your microwave decrement measurement. We were somewhat surprised by this outcome.

N. Bahcall: A576, one of the two clusters with significant ΔT diminution detection, is a rather low-luminosity X-ray cluster
and presumably of low density of hot gas.

Silk: I have observed Abell 576 with the IPC of the Einstein
Observatory. The X-ray flux is about 1/6 of the Uhuru
flux, but in agreement with other satellite measurements, making A576
a relatively low-luminosity X-ray cluster (9×10^{43} erg s^{-1} over 0.2 to
3 keV). However, the X-ray distribution is relatively uniform, spherically symmetric, and centrally condensed.

LARGE–ANGULAR–SCALE ANISOTROPY IN THE COSMIC BACKGROUND RADIATION

George F. Smoot
Space Sciences Laboratory and Lawrence Berkeley Laboratory
University of California, Berkeley, California 94720

ABSTRACT
       Measurements of the large–angular–scale anisotropy of the
cosmic background radiation made from the northern hemisphere are in
essential agreement with each other and indicate a first order spherical harmonic component with an amplitude of approximately 3 mK. New
data from the southern hemisphere support these previous results. This
first order anisotropy is interpreted as resulting from the motion of
the solar system relative to the cosmic background radiation. There
is no evidence of any higher order anisotropy to the level of 1 mK.

INTRODUCTION

       The cosmic microwave background radiation not only provides
the strongest evidence for a hot, compressed early universe – the Big
Bang – but also one of the most direct means for studying the early
universe. In particular, the large–angular–scale anisotropy of the
cosmic background radiation is a sensitive probe of several phenomena
of cosmological interest. These include the isotropy of the Hubble
expansion, possible rotation of the Universe, and the existence of very
long wavelength gravitational radiation. A large–scale anisotropy can
also arise from the non–uniform distribution of matter at the time of
decoupling.
       Ever since the discovery of the cosmic background, experimen-
ters have probed the isotropy of the radiation. The measurements on
the large–angular–scale anisotropy have been directed toward the twin
goals of finding an intrinsic anisotropy in the early universe and
detecting the solar motion through the radiation. Early measurements
of the anisotropy were hampered by atmospheric noise, Galactic radiation and limited sky coverage. In more recent experiments, low side-
lobe horn antennas, careful radiometer design, higher observing frequencies and measurements from high altitude (balloon and aircraft)
have minimized the systematic errors that plagued earlier measurements.
Anisotropy in the cosmic background radiation has now been clearly observed by groups at Princeton and Berkeley. For both measurements, the

G.O. Abell and P. J. E. Peebles (eds.), Objects of High Redshift, 321–328.
Copyright © 1980 by the IAU.

anisotropy is well described by a first order spherical harmonic
(dipole) distribution and is inconsistent with a quadrupole distribution. The new Berkeley measurements from the southern hemisphere are
in good agreement with northern hemisphere results.

REVIEW OF BERKELEY AND PRINCETON EXPERIMENTS

The Berkeley and Princeton experiments each detected a first
order anisotropy and set stringent upper limits on higher order components. The reported signals (3 mK) are extremely small when compared
to the ambient temperature (300 K) and the equivalent noise temperature
of the receivers. Maintaining the stability of a receiver to better
than one part in 10^5 while pointing the apparatus from one part of the
sky to the next would be exceedingly difficult; however, if the receiver is designed to have inputs from two identical antennas pointing at
different parts of the sky simultaneously, the stability requirements
are reduced typically to about a part in a thousand. The receiver must
be well regulated thermally to provide the necessary stability. Because
the inputs may not be precisely identical, the entire apparatus is also
rotated, interchanging the positions of the two antennas roughly once
per minute.

The results of earlier experiments have been limited by Galactic emissions -- primarily synchrotron radiation and thermal emission from HII regions at the low frequencies, and Galactic dust at very
high frequencies. However, there is a natural window in the Galactic
background in which to make anisotropy measurements (between 3 mm and
1.5 cm) where the Galactic contribution is lower than the reported cosmic background anisotropy.

Care must be taken to see that the two antennas are pointing
through nearly identical pathlengths of atmosphere, so that the atmospheric microwave emission is balanced. This is very difficult to do at
low altitudes (less than 12 km) because of the fluctuating atmospheric
water vapor. At higher altitudes, the water vapor is reduced substantially and the remaining atmospheric emission is due primarily to oxygen which is much more uniformly distributed.

Finally, the experiments must guard against stray emissions
from the earth, the apparatus itself, and other extraneous sources.
The receivers must be designed and constructed carefully to ensure good
performance and shielding against man-made interference. Rejecting the
unwanted stray radiation also requires the use of ground shields and
antennas with very low sidelobe response. All ferrite components are
magnetically shielded.

Table 1 lists some of the relevant experiment parameters.

Table 1 Experiment Parameters

	Berkeley	Princeton
Vehicle	U-2	Balloon
Altitude	20 km	27 km
Duration	4 hrs/flight	2 @ 8 hrs
Frequency	33 GHz	19 GHz
	(9 mm)	24.8 + 31.4 GHz
Opening Angle	60°	90°
Rotation	64 seconds	∿1 minute
Origin of Flights	Ames, Cal	Palestine, Texas
	lat 38°	lat 32°
Antenna	dual-mode	standard with ground
		shield
	corrugated horns	+ add flare on end.

The parameters of the best fit to a first order spherical harmonic distribution are:

	ΔT	α (hours)	δ	
Berkeley	3.5±0.5	11.2±0.5	16°±7°	(Smoot, Gorenstein & Muller)
Princeton	3.0±0.3	12.3±0.4	-1°±6°	(Cheng et al. 1979)

If these dipole distributions are due to the solar motion through the radiation, we can use the Doppler shift formula:

$T(\theta,\phi) = T_o \, \gamma/(1-\beta\cos\theta)$ where θ = angle between direction of motion and observation, β = velocity

$\simeq T_o + T_o \, \beta\cos\theta$ $\beta \ll 1$

The solar velocity is then found to be

$$V_\Theta = \Delta T \times 100 \text{ km/sec} \times \frac{3°K}{T_{CBR}}$$

where the direction of the velocity aligns with the direction of maximum anisotropy.
Using canonical values,

$$V_\Theta \simeq 300 \pm 40 \text{ km/sec}.$$

We can then calculate the velocity of the Galaxy and local group by subtracting the solar motion due to the rotation of the Galaxy (300 km/sec, $\ell=90°$, $b=0°$). The local group has a large net velocity of 500 km/sec towards $\alpha = 10.5^h$ and $\delta = -19.5°$ ($\ell=264°$, $b=33°$).

SOUTHERN HEMISPHERE MEASUREMENTS

        With a positive result from anisotropy measurements in the
northern hemisphere, our attention naturally turned toward possible
southern sky coverage. First, the unsurveyed southern hemisphere could
provide completely independent verification of the observed northern
hemisphere anisotropy. Second, the southern sky coverage (or at least
accurate data from a very different latitude) would be necessary to
distinguish unambiguously between the polar dipole and a polar-axially-
symmetric quadrupole component. The optimum location for separating
the spherical harmonic polar components of first and second order is
30° south latitude. Unfortunately, the Galactic background, geography,
and politics all conspire to make this a difficult latitude from which
to make measurements.

        On March 2-5, 1979, we were able to conduct a series of four
flights from Lima, Peru. The U-2 flew to southern Peru, and most of
the observations were conducted from 15° south latitude. The data from
the March 5 flight were taken from 10° latitude in order to utilize the
moon for an inflight calibration. Unfortunately, we were unable to
finalize the arrangements on schedule, and the four flights occurred
two weeks later than originally planned. This delay caused the first
data points (taken just after dark) to be one hour later in siderial
time than planned; the data-taking flight segments could not extend
beyond the original cut-off, because the Galactic plane would have produced a significant background signal. In addition, there were other
small time delays, and an error in which the pilot flew the flight plan
from the previous day. These combined difficulties resulted in a sky
coverage which was less than ideal, but which still provided much of
the information we sought. The sky coverage actually achieved is shown
in Figure 1.

        The general operating conditions in Peru were more severe
than California. In particular, atmospheric temperatures at altitude
were significantly colder. The heaters in the thermal controller for
the receiver and Dicke switch drew about one third more power, and the
temperature sensors imbedded in the antennas near the exposed ends
registered temperatures of less than -40°C, typically 10°C colder than
on flights from Ames. This extreme cold and the high humidity in Lima
combined to make the post-flight checkout a difficult task. As soon as
each flight was over, we would begin the checkout by covering each antenna with a piece of Saran Wrap to keep out dust and any moisture which
might condense. By the end of the checkout, the heaters would have
warmed the equipment to a higher than ambient temperature, removing
the possibility of condensation.

        We believe that the apparatus got very cold during the flight
down from California, and condensed water when it landed in Panama and/
or Lima. The condensation probably caused the atmospheric monitor (54
GHz radiometer) to fail before the first flight. We discovered the
failure in the checkout before the flight, but had insufficient time to
repair the 54 GHz radiometer (the equipment had arrived only four hours
before the scheduled flight). We found water in the Dicke switch wave-
guide when we disassembled and repaired it. Our previous and, as it

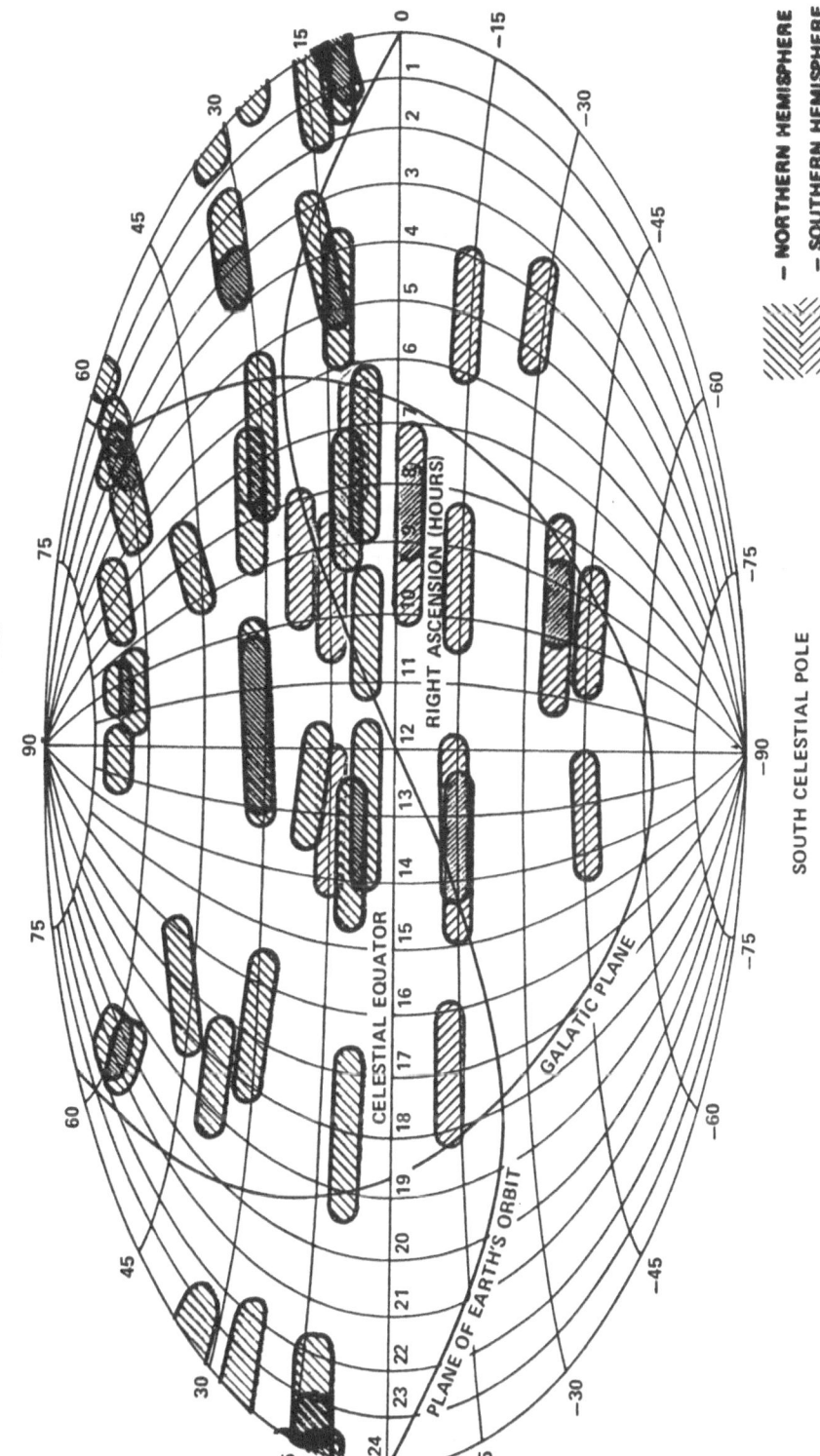

Figure 1. Plot of Sky Coverage in Celestial Coordinates. Sky coverage for the northern flights from NASA Ames in California and southern flights from Lima, Peru is indicated by the shaded regions. The width of each region is set by the 7° antenna beam widths, and the length is set by the rotation of the earth and the motion of the U2 back and forth along its flight path. The galactic and ecliptic planes are shown for reference.

turns out, succeeding experience verified that the U-2 flies sufficiently level to make the atmospheric corrections negligible, and the data
from the first flight appear to be satisfactory.

RESULTS

        Measurements from the southern hemisphere flight paths are
shown in Figure 2 along with the predicted response based on our
northern hemisphere data. The results are very encouraging, since they
show an anisotropy of roughly the same magnitude and direction as the
northern data. Fitting these southern data to a first order spherical
harmonic anisotropy produces:

	ΔT (m°K)	α(hours)	δ
Uncorrected	2.4±0.7	12.5±1	2°±13°
Corrected for Galactic Bkg	2.9±0.7	12.3±1	1°±13°

These results are in remarkable agreement with those of Princeton and
in reasonable agreement with our own northern hemisphere data.
        To obtain the best value for the first-order anisotropy and
limits on higher order spherical harmonics, we combined the northern
and southern hemisphere measurements. For the dipole solution, we find:

$$\Delta T = 3.1 \pm 0.4 \text{ m°K}, \ \alpha = 11.4 \pm 0.4 \text{ hours, and } \delta = 10° \pm 6°$$

The combined dipole and quadrupole best-fit parameters are:

$$\Delta T = 2.9 \pm 0.4 \text{ m°K}, \ \alpha = 11.1 \pm 0.5 \text{ and } \delta = -4° \pm 10°$$

so that 95% of the anisotropy remains in the three first-order components. The net quadrupole amplitude is approximately 0.5±0.5 mK, consistent with no net quadrupole term in a combined first and second
order fit to the level of 1 mK.

CONCLUSION

        The data from the southern hemisphere support and complement
the northern sky measurements. The combined data indicate a global
first order anisotropy at about the 3 mK level and limit any second or
higher order anisotropy to significantly lower levels. This conclusion,
in turn, supports the interpretation that the observed anisotropy arises
via a Doppler shift caused by the solar motion relative to the background radiation. A 3 mK anisotropy translates into a solar motion of
approximately 300 km/sec; when the rotation of the galaxy is taken into
account, we find a net velocity of approximately 500 km/sec for the
local group. This relatively high velocity disagress with the velocities inferred from the anisotropy in the Hubble diagram. Thus, the
southern measurements have highlighted the apparent inconsistency

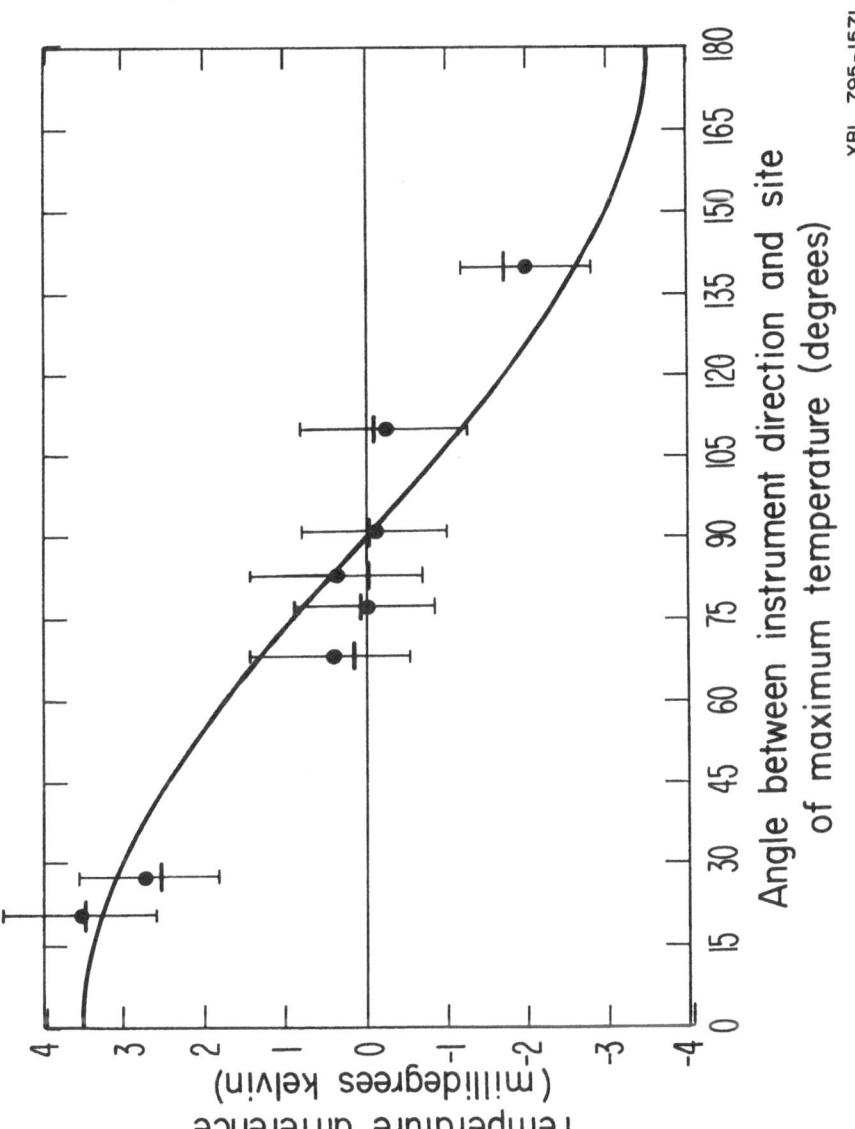

XBL 795-1571

Figure 2. Comparison of southern hemisphere data with predictions from northern measurements. The temperature difference, $\Delta T = T(\hat{\theta}_1) - T(\hat{\theta}_2)$, observed is plotted versus the angle between the instrument direction, $\hat{d} = \hat{\theta}_1 - \hat{\theta}_2$, and the direction of maximum temperature, $\hat{n} = \hat{x} = 11.2$ hours and $\delta = 16°$. The heavy horizontal bars represent the uncorrected data, while the dots show the data corrected for estimated galactic background.

between the background radiation measurements and the galaxy-magnitude-redshift studies.

The lack of observed higher-order anisotropy (no component with greater than 1 m°K amplitude) emphasizes the intrinsic large-scale isotropy of the universe and supports the Cosmological Principle. Apparently, the discovery of any intrinsic large-angular-scale anisotropies must await the next generation of experiments, or perhaps the full-sky coverage that will be achieved by the Cosmic Background Explorer (COBE) Satellite.

The COBE satellite will carry anisotropy-measuring radiometers operating at four frequencies, in order to distinguish between Galactic radiation and the intrinsic anisotropy in the cosmic background based on frequency dependence. The space craft will rotate at about 1 RPM to interchange the antennas, and will be in a polar orbit perpendicular to the sun-earth line. This orbit allows the radiometers to look away from the earth at all times, and to be shaded from the sun by a large conical shield. The radiometers will cover the entire sky in six months. The anisotropy measuring experiment should achieve a sensitivity of 0.1 mK on all angular scales larger than about 10°, and thus be able to detect higher order components which are approximately 1/30th the amplitude of the observed first-order anisotropy.

REFERENCES

Cheng, E.S., Saulson, P.R., Wilkinson, D.T., and Corey, B.E.: 1979, Astrophys.J.Letters, very soon.
Gorenstein, M.V., Muller, R.A., Smoot G.F., and Tyson, J.A.: 1978, Rev. of Sc. Ins. 49, 4, p. 440.
Gorenstein, M.V. and Smoot, G.F.: 1980, to be submitted to Astrophys.J.
Smoot, G.F., Gorenstein, M.V., and Muller, R.A.: 1977, P.R.L. 39, 898.
Smoot, G.F. and Lubin, P.M.: 1979, Astrophys.J.Lett. - soon.

DISCUSSION

Tyson: What is the magnitude of the left-right based anisotropy, and was it the same in the south?

Smoot: For the northern hemisphere data the equipment centered anisotropy revealed by reversing the aircraft flight path was about 2 mK. For the southern flights the aircraft centered offset appears to be about twice this value. The effect appears to be stable and cancels because we take our data in pairs of opposing direction flight paths. We do not yet understand why the offset is greater for the southern flights.

329

INDEX OF SUBJECTS